Geologic Hazards

A Field Guide for Geotechnical Engineers

Geologic Hazards

A Field Guide for Geotechnical Engineers

Roy E. Hunt, P.E., P.G.

CRC is an imprint of the Taylor & Francis Group,
an informa business

This material was previously published in *Geotechnical Engineering Investigations Handbook*, Second Edition © 2005 by CRC Press, LLC.

CRC Press
Taylor & Francis Group
6000 Broken Sound Parkway NW, Suite 300
Boca Raton, FL 33487-2742

Printed in the United States of America on acid-free paper
10 9 8 7 6 5 4 3 2 1

International Standard Book Number-10: 1-4200-5250-0 (Hardcover)
International Standard Book Number-13: 978-1-4200-5250-3 (Hardcover)

Library of Congress Cataloging-in-Publication Data

Hunt, Roy E.
Geologic hazards : a field guide for geotechnical engineers / by Roy E. Hunt.
p. cm.
Includes bibliographical references and index.
ISBN 1-4200-5250-0 (alk. paper)
1. Engineering geology--Handbooks, manuals, etc. 2. Natural disasters. I. Title.

TA705.H844 2006
363.34'9--dc22 2006052993

Visit the Taylor & Francis Web site at
http://www.taylorandfrancis.com

and the CRC Press Web site at
http://www.crcpress.com

Contents

Author

Now in private practice, Roy E. Hunt, P.E., P.G., has more than 50 years of experience in geo-technical and geological engineering. Hunt has been an adjunct professor of engineering geology at the Graduate School of Civil Engineering, Drexel University, and currently holds a similar position in the Geosciences Program at the University of Pennsylvania. He has been the consultant on two new nuclear power plants in Brazil; a toll road program in Indonesia and a new airbase in Israel; offshore mooring structures in the Philippines and Brazil; and on landslide studies in Bolivia, Brazil, Ecuador, Indonesia, Puerto Rico, and the continental United States. Assignments have also taken him to Barbados, England, France, the U.S. Virgin Islands, and locations throughout the continental United States. His past affiliations include Joseph S. Ward and Associates, where he was a partner, and Woodward-Clyde Consultants, where he was director of engineering in the Pennsylvania office.

His education includes an M.A. in soil mechanics and foundation engineering, Columbia University, New York (1956), and a B.S. in geology and physics, Upsala College, East Orange, New Jersey (1952). He is a registered professional engineer in New Jersey, New York, and Pennsylvania; a registered professional geologist in Delaware, Pennsylvania, and Brazil; and a certified professional geologist. His professional affiliations include the American Society of Civil Engineers (Life Member), Association of Engineering Geologists, and the American Institute of Professional Geologists. He has received the E.B. Burwell Jr. Memorial Award, Geologic Society of America, Engineering Geology Division, and the Claire P. Holdredge Award, Association of Engineering Geologists, for his book *Geotechnical Engineering Investigation Manual* (1984); and the Claire P. Holdredge Award, Association of Engineering Geologists, for his book *Geotechnical Engineering Techniques and Practices* (1986) — both books were published by McGraw-Hill, New York.

Introduction

Purpose and Scope

This section sets forth the basis for recognizing, understanding, and treating the geologic hazards to provide for safe and economical construction. It invokes general concepts rather than rigorous mathematical analyses.

Significance

Geologic hazards represent substantial danger to humans and their works. The hazards may exist as a consequence of natural events, but often they are the result of human activities.

Slope failures, such as landslides and avalanches, can occur in almost any hilly or mountainous terrain, or offshore, often with a very frequent incidence of occurrence, and can be very destructive, at times catastrophic. The potential for failure is identifiable, and therefore forewarning is possible, but the actual time of occurrence is not predictable. Most slopes can be stabilized, but under some conditions failure cannot be prevented by reasonable means.

Ground subsidence, collapse, and *expansion* usually are the result of human activities and range from minor to major hazards, although loss of life is seldom great as a consequence. Their potential for occurrence evaluated on the basis of geologic conditions is for the most part readily recognizable and they are therefore preventable or their consequences are avoidable.

Earthquakes represent the greatest hazard in terms of potential destruction and loss of life. They are the most difficult hazard to assess in terms of their probability of occurrence and magnitude as well as their vibrational characteristics, which must be known for aseismic design of structures. Recognition of the potential on the basis of geologic conditions and historical events provides the information for aseismic design.

Health hazards related to geologic conditions include asbestos, silica and radon, and the various minerals found in groundwater such as arsenic and mercury. Recently, mold has been added to the list of health hazards. Discussion of environmental concerns related to health from contaminants is beyond the scope of this book.

1

Landslides and Other Slope Failures

1.1 Introduction

1.1.1 General

Origins and Consequences of Slope Failures

Gravitational forces are always acting on a mass of soil or rock beneath a slope. As long as the strength of the mass is equal to or greater than the gravitational forces, the forces are in balance, the mass is in equilibrium, and movement does not occur. An imbalance of forces results in slope failure and movement in the forms of creep, falls, slides, avalanches, or flows.

Slope failures can range from being a temporary nuisance by partially closing a roadway, to destroying structures, to being catastrophic and even burying cities.

Failure Oddities

- *Prediction:* Some failures can be predicted, others cannot, although most hazardous conditions are recognizable.
- *Occurrence*: Some forms occur without warning; many other forms give warning, most commonly in the form of early surface cracks.
- *Movement velocities*: Some move slowly, others progressively or retrogressively, others at great velocities.
- *Movement distances*: Some move short distances; others can move for many miles.
- *Movement volume*: Some involve small blocks; others involve tremendous volumes.
- *Failure forms*: Some geologic formations have characteristic failure forms; others can fail in a variety of forms, often complex.
- *Mathematical analysis*: Some conditions can be analyzed mathematically, many cannot.
- *Treatments*: Some conditions cannot be treated to make them stable; they should be avoided.

Objectives

The objectives of this chapter are to provide the basis for:

- Prediction of slope failures through the recognition of the geologic and other factors that govern failure.
- Treatment of slopes that are potentially unstable and pose a danger to some existing development.
- Design and construction of stable cut slopes and side-hill fills.
- Stabilization of failed slopes.

1.1.2 Hazard Recognition

General

Slope failures occur in many forms. There is a wide range in their predictability, rapidity of occurrence and movement, and ground area affected, all of which relate directly to the consequences of failure. Recognition permits the selection of some slope treatment which will either *avoid, eliminate, or reduce the hazard.*

Hazard recognition and successful treatment require thorough understanding of a number of factors including:

- Types and forms of slope failures (classification)
- Relationship between geologic conditions and the potential failure form
- Significance of slope activity, or amount and rate of movement
- Elements of slope stability
- Characteristics of slope failure forms (see Section 1.2)
- Applicability of mathematical analysis (see Section 1.3)

Classification of Slope Failures

A classification of slope failures is given in Table 1.1. The most important classes are *falls, slides, avalanches*, and *flows*.

TABLE 1.1

A Classification of Slope Failures

Type	Form	Definition
Falls	Free fall	Sudden dislodgment of single or multiple blocks of soil or rock which fall in free descent.
	Topple	Overturning of a rock block about a pivot point located below its center of gravity.
Slides	Rotational or slump	Relatively slow movement of an essentially coherent block (or blocks) of soil, rock, or soil–rock mixtures along some well-defined arc-shaped failure surface.
	Planar or translational	Slow to rapid movement of an essentially coherent block (or blocks) of soil or rock along some well-defined planar failure surface.
	Subclasses	
	Block glide	A single block moving along a planar surface.
	Wedges	Block or blocks moving along intersecting planar surfaces.
	Lateral spreading	A number of intact blocks moving as separate units with differing displacements.
	Debris slide	Soil–rock mixtures moving along a planar rock surface.
Avalanches	Rock or debris	Rapid to very rapid movement of an incoherent mass of rock or soil–rock debris wherein the original structure of the formation is no longer discernible, occurring along an ill-defined surface.
Flows	Debris Sand Silt Mud Soil	Soil or soil–rock debris moving as a viscous fluid or slurry, usually terminating at distances far beyond the failure zone; resulting from excessive pore pressures (subclassed according to material type).
Creep		Slow, imperceptible downslope movement of soil or soil–rock mixtures.
Solifluction		Shallow portions of the regolith moving downslope at moderate to slow rates in Arctic to sub-Arctic climates during periods of thaw over a surface usually consisting of frozen ground.
Complex		Involves combinations of the above, usually occurring as a change from one form to another during failure with one form predominant.

Major factors of classification include:

- *Movement form*: Fall, slide, slide flow (avalanche), flow
- *Failure surface form*: Arc-shaped, planar, irregular, ill-defined
- *Mass coherency*: Coherent, with the original structure essentially intact although dislocated, or incoherent, with the original structure totally destroyed
- *Constitution*: Single or multiple blocks, or a heterogeneous mass without blocks, or a slurry
- *Failure cause*: Tensile strength or shear strength exceeded along a failure surface, or hydraulic excavation, or excessive seepage forces

Other factors to consider include:

- *Mass displacement*: Amount of displacement from the failure zone, which can vary from slight to small, to very large. Blocks can move together with similar displacements, or separately with varying displacements.
- *Material type*: Rock blocks or slabs, soil–rock mixtures (debris), sands, silts, blocks of overconsolidated clays, or mud (weak cohesive soils).
- *Rate of movement during failure*: Varies from extremely slow and barely perceptible to extremely rapid as given in Table 1.2.

TABLE 1.2
Velocity of Movement for Slope Failure Forms[a]

Velocity (m/s)	Movement	Rate	Stage	Classification
10^2, 10	Extremely rapid		Final stage (Planar slide-rock mass)	Avalanches and flows
		3 m/sec		
1, 10^{-1}, 10^{-2}	Very rapid		Final stage (Planar slide-rock mass)	Avalanches and flows
		0.3 m/min		
10^{-3}, 10^{-4}	Rapid		Final stage (Rotational slide)	Slides
		1.5 m/day		
10^{-5}, 10^{-6}	Moderate		Final stage (Rotational slide)	Slides
		1.5 m/month		
10^{-7}	Slow		Intermediate (Rotational slide)	Slides
		1.5 m/year		
10^{-8}	Very slow		Early stage	Creep
		0.3 m/5 year		
10^{-9}	Extremely slow		Early stage	Creep

[a] After Varnes, D. J., *Landslides and Engineering Practice*, Eckel, E. B., Ed., Highway Research Board Special Report No. 29, Washington, DC, 1958. Reprinted with permission of the Transportation Research Board.

Slope Failure Forms Related to Geologic Conditions

Anticipation of the form of slope failure often can be based on geologic conditions as summarized in Table 1.3. Detailed descriptions of the various forms are given in Section 1.2. Some forms of falls and slides in rock masses are illustrated in Figure 1.1. and Figure 1.2, slides in soil formations in Figure 1.3, and avalanches and flows in rock, soil, and mixtures in Figure 1.4.

Slope Activity

Slope activity relates to the amount and rate of slope movement that occur. Some failure forms occur suddenly on stable slopes without warning, although many forms occur slowly through a number of stages. Failure implies only that movement has occurred, but not necessarily that it has terminated; therefore, it is necessary to establish descriptive criteria for failure, or stability, in terms of stages. The amount and rate of movement vary with the failure stage for some failure forms.

Slide forms of failure may be classified by five stages of activity:

1. *Stable slope*: No movement has occurred in the past, or is occurring now.
2. *Early failure stage*: Creep occurs, with or without the development of tension cracks on the surface (see Figure 1.22). Slump form movement velocities are generally of the order of a few inches per year.

TABLE 1.3

Geologic Conditions and Typical Forms of Slope Failures

Geologic Condition	Typical Movement Forms
Rock masses: general	Falls and topples from support loss
	Wedge failure along joints, or joints, shears, and bedding
	Block glides along joints and shears
	Planar slide along joints and shears
	Multiplanar failure along joint sets
	Dry rock flow
Metamorphic rocks	Slides along foliations
Sedimentary rocks	Weathering degree has strong effect
Horizontal beds	Rotational, or a general wedge through joints and along bedding planes
Dipping beds	Planar along bedding contacts; block glides on beds from joint separation
Marine shales, clay shales	Rotational, general wedge, or progressive through joints and along mylonite seams
Residual and colluvial soils	Depends on stratum thickness
Thick deposit	Rotational, often progressive
Thin deposit over rock	Debris slide, planar; debris avalanche or flow
Alluvial soils	Depends on soil type and structure
Cohesionless	Runs and flows
Cohesive	Rotational or planar wedge
Stratified	Rotational or wedges, becoming lateral spreading in fine-grained soils
Aeolian deposits	Variable
Sand dunes or sheets	Runs and flows
Loess	Block glides: flows during earthquakes
Glacial deposits	Variable
Till	Rotational
Stratified drift	Rotational
Lacustrine	Rotational becoming progressive
Marine	Rotational to progressive: rotational becoming lateral spreading: flows

(a)

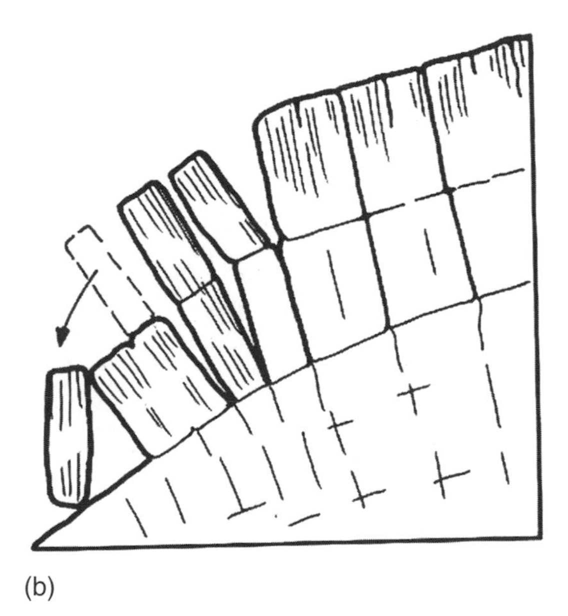

(b)

FIGURE 1.1
Forms of falls in rock masses: (a) free fall; (b) toppling by overturning.

3. *Intermediate failure stage*: Progressive slumps and scarps begin to form during rotational slides, and blocks begin to separate during planar slides, as tension cracks grow in width and depth. Movement velocities may range up to about 2 in./day (5 cm/day), accelerating during rainy seasons and storms and diminishing during dry periods. Movement is affected also by flooding, high tides, and earthquake forces. The slope is essentially intact and may remain in this condition for many years (see Figure 1.89).
4. *Partial total failure*: A major block or portion of the unstable mass has moved to a temporary location leaving a large scarp on the slope (Figure 1.25a).
5. *Complete failure*: The entire unstable mass has displaced to its final location (see Figure 1.93), moving rapidly at rates of about 3 ft/min (1m/min) for the case of

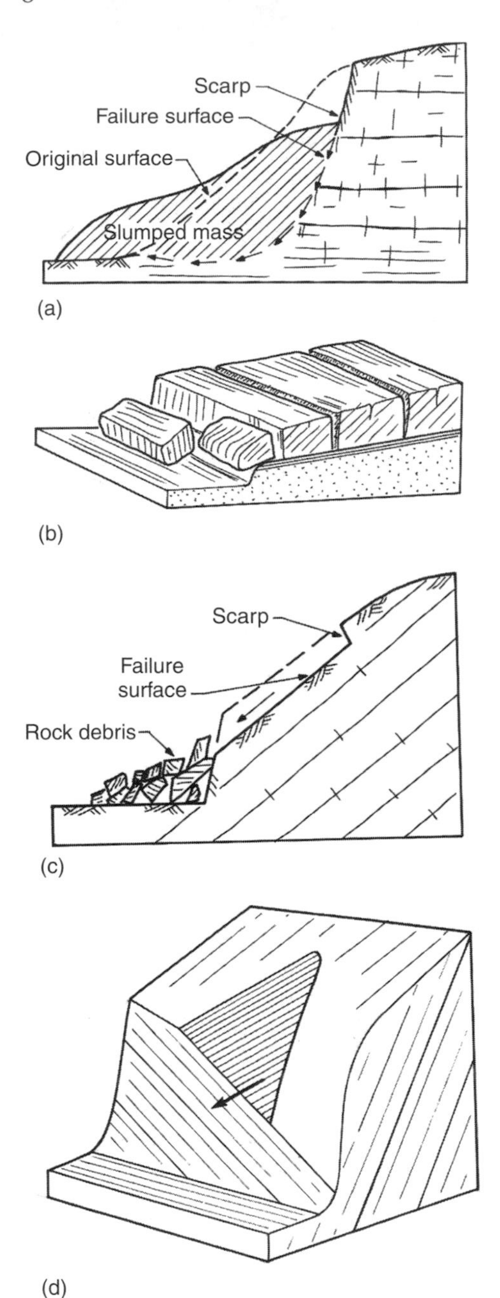

FIGURE 1.2
Slide forms in rock masses. (a) Rotational slide failed through joints and weak basal horizontal bed. (b) Translational sliding of blocks along a weak planar surface such as shale. (c) Planar slide failed along steeply dipping beds after cutting along lower slope. (d) Wedge failure scar. Failure occurred along intersecting joints and bedding planes when cut was made in obliquely dipping beds (see Sections 1.2.3 and 1.2.4).

rotational slides (Table 1.2). Planar slides in rock masses commonly reach velocities of 10 to 50 m/h (Banks and Strohn, 1974). Large planar slides in rock masses can achieve tremendous velocities, at times of the order of 200 m/h, as has been computed for the Vaiont slide (see Section 1.2.3). Habib (1975) considers these high velocities to be the result of movement of the rock mass over a cushion of water that negates all frictional resistance. The cushion is caused by heat, generated by shearing forces, which vaporizes the pore water. Such velocities are the major reason for the often disastrous effects of planar rock slides. Slide failures are usually progressive, and can develop into failure by lateral spreading, as well as into avalanches and flows.

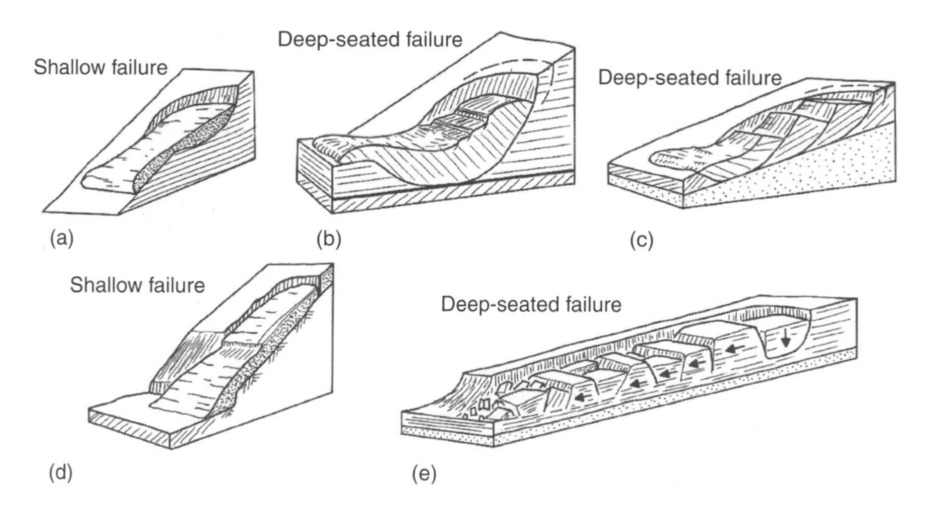

FIGURE 1.3
Slide forms in soil formations. (a) Single block failed along slope as a result of high groundwater level or strength increase with depth in cohesive soils. (b) Single block in homogeneous cohesive soils failed below toe of slope because of either a stronger or a weaker soil boundary at base. (c) Failure of multiple blocks along the contact with strong material. (d) Planar slide or slump in thin soil layer over rock. Often called debris slides. Common in colluvium and develop readily into flows. (e) Failure by lateral spreading. Occurs in glaciomarine or glaciolacustrine soils (parts a–c are rotational forms; parts d and e are planar or translational forms).

Avalanches and flows may develop from slide forms as mentioned above, or may undergo an early stage, but total final failure often occurs suddenly without warning on a previously stable slope as the result of some major event such as a very large rainfall or an earthquake. Velocities are usually very rapid to extremely rapid as given in Table 1.2.

Falls may occur suddenly, but often go through an early stage evidenced by the opening of tension cracks.

Deposition

Talus is rock debris at slope toes resulting from falling blocks. *Colluvium* is the residue of soil materials composing the soil mass, generally resulting from complete failure.

1.1.3 Rating the Hazard and the Risk

Significance

An existing or potential slope failure must be evaluated in terms of the degree of the hazard and the risk when plans for the treatment are formulated (see Section 1.4). Some conditions cannot be improved and should be avoided; in most, however, the hazard can be eliminated or reduced.

Hazard refers to the slope failure itself in terms of its potential magnitude and probability of occurrence.

Risk refers to the consequences of failure on human activities.

Hazard Degree

The rating basis for hazard is the potential magnitude and probability of failure.

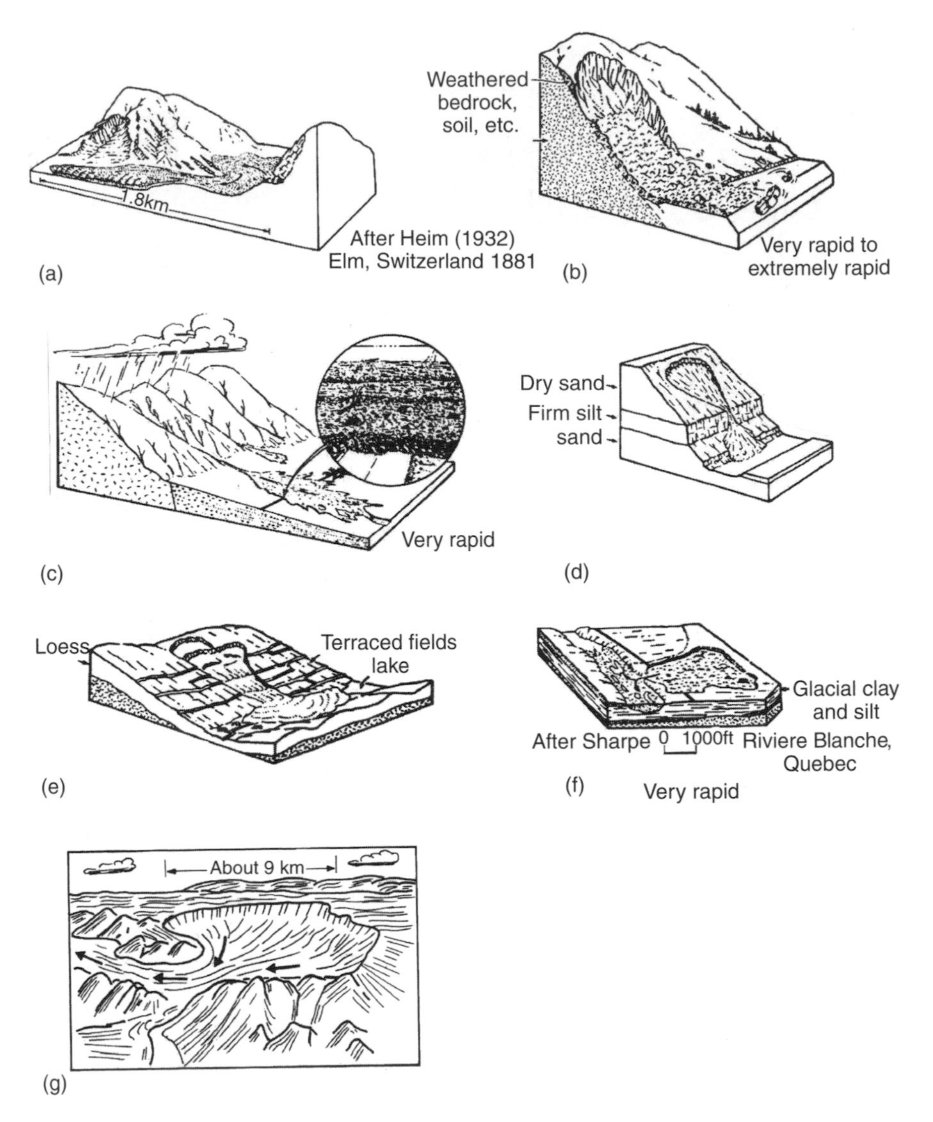

FIGURE 1.4
Avalanches and flows in rock, debris, and soil. (a) Rock fragment flow or rockfall avalanche. (This type of movement occurs only when large rockfalls and rockslides attain unusual velocity. Extremely rapid [more than 130 ft/sec] at Elm, Switzerland.) (b) Debris avalanche. (c) Debris flow. (d) Sand run: rapid to very rapid. (e) Dry loess flow caused by earthquake (Kansu Province, China, 1920). Extremely rapid movement. (f) Soil or mud flow. (g) Achacolla mud flow (La Paz, Bolivia). Huge mass of lacustrine soils slipped off the altiplano and flowed downstream for 25 km (see Figure 1.57). (Parts a–f from Varnes, D. J., *Landslides and Engineering Practice*, Eckel, E. B., Ed., Highway Research Board, Washington, DC, 1958. Reprinted with permission of the Transportation Research Board.)

- *Magnitude* refers to the volume of material which may fail, the velocity of movement during failure, and the land area which may be affected. It depends very much on the form of failure as related to geology, topography, and weather conditions.
- *Probability* is related in a general manner to weather, seismic activity, changes in slope inclinations, and other transient factors.

No hazard: A slope is not likely to undergo failure under any foreseeable circumstances.

Low hazard: A slope may undergo total failure (as compared with partial failure) under extremely adverse conditions which have a low probability of occurrence (for example, a 500 year storm, or a high-magnitude earthquake in an area of low seismicity), or the potential failure volume and area affected are small even though the probability of occurrence is high.

Moderate hazard: A slope probably will fail under severe conditions that can be expected to occur at some future time, and a relatively large volume of material is likely to be involved. Movement will be relatively slow and the area affected will include the failure zone and a limited zone downslope (moderate displacement).

High hazard: A slope is almost certain to undergo total failure in the near future under normal adverse conditions and will involve a large to very large volume of materials; or, a slope may fail under severe conditions (moderate probability), but the potential volume and area affected are enormous, and the velocity of movement very high.

Risk Degree

The rating basis for risk is the type of project and the consequences of failure.

No risk: The slope failure will not affect human activities.

Low risk: An inconvenience easily corrected, not directly endangering lives or property, such as a single block of rock of small size causing blockage of a small portion of roadway and easily avoided and removed.

Moderate risk: A more severe inconvenience, corrected with some effort, but not usually directly endangering lives or structures when it occurs, such as a debris slide entering one lane of a roadway and causing partial closure for a brief period until it is removed. Figure 1.5 illustrates a debris avalanche that closed a roadway for some days.

High risk: Complete or partial loss of a roadway or important structure, or complete closure of a roadway for some period of time, but lives are not necessarily endangered during the failure. Figure 1.6 illustrates a partial loss of a roadway. If failure continues it will result in total loss of the roadway and will become a *very high risk* for traffic.

Very high risk: Lives are endangered at the time of failure by, for example, the destruction of inhabited structures or a railroad when there is no time for a warning. The scars on the steep slopes in Figure 1.7 are the result of debris slides and avalanches resulting from roadway cuts upslope. The town shown on the lower right of the photo is located on the banks of a river. Concerns were from debris avalanches (1) filling and damming the river resulting in flooding of the town, and (2) falling on the town. Studies showed that the width, depth, and flow velocity of the river would remove any foreseeable volume of debris, and damming would not be expected. As long as the vegetation upslope of the town remained, the slope would be stable. Treatments were recommended to stabilize the areas upslope where failures had occurred. Therefore, the possible very high risk was reduced to low.

1.1.4 Elements of Slope Stability

General

Dependent Variables

Stated simply, slope failures are the result of gravitational forces acting on a mass which can creep slowly, fall freely, slide along some failure surface, or flow as a slurry. Stability can depend on a number of complex variables, which can be placed into four general categories as follows:

1. Topography — in terms of slope inclination and height
2. Geology — in terms of material structure and strength
3. Weather — in terms of seepage forces and run-off quantity and velocity
4. Seismic activity — as it affects inertial and seepage forces

FIGURE 1.5
Debris avalanche closes a roadway in Ecuador for some days. A temporary bypass was constructed and used until the debris was removed.

It is important to note that, although topography and geology are usually constant factors, there are situations where they are transient.

Mechanics of Sliding Masses

Masses that fail by sliding along some well-defined surface, moving as a single unit (as opposed to progressive failure or failure by lateral spreading), are the only slope failure form that can be analyzed mathematically in the present state of the art (see Section 1.3).

FIGURE 1.6
Partial loss of mountain roadway in Ecuador. Failure resulted from discharge from roadway drains causing downslope erosion. Additional failure will result in total loss of roadway and closure probably for months.

The diagrams given in Figure 1.8 illustrate the concept of failure that occurs when driving forces exceed resisting forces.

In the figure, the weight of mass *W* bounded by slice abc (in [a] acted on by the lever arm E; in [b] a function of the inclination of the failure surface) causing the driving force, is resisted by the shear strength *s* mobilized along the failure surface of length *L* (in case [a] acted on at "a" by lever arm R). The expression for factor of safety (FS) given in the figure is commonly encountered but is generally considered unsatisfactory because the resisting moment and the driving moment in (a) are ambiguous. For example, the portion of the rotating mass to the left of the center of rotation could be considered as part of the resisting moment. For this reason, FS is usually defined as

$$\text{FS} = \frac{\text{shearing strength available along sliding surface}}{\text{shearing stresses tending to produce failure along surface}}$$

The four major factors influencing slope stability are illustrated in Figure 1.9 and described in the following sections.

Slope Geometry (Figure 1.9a)

Significance

Driving forces and *runoff* are increased as slope inclination and height increase. Runoff quantity and velocity are related directly to amount of erosion, and under severe conditions cause "hydraulic excavation," resulting in avalanches and flows (see discussion of runoff below).

Inclination

Geologic formations often have characteristic inclinations at which they are barely stable in the natural state, for examples, residual soils at 30 to 40°, colluvium at 10 to 20°, clay shales at 8 to 15°, and loess, which often stands vertical to substantial heights.

FIGURE 1.7
Debris avalanches resulted from roadway construction on a steep slope in the Bolivian Andes. The small town of Pacallo in the photo lower right is located adjacent to a fast-flowing mountain stream. The major concern was with future avalanches damming the stream with debris resulting in flooding of the town.

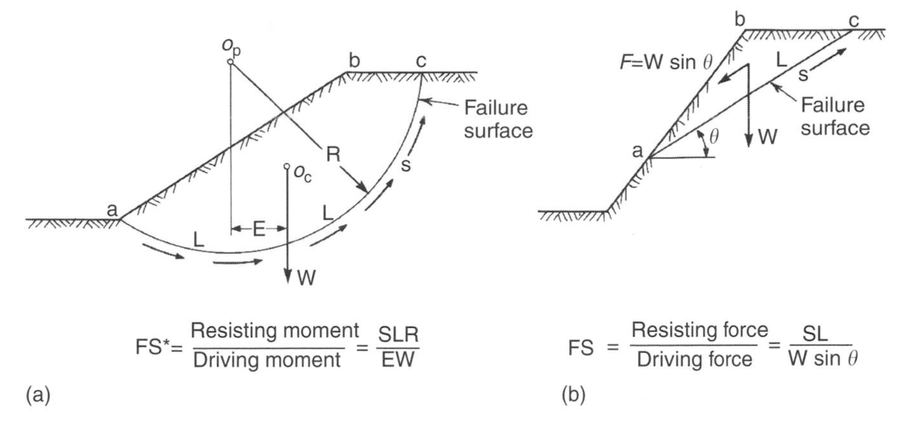

FIGURE 1.8
Forces acting on cylindrical and planar failure surfaces. (a) Rotational cylindrical failure surface with length L. Safety factor against sliding, FS. (b) Simple wedge failure on planar surface with length L. (*Note that the expression for FS is generally considered unsatisfactory; see text.)

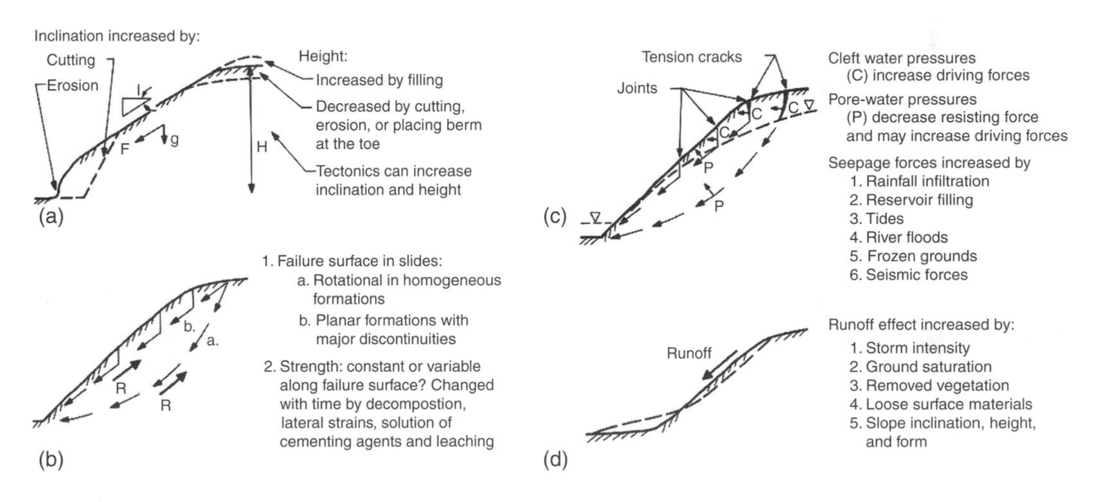

FIGURE 1.9
The major factors influencing slope stability: (a) increasing slope inclination and height increases the driving forces F; (b) geologic structure influences form and location of failure surface, material strength provides the resisting force R; (c) seepage forces reduce resisting forces along failure surface and increase driving forces in joints and tension cracks; (d) runoff quantity and velocity are major factors in erosion, avalanches, and flows.

Inclination is increased by:

- Cutting during construction, which should be controlled by analysis and judgment.
- Erosion, as a result of undercutting at the slope toe by wave or stream activity, of seepage exiting from the slope face, or of removal of materials by downslope runoff. All these are significant natural events.
- Tectonic movements in mountainous terrain, a very subtle and long-term activity which provides a possible explanation for the very large failures that occur from time to time and for which no other single explanation appears reasonable. An example is the disastrous rock slide at Goldau, Switzerland (see Section 1.2.3).

Slope Height

Slope height is increased by filling at the top, erosion below the toe, or tectonic activity. It is decreased by excavation and erosion at the top, or by placing a berm at the bottom. The driving forces are affected in failure forms where the limited slope condition applies (see Figure 1.8).

Material Structure (Figure 1.9b)

Significance

Material structure influences the failure form and the location and shape of the potential failure surface, and can be considered in two broad categories: uniform and nonuniform.

Uniform Materials

Uniform materials consist of a single type of soil or rock, essentially intact and free of discontinuities. From the aspect of slope stability, they are restricted to certain soil formations. Rotational failure is normal; the depth of the failure surface depends on the location of the phreatic surface and on the variation of strength with depth. Progressive failures are common, and falls and flows possible; flows are common in fine-grained granular soils.

Nonuniform Materials

Formations containing strata of various materials, and discontinuities represented by bedding, joints, shears, faults, foliations, and slickensides are considered nonuniform. The controlling factor for stability is the orientation and strength of the discontinuities, which represent surfaces of weakness in the slope.

Planar slides occur along the contacts of dipping beds of sedimentary rock and along joints, fault and other shear zones, slickensides, and foliations. Where a relatively thin deposit of soil overlies a sloping rock surface, progressive failure is likely and may develop into a debris avalanche. Along relatively flat-lying strata of weak material, failure can develop progressively in the form of lateral spreading, and can develop into a flow.

Rotational slides occur in horizontally bedded soil formations, and in certain rock formations such as clay shales and horizontally bedded sedimentary rocks.

Falls occur from lack of tensile strength across joints in overhanging or vertical rock masses. Changes in the orientation of the discontinuities with respect to the slope face occur normally as a result of excavation, but can also be caused by tectonic activity. Joint intensity can be affected by construction blasting.

Material Strength (Figure 1.9b)

Significance

Material strength provides the resisting forces along a surface of sliding. It is often neither the value determined by testing, nor the constant value assumed in analysis.

Variations along the Failure Surface

Slopes normally fail at a range of strengths, varying from peak to residual, distributed along the failure surface as a function of the strains. Slopes that have undergone failure in the past will have strengths at or near residual, depending upon the time for restitution available since failure.

Changes with Time

Chemical weathering is significant in residual soils and along discontinuities in rock masses in humid climates, and provides another possible explanation for the sudden failure of rock-mass slopes that have remained stable for a very long period of time under a variety of weather and seismic conditions.

Lateral strains in a slope tend to reduce the peak strength toward the residual, a significant factor in the failure of slopes in clay shales and some overconsolidated clays containing recoverable strain energy (Bjerrum, 1966), as well as in materials where slope movements have occurred.

Solution of Cementing Agents Reduces Strength.

Leaching of salts from marine clays increases their sensitivity and, therefore, their susceptibility to liquefaction and flow (Bjerrum et al., 1969).

Seepage Forces (Figure 1.9c)

Significance

Seepage forces may reduce the resisting forces along the failure surface or increase the driving forces.

Factors Causing Increased Seepage Forces

In general, seepage forces are increased by rainfall infiltration or reservoir filling, which raises the water table or some other phreatic surface (perched water level); sudden drawdown of a flooded stream or an exceptionally high tide; melting of a frozen slope that had blocked seepage flow; and earthquake forces.

Rising groundwater level is a common cause. Variables affecting such a rise include rainfall accumulation and increase in ground saturation for a given period, the intensity of a particular storm, the type and density of ground vegetation, drainage characteristics of the geologic materials, and the slope inclination and other features of topographic expression. Vegetation, geology, and topography influence the amount of infiltration that can occur, and careful evaluation of these factors often can provide the reasons for failure to occur at a particular location along a slope rather than at some other position during a given storm or weather occurrence.

Earthquake forces can cause an increase in pore-air pressures, as well as porewater pressures. Such an increase is believed to be the cause of the devastating extent of the massive landslides in loess during the 1920 earthquake in Kansu, China, which left 200,000 or more dead.

Runoff (Figure 1.9d)

Significance

The quantity and velocity of runoff are major factors in erosion, and are a cause of avalanches and flows. Storm intensity, ground saturation, vegetation, frozen ground, the nature of the surficial geologic materials, and slope inclination and other topographic features affect runoff.

Hydraulic Excavation

Many avalanches and flows are caused by hydraulic excavation during intense storms, a common event in tropical and semiarid climates. Water moving downslope picks up soils loosened by seepage forces, and as the volume and velocity increase, the capacity to remove more soil and even boulders increases, eventually resulting in a heavy slurry which removes everything loose in its path as it flows violently downslope. The scar of a debris avalanche is illustrated in Figure 1.10. Failure could hardly have been foreseen at that particular location along the slope, since conditions were relatively uniform.

FIGURE 1.10
Exposed rock surface remaining after runoff from torrential rains removed all vegetation, soil, and loose rock, depositing the debris mass at the toe of the slope (BR 116, km 56, Teresopolis, R. J., Brazil).

1.2 Slope Failure Form Characteristics

1.2.1 Creep

General

Creep is the slow, imperceptible deformation of slope materials under low stress levels, which normally affects only the shallow portion of the slope, but can be deep-seated where a weak zone exists. It results from gravitational and seepage forces, and is indicative of conditions favorable for sliding.

Recognition

Creep is characteristic of cohesive materials and soft rock masses on moderately steep to steep slopes. Its major surface features are parallel transverse slope ridges ("cow paths") as illustrated in Figure 1.11, and tilted fence posts, poles, and tree trunks. Straight tilted tree trunks indicate recent movement (Figure 1.12), whereas bent tree trunks indicate old continuing movement (Figure 1.13) (see Section 1.5.2, Dating Relict Slide Movements).

1.2.2 Falls

General

Falls are the sudden failures of vertical or near-vertical slopes involving single or multiple blocks wherein the material descends essentially in free fall. Toppling, or overturning of rock blocks, often results in a fall.

In soils, falls are caused by the undercutting of slopes due to stream or wave erosion, usually assisted by seepage forces. In rock masses, falls result from undercutting by erosion or human excavation; increased pressures in joints from frost, water, or expanding materials; weathering along joints combined with seepage forces; and differential weathering wherein less-resistant beds remove support from stronger beds.

Their engineering significance lies normally in the occurrence of a single or a few blocks falling on a roadway, or occasionally encountering structures on slopes. At times, however, they can be massive and very destructive as shown in Figure 1.14.

Recognition

Falls are characteristic of vertical to near-vertical slopes in weak to moderately strong soils and jointed rock masses. Before total failure some displacement often occurs, as indicated by tension cracks; after total failure, a fresh rock surface remains and talus debris accumulates at the toe.

FIGURE 1.11
Creep ridges and erosion in residual soils after removal of vegetation (state of Rio de Janeiro, Brazil).

FIGURE 1.12
Trees bent in the lower portions and then growing straight up indicate long-term slope movements. The scarp in the photo is the head of a progressive failure in marine shales extending downslope for over a kilometer near Bandung, Java.

1.2.3 Planar Slides in Rock Masses

General

Forms of planar slides in rock masses include:

- Block glide involving a single unit of relatively small size (photo, Figure 1.15).
- Slab glide involving a single unit of relatively small to large size (photo, Figure 1.16).
- Wedge failures along intersecting planes involving single to multiple units, small to very large in size (Figure 1.2d). A small wedge failure is illustrated in Figure 1.81.
- Translational slide: Sliding as a unit, or multiple units, downslope along one or more planar surfaces (Figure 1.2b). Failure often is progressive (Section 1.2.6).
- Massive rock slide involving multiple units, small to very large in size, often with very high velocities (Figure 1.17).

Block and slab slides can be destructive, but massive rock slides are often disastrous in mountainous regions and in many cases cannot be prevented, only avoided.

FIGURE 1.13
Tilting tree trunks on a creeping hillside of varved clays indicate relatively recent movement (Tompkins Cove, New York).

FIGURE 1.14
Rockfall destroyed a powerhouse (Niagara Falls, New York). Failure may actually be in the form of a huge topple. (Photo by B. Benedict, 1956.)

FIGURE 1.15
Small granite block glide (Rio de Janeiro, Brazil).

FIGURE 1.16
Exfoliation loosening granite slabs. Impact wall on right was constructed to deflect falling and sliding blocks from buildings on lower slopes. Damage from falls and slab slides is a serious problem in Rio de Janeiro.

FIGURE 1.17
The scar of the Gros Ventre slide as seen from the Gros Ventre River, Wyoming, in August 1977.

Recognition

Planar slides are characteristic of:

- Bedded formations of sedimentary rocks dipping downslope at an inclination similar to, or less than, the slope face. They result in block glides or massive rock slides (see Examples below).
- Faults, foliations, shears or joints forming long, continuous planes of weakness that intersect the face of the slope.
- Intersecting joints result in wedge failures, which can be very large in open-pit mines.
- Jointed hard rock results in block glides.
- Exfoliation in granite masses results in slab glides.

Surface features:

- Before total failure, tension cracks often form during slight initial displacement.
- After total failure, blocks and slabs leave fresh scarps. Massive rock slides leave a long fresh surface denuded of vegetation, varying in width from narrow to wide and with a large debris mass at the toe of the slope and beyond. Since they can achieve very high velocities, they can terminate far beyond the toe.

Examples of Major Failures

Goldau, Switzerland

In September 1806, a massive slab 1600 m long, 330 m wide, and 30 m thick broke loose and slid downslope during a heavy rainstorm, destroying a village and killing 457 persons. The slab consisted of Tertiary conglomerate with a calcareous binder resting on a 30° slope. At its interface with the underlying rock was a porous layer of weathered rock.

Three possible causes were offered by Terzaghi (1950):

1. The slope inclination gradually increased from tectonic movements.
2. The shearing resistance at the slab interface gradually decreased because of progressive weathering or from removal of cementing material.

3. The piezometric head reached an unprecedented value during the rainstorm. Terzaghi was hesitant to accept this as the only cause, since he considered it unlikely that in the entire geologic history of the region, there had not been a more severe storm. Therefore, he concluded that the slide resulted from two or more changing conditions.

Gros Ventre, Wyoming

On June 23, 1925, following heavy rains and melting snow, approximately 50 million yd^3 slid in a few minutes down the mountainside along the Gros Ventre River near Grand Teton National Park in Wyoming. The debris formed a natural dam as high as 250 ft which blocked the river, and resulted in a lake 3 mi long. Almost 2 years later, in May 1927, water from heavy rains and melting snow filled the reservoir, over-topped the natural dam, eroded a large channel, and released flood waters which resulted in a number of deaths.

The slide scar which is still evident in 1977, 52 years later, is illustrated in Figure 1.17. A geologic section is given in Figure 1.18. Failure occurred along clay layers in the carbonaceous Amsden formation, dipping downslope. It appears that water entered the joints and pores of the Tensleep sandstone saturated the clay seams, and reduced or eliminated the normal stresses.

Vaiont, Italy

On October 9, 1963, the worst dam disaster in history occurred when more than 300 million m^3 of rock slid into the reservoir formed by the world's highest thin-arch concrete dam causing a tremendous flood which overtopped the dam and flowed into the Piave River valley, taking some 2600 lives. The slide involved an area on the south side of the valley roughly 2.3 km in width and 1.3 km in length, as shown in Figure 1.19. The natural slope was of the order of 20 to 30°.

A geologic section is given in Figure 1.20. The valley had formed in the trough of a syncline, and the beds forming the limbs dipped downslope at inclinations a few degrees steeper than the slope. The south slope consisted of Jurassic sedimentary rocks, primarily limestones and marls occasionally interbedded with clay seams (bentonite clay at residual strength; Patton, F. D. and Hendron, A. J., unpublished). Tectonic activity had caused regional folding, faulting, and fracturing of strata, and some of the tectonic stresses

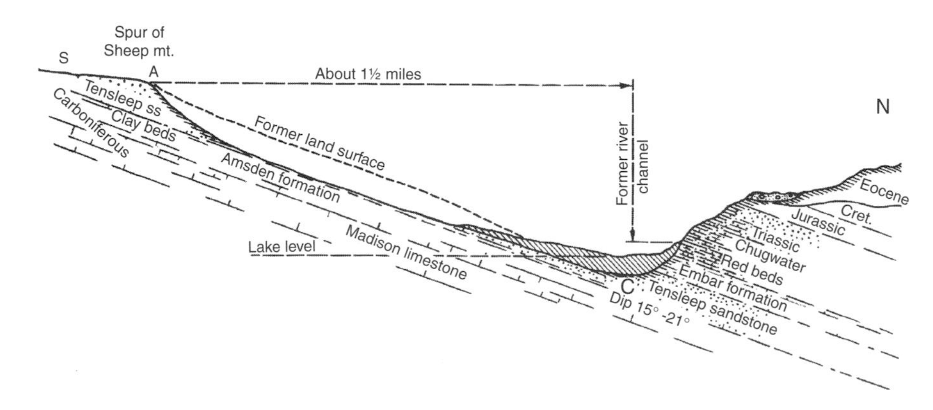

FIGURE 1.18

Section showing geologic conditions after the Gros Ventre landslide. The landslide dammed the Gros Ventre River. (From Alden, W. C., in *Focus on Environmental Geology*, Tank, R., Ed., Oxford University Press, New York [1973], 1928, pp. 146–153. With permission.)

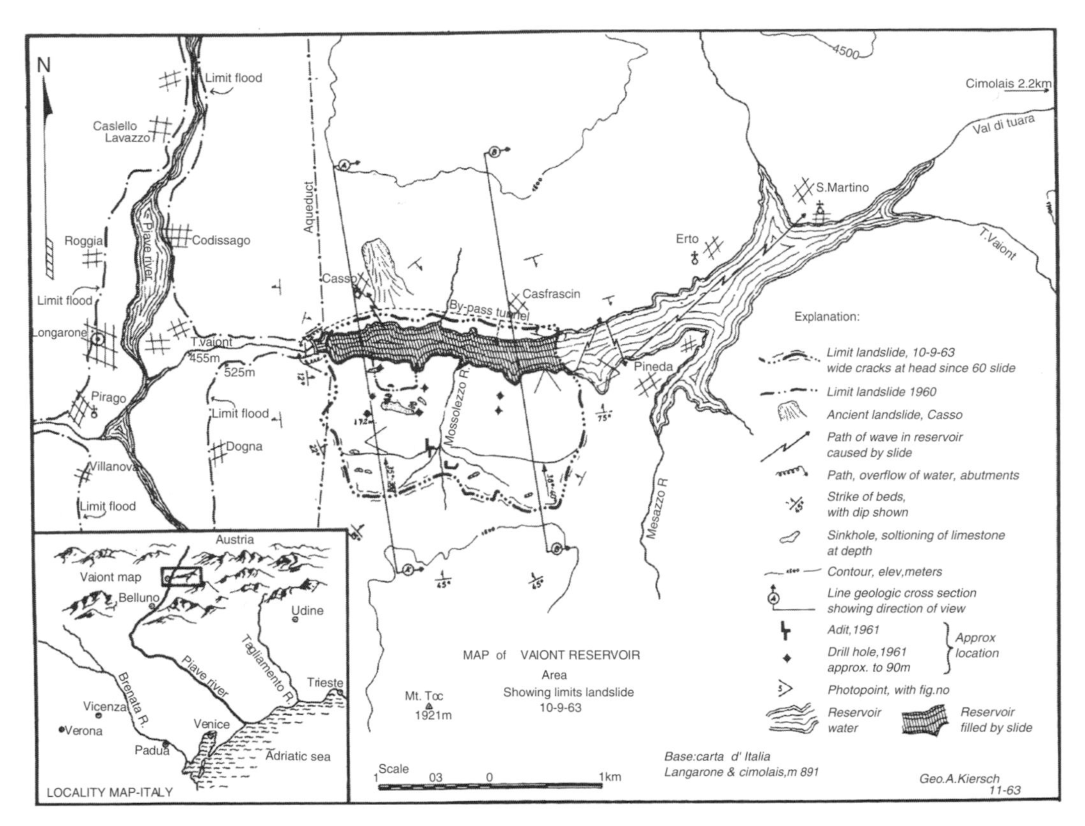

FIGURE 1.19
Map of Vaiont Reservoir slide. (From Kiersch, G. A., in *Focus on Environmental Geology*, Tank, R., Ed., Oxford University Press, New York [1973], Section 17, 1965, pp. 153–164. With permission.)

probably remained as residual stresses in the mass. Erosion of the valley caused some stress relief of the valley walls, resulting in numerous rebound joints that produced blocky masses. In addition, groundwater had attacked the limestone, leaving cavities and contributing to the generally unstable conditions (Kiersch, 1965).

The slide history is given by Kiersch (1965). Large-scale slides had been common on the Vaiont valley slopes, and evidence of creep had been observed near the dam as early as 1960, when the dam was completed at its final height of 267 m. During the spring and summer of 1963, the slide area was creeping at the rate of 1 cm/week. Heavy rains occurred during August and September and movement accelerated to 1 cm/day. In mid-September, movement accelerated to 20 to 30 cm/day, and on the day of failure, 3 weeks later, it was 80 cm/day. Since completion of the dam, the pool had been filled gradually and the elevation maintained at about 50 m below the crest or lower. During September, the pool rose at least 20 m higher, submerged the toe of the sliding mass, and caused the groundwater level to rise in the sliding mass. Collapse was sudden and the entire mass to a depth of 200 m broke loose and slid to the valley floor in 30 to 60 sec, displacing the reservoir and causing a wave that rose as much as 140 m above reservoir level. The dam itself was only slightly damaged by the wall of water but was rendered useless.

Sliding was apparently occurring along the clay seams, but the actual collapse is believed to have been triggered by artesian pressures and the rising groundwater levels that decreased the effective weight of the sliding mass and, thereby, the resisting force at the toe.

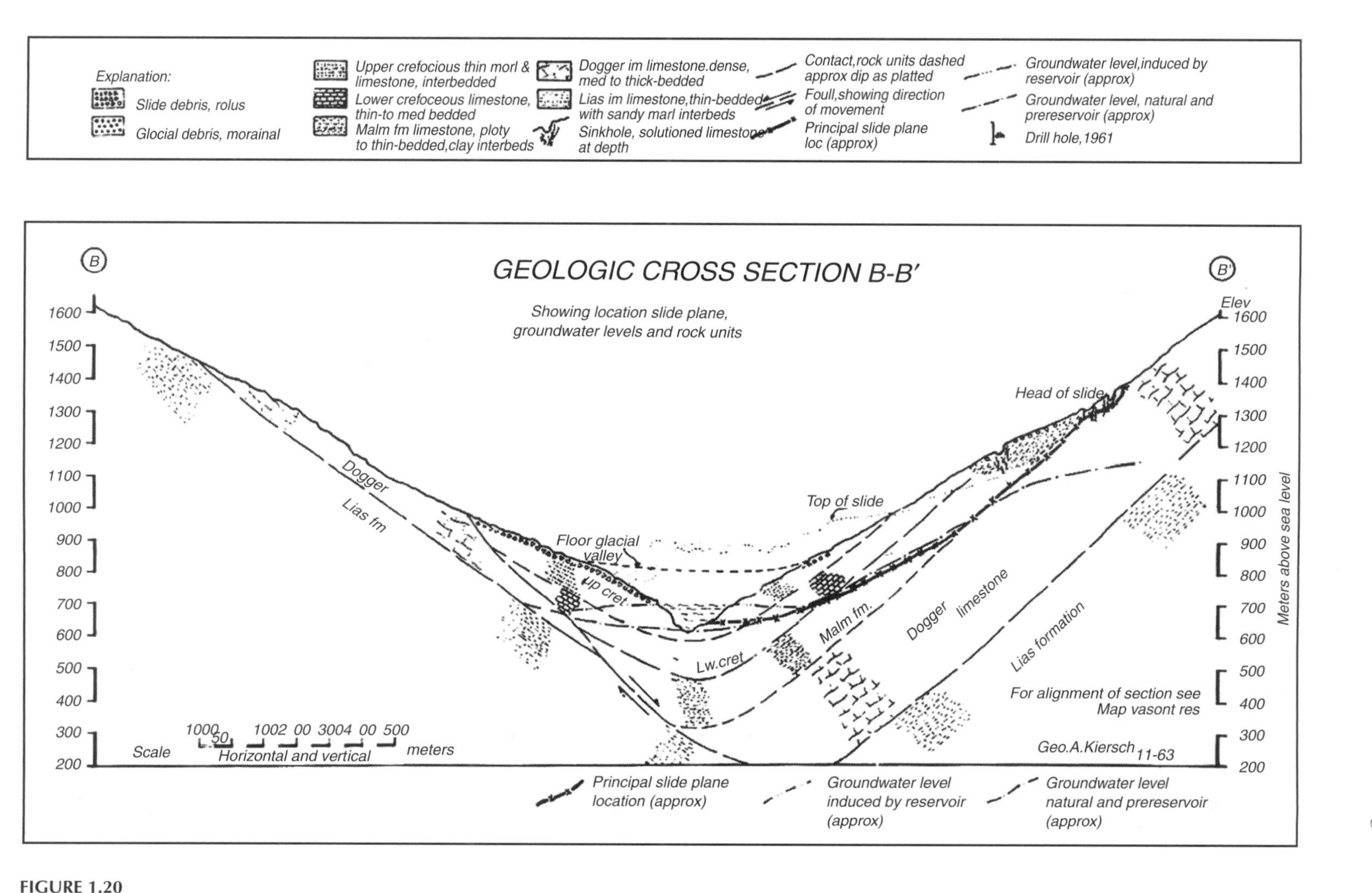

FIGURE 1.20
Geologic section through the Vaiont Reservoir slide. (From Kiersch, G. A., *Focus on Environmental Geology*, Tank, R., Ed., Oxford University Press, New York [1973], Section 17, 1965, pp. 153–164. With permission.)

1.2.4 Rotational Slides in Rock

General

In the rotational slide form, a spoon-shaped mass begins failure by rotation along a cylindrical rupture surface; cracks appear at the head of the unstable area, and bulging appears at the toe as the mass slumps (Figure 1.2a). At final failure, the mass has displaced substantially, and a scarp remains at the head (see Section 1.2.5 for nomenclature). The major causes are an increase in slope inclination, weathering, and seepage forces.

Recognition

Rotational slides are essentially unknown in hard-rock formations, but are common in marine shales and other soft rocks, and in heavily jointed stratified sedimentary rocks with weak beds.

Marine shales, with their characteristic expansive properties and highly fractured structure, are very susceptible to slump failures, and their wide geographic distribution makes such failures common. Natural slope angles are low, about 8 to 15°, and stabilization is often difficult. Failure is often progressive and can develop into large moving masses (see Section 1.2.6).

Stratified sedimentary rocks can on occasion result in large slides, and in humid climates slope failures can be common (Hamel, 1980) (see Example below).

Surface features before total failure are tension cracks; after total failure, a head scarp remains along with spoon-shaped slump topography (see Section 1.2.5).

Example of Major Failure

Event

At the Brilliant cut, Pittsburgh, Pennsylvania, on March 20, 1941, a rotational slide involving 120,000 yd^3 of material displaced three sets of railroad tracks and caused a train to be derailed (Hamel, 1972). A plan of the slide area is given in Figure 1.21b.

Geological conditions are illustrated on the section given in Figure 1.21a. The basal stratum, Zone 1, is described as "soft clay shale and indurated clay (a massive slickensided claystone)." The Birmingham shale of Zone 4 is heavily jointed vertically.

Slide History

In the 1930s, a large tension crack opened at the top of the slope. Sealing with concrete to prevent infiltration was unsuccessful in stopping movement and the crack continued to open over a period of several years. The rainfall that entered the slope through the vertical fractures normally drained from the slope along pervious horizontal beds. On the day of failure, which followed a week of rainfall, the horizontal passages were blocked with ice. Hamel (1972) concluded that final failure was caused by water pressure in the mass, and the failure surface was largely defined by the existing crack at the top of the slope and the weak basal stratum.

1.2.5 Rotational Slides in Soils

General

A common form of sliding in soil formations is the rotation about some axis of one or more blocks bounded by a more or less cylindrical failure surface (Figures 1.3a–c). The major causes are seepage forces and increased slope inclination, and relict structures in residual soils.

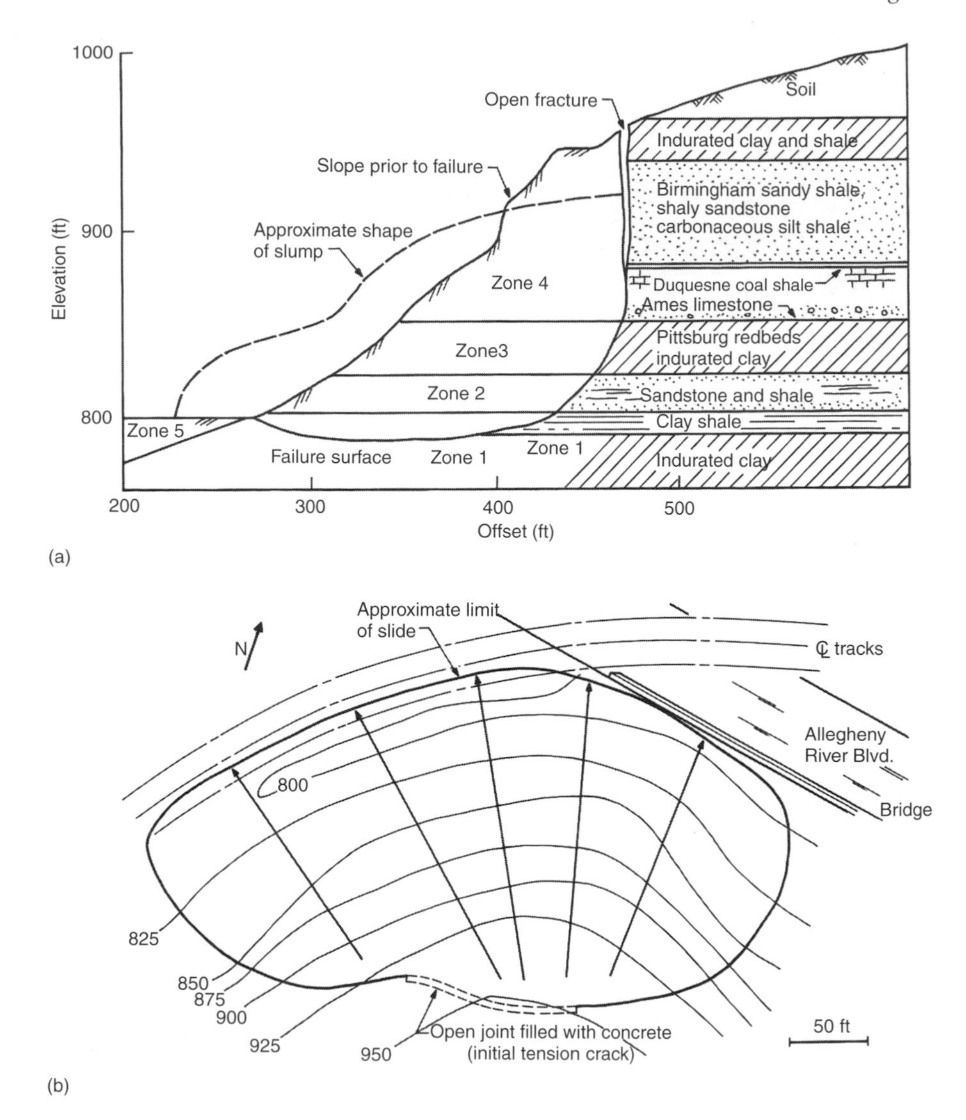

FIGURE 1.21
Rotational slide in rock at Brilliant Cut (Pittsburgh, Pennsylvania, March 28, 1941). (a) Generalized section through Brilliant Cut; (b) plan of slide area. (From Hamel, J. V., *Proceedings of the ASCE, 13th Symposium on Rock Mechanics*, Urbana, Illinois [1971], 1972, pp. 487–572. With permission.)

Usually, neither the volume of mass involved nor the distance moved is great; therefore, the consequences are seldom catastrophic although slump slides cause substantial damage to structures. If their warning signs are recognized they usually can be stabilized or corrected.

Recognition

Occurrence

Slump or rotation slides are characteristic of relatively thick deposits of cohesive soils without a major weakness plane to cause a planar failure. The depth of the failure surface varies with geology.

Deep-seated failure surfaces are common in soft to firm clays and glaciolacustrine, and glaciomarine soils. Deep to shallow failure surfaces are common in residual soils, depending on the strength increase with depth and relict rock defects. Relatively shallow failure surfaces are characteristic of colluvial soils.

Surface Features

During *early failure stages* tension cracks begin to form as shown in Figure 1.22 and Figure 1.23. After *partial failure*, in a progressive mode, the slope exists as a series of small slumps and scarps with a toe bulge as shown in Figure 1.24, or it may rest with a single large scarp and a toe bulge as illustrated in Figure 1.25(a). After *total failure*, surface features include a large head scarp and a mass of incoherent material at the toe as shown in Figure 1.25(b) and Figure 1.93.

Slump landforms remaining after total failures provide forewarning of generally unstable slope conditions. They include spoon-shaped irregular landforms, as seen from the air (Figure 1.26 and Figure 1.27), cylindrical scarps along terraces and water courses, and hummocky and irregular surfaces, as seen from the ground (Figure 1.28 and Figure 1.29). In the stereo-pair of aerial photos shown in Figure 1.26, the slump failure mass has stabilized temporarily but probably will reactivate when higher than normal seasonal rainfall arrives. A small recent failure scar exists along the road in the center of the slide mass. The rounded features of the mass, resulting from weathering, and vegetation growth indicate that the slide is probably 10 to 15 years old, or more. In the photo, it can be seen that the steep highway cut on the opposite side of the valley appears stable, indicative of different geologic conditions. In general, the geology consists of residual soils derived from metamorphic rocks in a subtropical climate. In Figure 1.27, an old slump scar in residual soils, weathering has strongly modified the features. In the photo, the tongue lobe at the intersection of the trails and the creep ridges are to be noted. The location is near the slide of Figure 1.26.

1.2.6 Lateral Spreading and Progressive Failure

General

Failure by lateral spreading is a form of planar failure which occurs in both soil and rock masses. In general, the mass strains along a planar surface, such as shown in Figure 1.3e, represent a weak zone. Eventually, blocks progressively break free as movement retrogresses toward the head. The major causes are seepage forces, increased slope inclination and height, and erosion at the toe.

Failure in this mode is essentially unpredictable by mathematical analysis, since one cannot know at what point the first tension crack will appear, forming the first block. Nevertheless, the conditions for potential instability are recognizable, since they are characteristic of certain soil and rock formations. Failure usually develops gradually, involving large volumes, but can be sudden and disastrous. Under certain conditions, it is unavoidable and uncontrollable from the practical viewpoint, and under other conditions control is difficult at best.

Recognition

The failure mode is common in river valleys, particularly where erosion removes material from the river banks. Characteristically, occurrence is in stiff fissured clays, in clay shales, and in horizontal or slightly dipping strata with a continuous weak zone such as those that occur in glaciolacustrine and glaciomarine soils. Colluvium over gently sloping residual soils or rock also fails progressively in a form of lateral spreading.

Surface features are characterized during the early stages by tension cracks, although failure can be sudden under certain conditions such as earthquake loadings. During the

(a)

(b)

FIGURE 1.22
(a) Small scarp along tension crack appears in photo (middle right). Small highway cut is far below to the left. Scarp appeared after soil was removed from small slump failure at the toe (BR 101, Santa Catarina, Brazil). Movement is in residual soil. If uncorrected, a very large failure will develop. (b) Tension crack in the same slope found in another location.

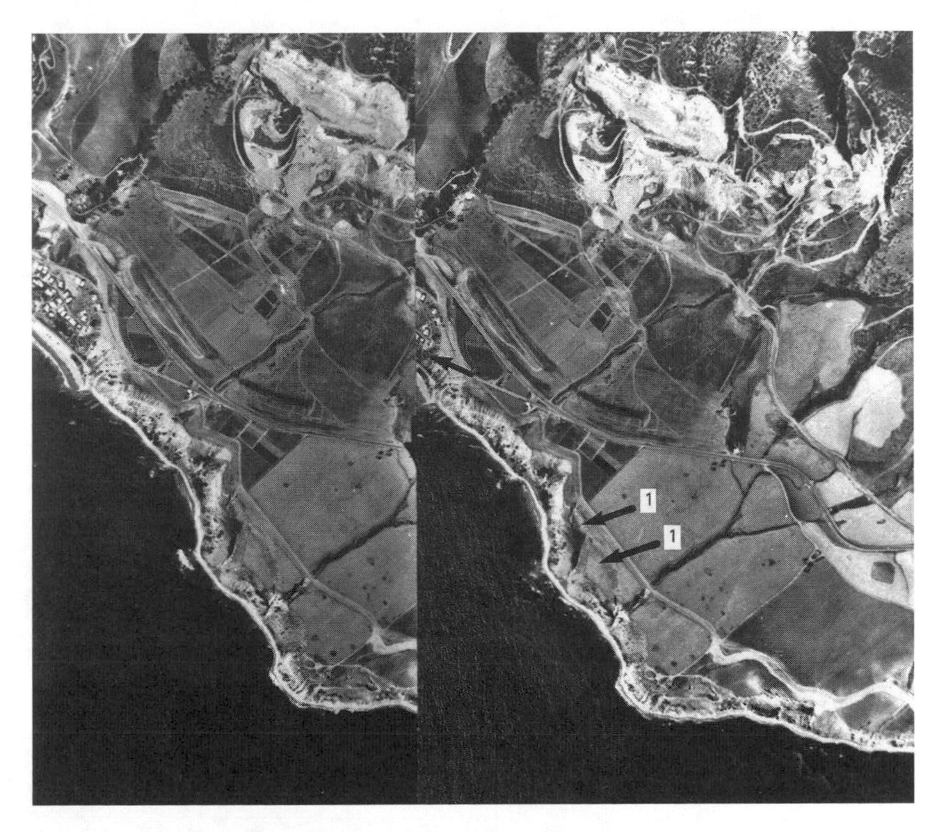

FIGURE 1.23
Stereo-pair of aerial photos showing tension cracks of incipient slides, such as at (1) along the California coast, a short distance from Portuguese Bend.

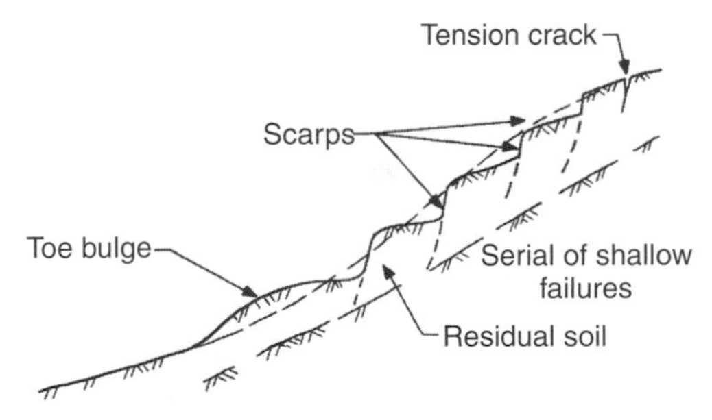

FIGURE 1.24
During the intermediate stage (during partial failure), residual soils often fail progressively, forming a series of slumps in tropical climates. Blocks move downhill during rainy periods and stabilize during dry periods.

progressive failure tension cracks open and scarps form, separating large blocks. The cracks can extend far beyond the slope face when a large mass goes into tension, even affecting surface structures as shown in Figure 1.30. Final failure may not develop for many years, and when it occurs it may be in a form resembling a large slump slide, or it may develop into a flow with individual blocks floating in a highly disturbed mass, depending upon natural conditions as described in the examples below.

Failure Examples

Marine Shales: General

Clay shales, particularly those of marine origin, are susceptible to several modes of slope failures as shown in Figure 1.31, of which progressive failure involves the largest volumes

FIGURE 1.25
Slump failure occurred after cutting in fine-grained glacial till (Mountainside, New Jersey). (a) Head scarp, toe bulge, and seepage at the toe. (b) Total failure some weeks later. Slope stabilized by benching, installation of trench drain, and counterberm along the toe.

and can be the most serious from the engineering viewpoint. Their most significant characteristics are their content of montmorillonite and their high degree of overstress. Excavation, either natural or human, results in lateral strains causing the strength along

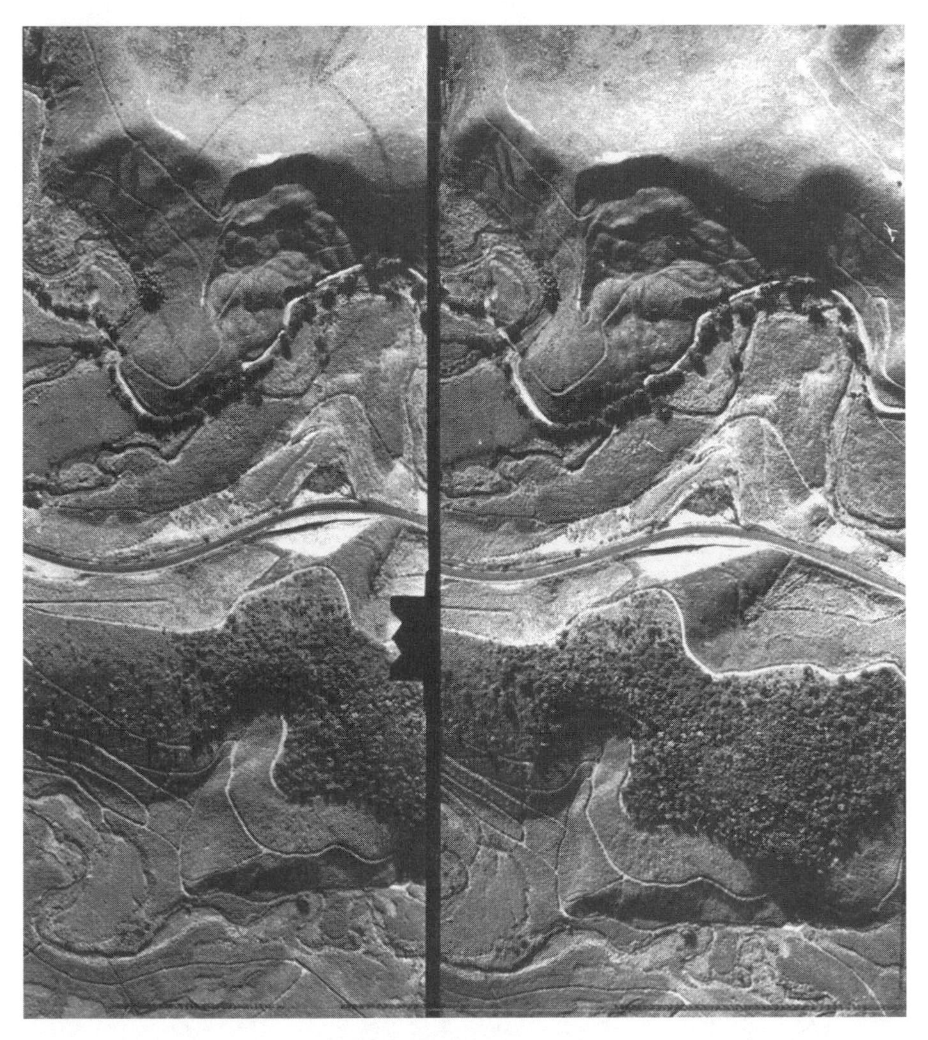

FIGURE 1.26
Stereo-pair of aerial photos showing *slump failure landform* (scale 1:8000). (From Hunt, R. E. and Santiago, W. B., *Proceedings of the 1st Congress Brasileiro de Geologia de Engenharia*, Rio de Janeiro, August, Vol. 1, 1976, pp. 79–98. With permission.)

certain planes to be reduced to residual values. Water entering the mass through open tension cracks and fractures assists in the development of failure conditions.

Marine Shale: Forest City Landslide

The Forest City Landslide, located on the banks of the Oahe Reservoir in South Dakota, includes an area of about 700 acres (Hunt et al., 1993). The hummocky landform, typical of marine shales and a large head scarp, is shown in the aerial oblique of Figure 1.32. Movements toward the reservoir, of the order of several inches or more per year, were causing distress in a large bridge structure. Investigations, including test borings and inclinometer data, identified the main failure surface at depths of the order of 100 ft, extending upslope to the head scarp, a distance of 2200 ft. The approach roadway embankment was moving laterally on shallower failure surfaces. A geologic section is given in Figure 1.33.

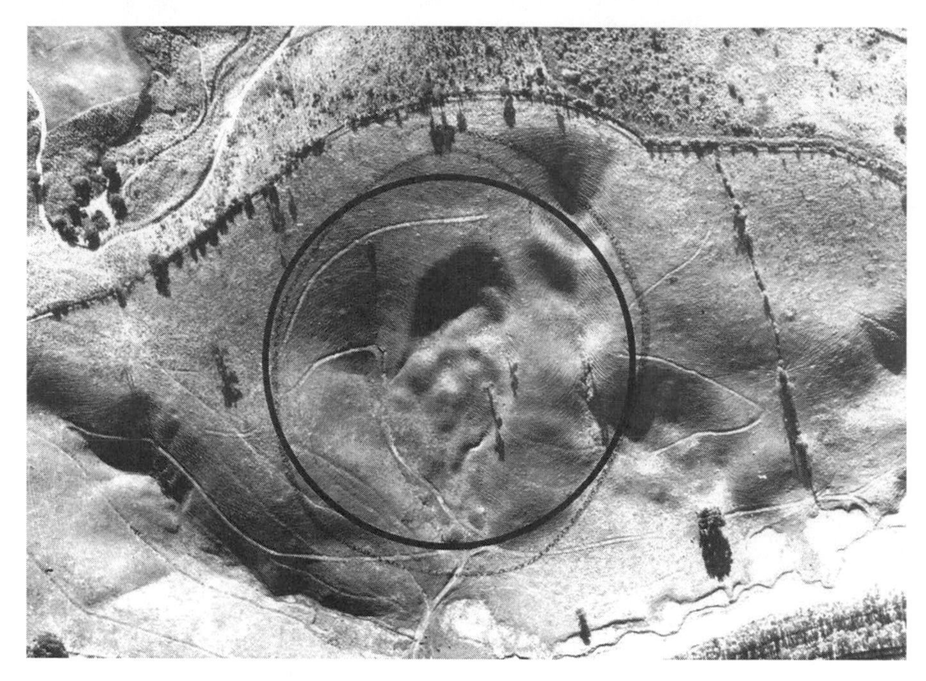

FIGURE 1.27
Old slump slide in residual soils located near slide in Figure 1.26.

FIGURE 1.28
Slump landform in glaciolacustrine soils showing shallow slopes, creep ridges, and seepage (Barton River, Vermont). Trees in upper left are growing on slide area. Slope failures are common in this region in the spring when the ground thaws and rains arrive.

FIGURE 1.29
Slump-slide landform (valley of the Rio Choqueyapa, La Paz, Bolivia). High center scarp in strong sands and gravels remains after failure of underlying lacustrine soils. Slopes were extremely unstable prior to channelization of the river, because of river erosion and flood stages. Grading of old slide in upper left is not arresting slope movements as evidenced by cracks in new highway retaining wall (not apparent in photo). Slope failures continue to occur from time to time throughout the valley (photo taken in 1973).

FIGURE 1.30
One-year-old church being split into half from slope movements although located over 1 km from the slope shown in Figure 1.29 (La Paz, Bolivia, 1972).

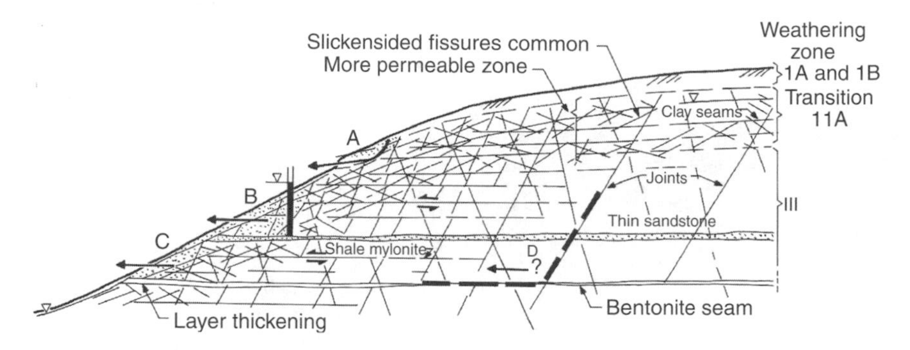

FIGURE 1.31
Failure forms in weathered clay shales: (A) surface slump in shallow weathered zone; (B) wedge failure along joints and sandstone seam; (C) wedge failure along thin bentonite seam may develop into large progressive failure to (D) or beyond. (From Deere, D. U., and Patton, F. D., *Proceedings of ASCE, 4th Pan American Conference on Soil Mechanics and Foundation Engineering*, San Juan, P. R., 1971, pp. 87–170. With permission.)

FIGURE 1.32
Aerial oblique, Forest City Landslide, South Dakota. Note the head scarp and hummocky landform. (Photo by Vermon Bump, SDDOT.)

Slope failures probably began in early postglacial times when the Missouri River incised its channel. Modern reactivation was caused by the filling of the valley with the reservoir, and subsequent relatively rapid changes in reservoir water levels. The failing mass consisted of a number of blocks, evidenced by surface tension cracks.

Stabilization of the overall sliding mass was essentially achieved by excavating a large cut at the escarpment at the head and relocating the approach roadway into the cut. The approach embankment, failing separately, was remediated by the installation of reinforced concrete "dowels."

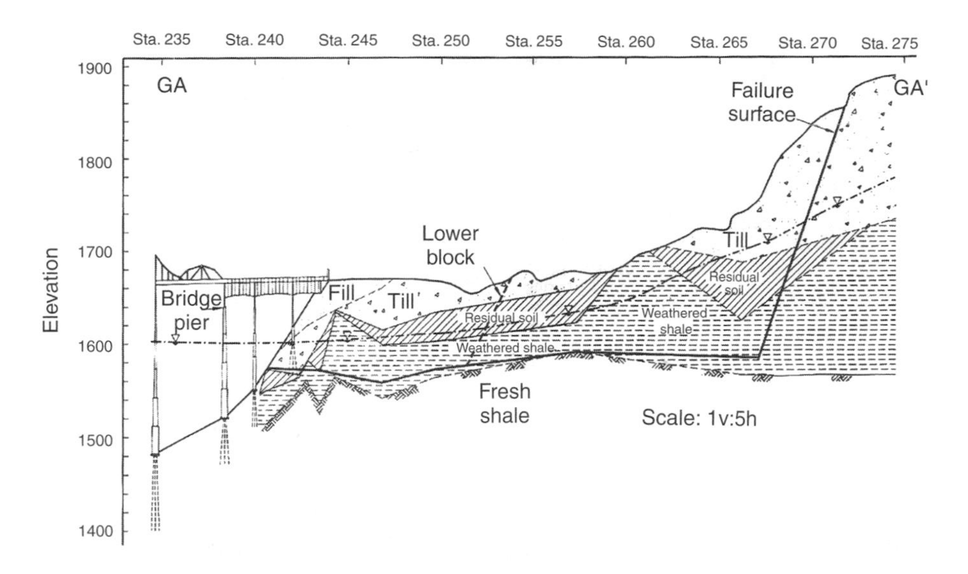

FIGURE 1.33
Geologic section, Forest City Landslide, South Dakota (From Hunt, R. E. et al., *3rd International Conference, Case Histories in Geotechnical Engineering*, St. Louis, MD, 1993. With permission.)

Marine Shale: Panama Canal Slides

Event: Massive slides occurred during 1907 and 1915 in the excavation for the Panama Canal in the Culebra Cut (Binger, 1948; Banks, 1972).

Geology: On the plan view of the slide areas (Figure 1.34), the irregular to gentle topography of the Cucaracha formation (Tertiary) is apparent. The Cucaracha is a montmorillonitic shale with minor interbedding of sandstone and siltstone more or less horizontally bedded but occasionally dipping and emerging from natural slopes. It is heavily jointed and slickensided, and some fractures show secondary mineral fillings. Natural slopes in the valley were relatively gentle, as shown on the geologic section given in Figure 1.35, generally about 20° or less. Laboratory consolidation tests gave values for preconsolidation pressure as high as 200 tsf.

Slide history: Excavations of the order of 300 ft in depth were required in the Cucaracha formation. Some minor sliding occurred as the initial excavations were made on slopes of 1:1 through the upper weathered zones to depths of about 50 ft. The famous slides began to occur when excavations reached about 100 ft. They were characterized by a buckling and heaving of the excavation floor, at times as great as 50 ft; a lowering of the adjacent ground surface upslope; and substantial slope movements. Continued excavation resulted in progressive sliding on a failure surface extending back from the cut as far as 1000 ft. The causes of the sliding are believed to be stress relief in the horizontal direction, followed by the expansion of the shale, and finally rupture along a shallow arc surface (Binger, 1948).

Analysis: Banks (1972) found that at initial failure conditions, the effective strength envelope yielded $\phi = 19°$ and $c' \approx 0$. For the case of an infinite slope (see Section 1.3.2) without slope seepage these values would produce a stable slope angle of 19°, or for the case of seepage parallel and coincident with the slope face, $1/2\phi$, or 1.5°. Since movements had occurred, the value 9.5° is considered to be the residual strength.

Solution: The slides were finally arrested by massive excavation and cutting the slopes back to 9.5° ($1/2\phi_r$), which is flatter than the natural slopes. Banks reported that measurements with slope inclinometers indicated that movement was still occurring in 1969, and that the depth of sliding was at an elevation near the canal bottom.

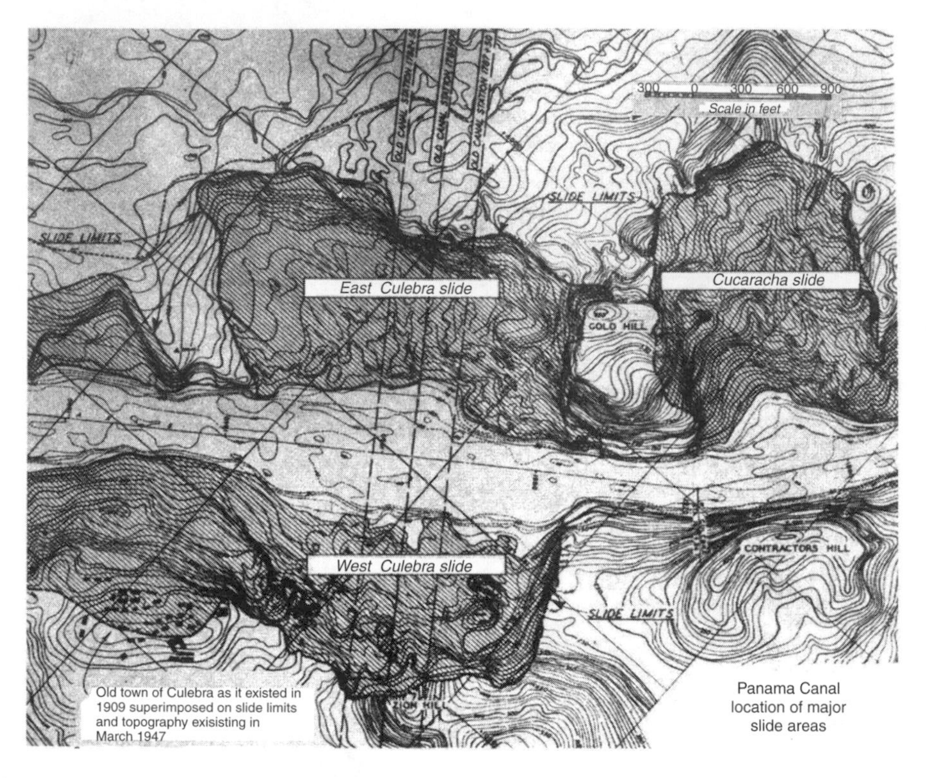

FIGURE 1.34
Plan view of slides and topography, Culebra Cut, Panama Canal. (From Binger, W. V., *Proceedings of the 2nd International Conference on Soil Mechanics and Foundation Engineering*, Rotterdam, Vol. 2, 1948, pp. 54–60. With permission.)

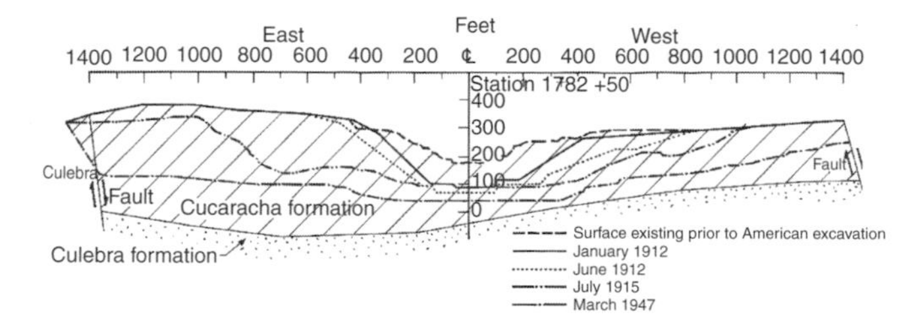

FIGURE 1.35
Sketch of east and west Culebra slides (Panama Canal) showing progress of slide movement: Cucaracha tuffaceous shale and Culebra tuffaceous shale, siltstone, and sandstone. (From Binger, W. V., *Proceedings of the 2nd International Conference on Soil Mechanics and Foundation Engineering*, Rotterdom, Vol. 2, 1948, pp. 54–60. With permission.)

Coastal Plain Sediments: Portuguese Bend Slide

Event: At Portuguese Bend, Palos Verdes Hills, California (see Figure 2.3 for location), a slide complex with a maximum width of roughly 4000 ft and a head-to-toe length of about 4600 ft began moving significantly in 1956 and, as of 1984, was still moving. Coastal plain sediments are involved, primarily marine shales. This slide may be classified as progressive block glides or failure by lateral spreading. It is one of the most studied active slides in the United States (Jahns and Vonder Linden, 1973).

Physiography: The limit of the slide area is shown in Figure 1.36, and the irregular hummocky topography is shown on the stereo-pair of aerial photos in Figure 1.37. In the slide

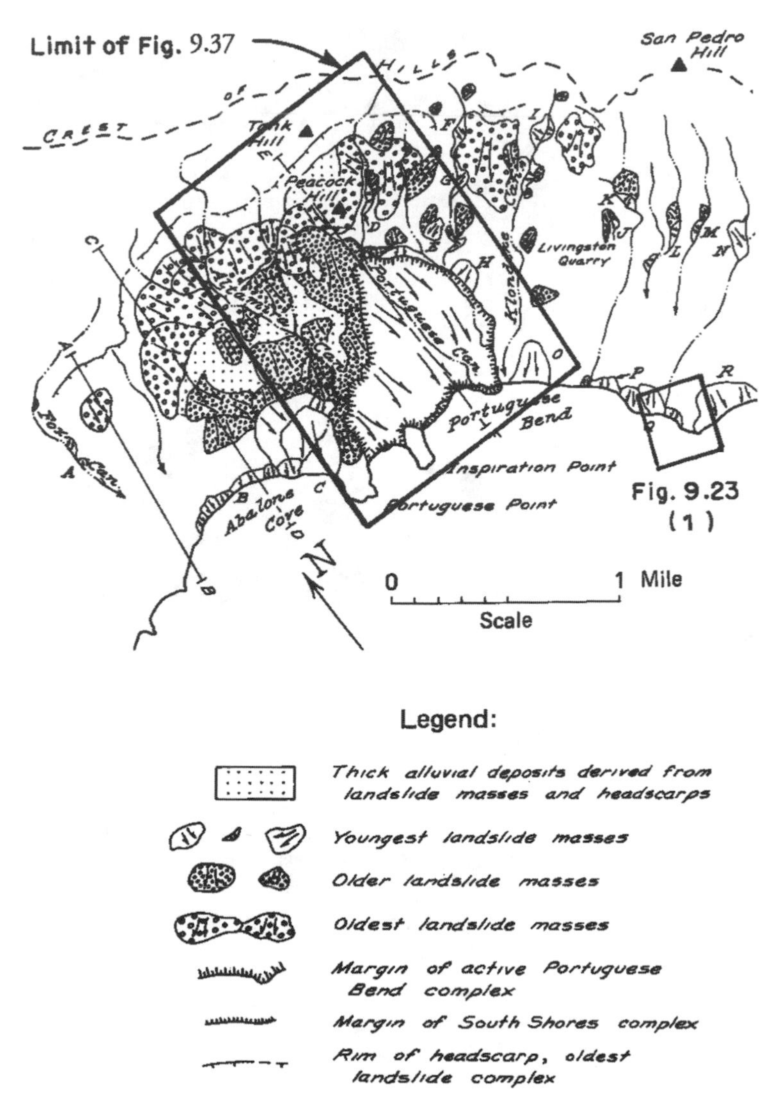

FIGURE 1.36
Distribution of principal landslides and landslide complexes in Palos Verdes Hills, California. (From Jahns, R. H. and Vonder Linden, C., *Geology, Seismicity and Environmental Impact*, Special Publication Association Engineering Geology, Los Angeles, 1973, pp. 123–138. With permission.)

area, the land rises from the sea in a series of gently rolling hills and terraces to more than 600 ft above sea level. The hills beyond the slide area rise to elevations above 1200 ft and the cliffs along the oceanfront are roughly 150 ft above the sea. A panoramic view of the slide is given in Figure 1.38.

Geology: The slide zone occurs in Miocene sediments of heavily tuffaceous and sandy clays interbedded with relatively thin strata of bentonitic clays. When undisturbed, the beds dip seaward at about 10 to 20°, which more or less conforms with the land surface as illustrated on the section (Figure 1.39). A badly crushed zone of indurated clayey silt forming a soil "breccia" (Figure 1.40) is found in the lower portions of the slide area. Present movement of the slide appears to be seated at a depth of about 100 ft below the surface in the "Portuguese tuff," originally deposited as a marine ash flow.

FIGURE 1.37
Stereo-pair of aerial photos of the Portuguese Bend area of the Palos Verdes Hills (January 14, 1973, scale 1:24,000).

Slide history: The area has been identified as one with ancient slide activity (see Figure 1.36). Using radiometric techniques, colluvium older than 250,000 years has been dated, and intermediate activity dated at 95,000 years ago. In recent times some block movement was noted in 1929, but during the 1950s, when housing development began on the present slide surface, the slide was considered as inactive. Significant modern movement began in 1956, apparently triggered by loading the headward area of the slide with construction fill for a roadway.

Slide movements: The mass began moving initially during 1956 and 1957, at rates of 2 to 5 in./year, continuing at rates varying from 6 to 24 in./year during 1958, then 3 to 10 ft/year during 1961 to 1968. After 1968, a significant increase in movement occurred. Eventually, 120 houses were destroyed over a 300 acre area. Studies have correlated acceleration in rate of advance with earthquake activity, abnormally high tides, and rainfall (Easton, 1973). The average movement in 1973 ranged from about 3 in./day during the dry season, to 4 in./day

FIGURE 1.38
Panorama of the Portuguese Bend landslide looking south. The highway on the left is continually moving, and the old abandoned road appears in the photo center. The broken ground on the right is the head scarp of rotational slides in the frontal lobe of the unstable mass (photo taken in 1973).

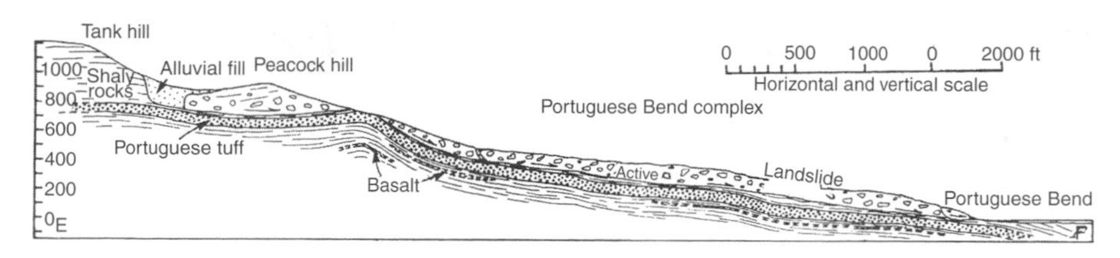

FIGURE 1.39
Geologic section through the Portuguese Bend landslide. For location see Figure 1.36, section along lines e–f. (From Jahns, R. H. and Vonder Linden, C., *Geology, Seismicity and Environmental Impact*, Special Publication Association Engineering Geology, Los Angeles, 1973, pp. 123–138. With permission.)

during the rainy season, to peaks of 6 in./day during heavy rains. Rainfall penetrating deeply into the mass through the many large tension cracks builds up considerable hydrostatic head to act as a driving force on the unstable blocks supported by material undoubtedly at residual strength. The maximum horizontal displacement between 1968 and 1970 was about 130 ft and the maximum vertical displacement about 40 ft. An interesting feature of the slide is the gradual and continuous movement without the event of total collapse.

Stabilization: Because of the large area involved and the geologic and other natural conditions, there appears to be no practical method of arresting slide movements. The cracks on the surface are too extensive to consider sealing to prevent rainwater infiltration, and the strength of the tuff layer is now inadequate to restrain gravity movement even during the dry season. A possible solution to provide stability might be to increase the shearing resistance of the tuff by chemical injection. Since this would be extremely costly, it appears prudent to leave the unstable area as open space although continuous maintenance of the roadway in Figure 1.38 has been necessary.

Glaciolacustrine Soils

Glaciolacustrine soils composing slopes above river valleys normally are heavily overconsolidated. Shear strengths, as measured in the laboratory, are often high, with cohesion ranging from 1 to 4 tsf. Therefore, these soils would not usually be expected to be slide-prone on moderately shallow slopes, and normal stability analysis would yield an adequate factor of safety against sliding (Bjerrum, 1966). Sliding is common, however, and often large in scale, even on shallow slopes.

FIGURE 1.40
Soil "breccia" of fragments of indurated clayey silt in a crushed uplifted zone at Portuguese Bend.

In the Seattle Freeway slides (Figure 1.41), failure occurred along old bedding plane shears associated with lateral expansion of the mass toward the slopes when the glacial ice in the valley against the slopes disappeared (Palladino and Peck, 1972). Similar conditions probably existed at the site of the slide occurring at Kingston, New York, in the Hudson River Valley in August 1915 (Terzaghi, 1950). The Kingston slide was preceded by a period of unusually heavy rainfall. Factors contributing to failure, as postulated by Terzaghi, were the accumulation of stockpiles of crushed rock along the upper edge of the slope and perhaps the deforestation of outcrops of the aquifer underlying the varved clays which permitted an increase in pore-water pressures along the failure surface.

Glaciomarine Soils: South Nation River Slide

Event: The South Nation River slide in Casselman, Ontario, of May 16, 1971, is typical of many slides occurring in the sensitive Champlain clays of glaciomarine origin, in Quebec province, Canada (Eden et al., 1971). These clays are distributed in a broad belt along the St. Lawrence River and up the reaches of the Saguenay River. Most of the slides occur along riverbanks, commencing as either a slump or block glide and retrogressing through either slumping or lateral spreading. At times, the frontal lobes of the slides liquefy and become flows (Figure 1.4f) (see Section 1.2.11).

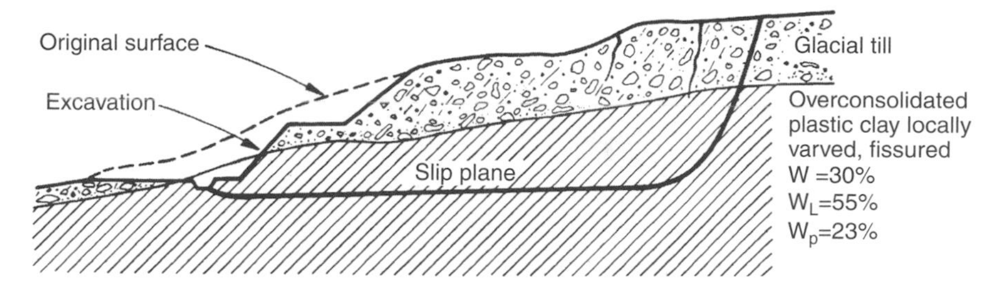

FIGURE 1.41
Failure surface in overconsolidated, fissured clays, undergoing progressive failure as determined by slope inclinometer measurements along the Seattle Freeway. (From Bjerrum, L., *Terzaghi Lectures 1963–1972, ASCE* [1974], 1966, pp. 139–189. With permission.)

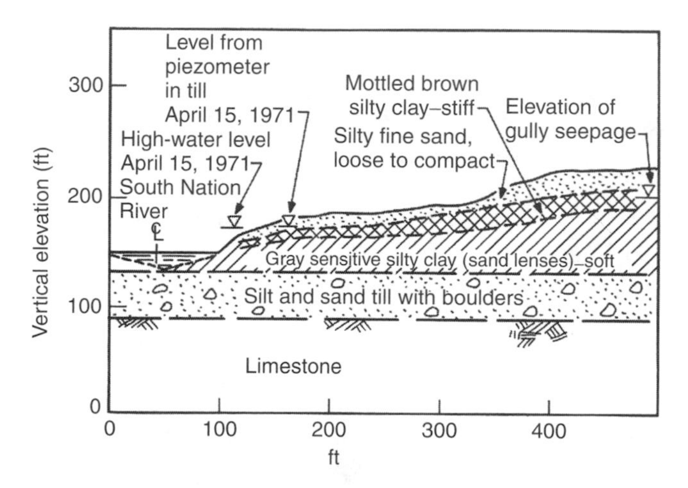

FIGURE 1.42
Stratigraphy prior to the South Nation River slide. An overconsolidated crust over soft, sensitive clays. (From Eden, W. J. et al., *Can. Geotech. J.*, 8, 1971. Reprinted with permission of the National Research Council of Canada.)

Geology: The stratigraphy at the South Nation River slide prior to failure consisted of 6 to 23 ft of stratified silty fine sands overlying the Champlain clay (Leda clay) as shown in Figure 1.42. The undrained strength of the clay was about 0.5 tsf, its sensitivity ranged from 10 to 100, the average plastic limit was 30%, and liquid limit 70%, and the natural water content was at the liquid limit.

Slide history: An all-time record snowfall of 170 in. (432 cm) occurred during the 1970 to 1971 winter and gradual melting resulted in saturation of the upper clays. The slide occurred at the end of the snow-melting season during a heavy rainstorm. A contributing factor was the river level at the slide toe area. It had risen as much as 30 ft during spring floods, remained at that level for a week, then dropped back rapidly to preflood levels. At the time of the slide, groundwater at the lower part of the slope was observed to be nearly coincident with the surface. From the appearance of the ground after failure, it appears that the slide retrogressed as a series of slumps as shown in Figure 1.43.

Glaciomarine Soils: Turnagain Heights Slide

Event: Much of the damage to the Anchorage, Alaska, area from the March 1964 earthquake was caused by landslides induced by seismic forces. The slides occurred in the city in the Ship Creek area and along the waterfront formed by the Knik Arm of Cook Inlet. The largest slide occurred at Turnagain Heights, a bluff some 60 ft high overlooking Knik Arm. Many homes were destroyed in the slide area of 125 acres as illustrated in Figure 1.44. The slide at Turnagain Heights is an example of sliding along horizontal strata. It was planar and

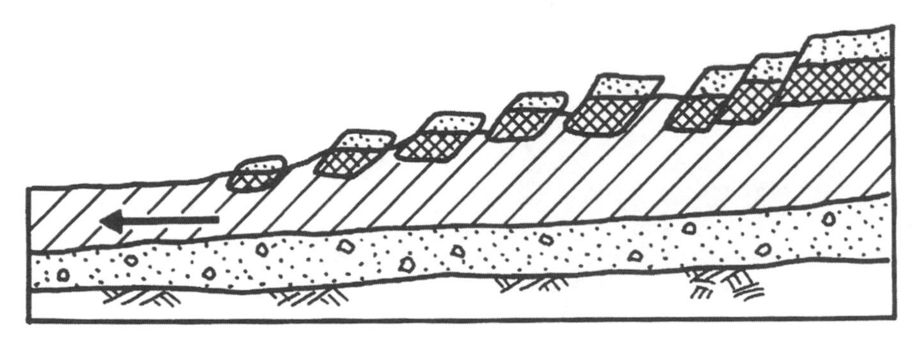

FIGURE 1.43
Stratigraphy after slide at South Nation River. Blocks broke loose and moved by lateral spreading. (After Mollard, J. D., *Reviews in Engineering Geology*, Vol. III, *Landslides*, Geological Society of America, 1977, pp. 29–56.)

FIGURE 1.44
Failure by block gliding and lateral spreading resulting from the 1964 earthquake, Turnagain Heights, Anchorage, Alaska. (Photo courtesy of U.S. Geological Survey, Anchorage.)

evolved by block gliding or slump failure at the bluff, followed by lateral spreading of the mass for a width of 8000 ft, and extending as much as 900 ft inland.

Geology: Anchorage and the surrounding area are underlain by the Bootlegger Cove clay of glaciomarine origin. Soil stratigraphy at the bluff consisted of a thin layer of sand and gravel overlying a clay stratum over 100 ft thick as shown in Figure 1.45a. The consistency of the upper portions of the clay was stiff to medium, becoming soft at a depth of about 50 ft. The soft zone extended to a depth of about 23 ft below sea level. Layers of silt and fine sand were present at depths of a several feet or so above sea level.

Slide history: Seed and Wilson (1967) postulated that cyclic loading induced by the earthquake caused liquefaction of the silt and fine sand lenses resulting in instability and block gliding along the bluff. Blocks continued to break loose and glide retrogressively, resulting

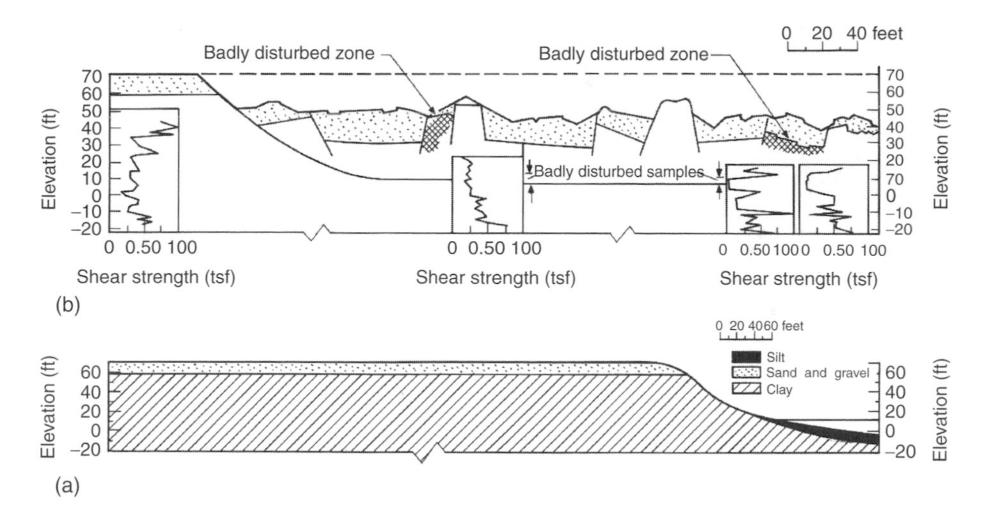

FIGURE 1.45
Soil profiles through east end of slide area at Turnagain Heights (a) before and (b) after failure. (From Seed, H.B. and Wilson, S.D., *Proceedings of ASCE, J. Soil Mechanics Foundation Engineering Division,* Vol. 93, 1967, pp. 325–353. With permission.)

in lateral spreading of the mass which came to rest with a profile more or less as illustrated in Figure 1.45b. The movement continued for the duration of the earthquake (more than 3 min), but essentially stopped once strong ground motion ceased.

Conclusions: The magnitude of the 1964 event was 8.5 (Richter) with an epicenter 80 mi east of Anchorage. Previous earthquakes of slightly lower magnitudes but closer epicenters had occurred, but the Turnagain Heights area had not been affected (see Section 3.3.4). Seed and Wilson (1967) concluded that, in light of previous earthquake history, the slide was the result of a continuous increase in pore pressures caused by the long duration of the 1964 event; and, that it is extremely unlikely that any analysis would have anticipated the extent of inland transgression of the failure. Considering the local stratigraphy and seismic activity, however, the area certainly should be considered as one with a high slope failure hazard.

1.2.7 Debris Slides

General

Debris slides involve a mass of soil, or soil and rock fragments, moving as a unit or a number of units along a steeply dipping planar surface. They often occur progressively and can develop into avalanches or flows. Major causes are increased seepage forces and slope inclination, and the incidence is increased substantially by stripping vegetation. Very large masses can be involved, with gradually developing progressive movements, but at times total failure of a single block can occur suddenly.

Recognition

Occurrence is common in colluvial or residual soils overlying a relatively shallow, dipping rock surface. During the initial stages of development, tension cracks are commonly formed. After partial failure, the tension cracks widen and the complete dislodgment of one or more blocks may occur, often leaving a clean rock surface and an elliptical failure scar as shown in Figure 1.46. Total failure can be said to have occurred when the failure

FIGURE 1.46
Small debris-slide scar along the Rio Santos Highway, Brazil. Note seepage along the rock surface, the sliding plane. Failure involved colluvium and the part of the underlying rock. The area subsequently was stabilized by a concrete wall.

surface reaches to the crest of the hill. If uncorrected, failure often progresses upslope as blocks break loose.

Examples of Major Failures

Pipe Organ Slide

Description: The Pipe Organ slide in Montana (Noble, 1973) triggered by a railroad cut made at the toe involved about 9 million yd^3 of earth. The failure surface developed at depths below the surface of 120 to 160 ft in a Tertiary colluvium of stiff clay containing rock fragments, which overlay a porous limestone formation.

Slide movements: During sliding, movement continued for a year at an average rate of 2 in./week, developing in a progressive mode. Total movement was about 12 ft and the length of the sliding mass was 2000 ft along the flatter portions of a mountain slope.

Stabilization: Movement was arrested by the installation of pumped wells, which were drilled into the porous limestone and later converted into gravity drains. The water perched on the sliding surface near the interface between the colluvium and the limestone drained readily into the limestone and the hydrostatic pressures were relieved.

Golden Slide

Description: A railroad cut into colluvium of about 50 ft thick caused a large slide near Golden, Colorado (Noble, 1973). The colluvium is an overconsolidated clay with fragments of clay shale and basalt overlying a very hard blue-gray clayey siltstone. The water table was midway between the ground surface and the failure surface, and there was evidence of artesian pressure at the head of the slide.

Slide movements: Movement apparently began with a heavy rainfall and was about 1 in./day. Tension cracks developed in the surface and progressed upslope with time. About 500,000 yd^3 of material were moving within a length of about 1000 ft.

Stabilization: Cutting material from the head of the slide and placing approximately 100,000 yd^3 against a retaining wall at the toe that penetrated into underlying sound rock was unsuccessful. Movement was finally arrested by the installation of horizontal drains as long as 400 ft, and vertical wells which were being pumped daily at the time that Noble (1973) prepared his paper.

Colluvium on Shale Slopes

The Pennington shale of the Cumberland plateau in Tennessee and the sedimentary strata in the Appalachian plateau of western Pennsylvania develop thick colluvial overburden, which is the source of many slide problems in cuts and side-hill fills. The geology, nature of slope problems, and solutions are described in detail by Royster (1973, 1979) and Hamel (1980).

1.2.8 Debris Avalanches

General

Debris avalanches are very rapid movements of soil and rock debris which may, or may not, begin with rupture along a failure surface. All vegetation and loose soil and rock material may be scoured from a rock surface as shown in Figure 1.7. Major causes are high seepage forces, heavy rains, snowmelts, snowslides, earthquakes, and the creep and gradual yielding of rock strata.

Failure is sudden and without warning, and essentially unpredictable except for the recognition that the hazard exists. Effects can be disastrous in built-up areas at the toes of high steep slopes under suitable geologic conditions (see Examples below).

Occurence

Debris avalanches are characteristic of mountainous terrain with steep slopes of residual soils where topography causes runoff concentration (see Figure 1.86), or badly fractured rock such as illustrated in Figure 1.47.

There is usually no initial stage, although occasionally tension cracks may be apparent under some conditions. Total failure occurs suddenly either by a rock mass breaking loose or by "hydraulic excavation" which erodes deep gullies in soil slopes during torrential rains as shown in Figure 1.48. All debris may be scoured from the rock surface and deposited as a terminal lobe at a substantial distance from the slope. As shown in Figure 1.49, the force is adequate to move large boulders, and erosion can cause the failure area to progress laterally to affect substantial areas.

In the Andes Mountains of South America, debris avalanches are the most common form of slope failures. They occur occasionally in natural slopes, not impacted on by construction, but normally are caused by roadway cuts. Typical geologic conditions are shown in Figure 1.50. Illustrated in Figure 1.51 are the steep slopes and irregular topography that result in the necessity for numerous cuts for roadway construction. The slope shown in the photo, Figure 1.52, taken in 2002, was free of slope failures in 1995. Investigation and treatments of slopes in the Andes is discussed in Section 1.5.

Examples of Major Failures

Rio de Janeiro, Brazil

Event: A debris avalanche occurred during torrential rains in 1967 in the Laranjeiras section of Rio de Janeiro, which destroyed houses and two apartment buildings, causing the death of more than 130 persons. The avalanche scar and a new retaining wall are shown in Figure 1.53.

FIGURE 1.47
Scarred surface remaining after a rock and debris avalanche in a limestone quarry. The rock is heavily jointed with sets oriented more or less parallel to the slope and across the bedding plane. Failure was induced by wedging from water and ice pressures, and occurred in the early spring.

Climatic conditions: Hundreds of avalanches and slides occurred in Rio de Janeiro and the nearby mountains during the unusually heavy rains of 1966 and 1967 when intensities as high as 200 mm/h (8 in./h) were recorded (Jones, 1973).

During a 3-day storm beginning on January 10, 1966, a gaging station at Alto da Boa Vista, in the mountains a few kilometers from the city, recorded 675 mm (26.2 in.) of rainfall. It was an unprecedented amount. Although heavy rains occur each year during the summer months of January and February, with rainfall averaging 171 mm (6.7 in.) during January, most of the rainfalls during intense storms. The potential for slope failures is very much dependent upon the accumulated rainfall and associated water-table conditions for a given rainy season (see Section 1.3.4).

Local geology: Typical profiles in the residual soils (see Figure 1.69) along the coastal mountains of Brazil show that these soils are most impervious near the surface and that permeability increases downward through the soil profile into the underlying decomposed and fractured crystalline rocks. Fissures in the outer portions of the residual and colluvial soils close during rainfall; they thereby block drainage and cause a rapid increase in pore pressures, resulting in sudden failure, which combined with high runoff develops into an avalanche or even a flow. To minimize the slope-failure hazard, the city of Rio has zoned some areas of high steep slopes to prohibit construction, and has undertaken the construction of numerous stabilization works throughout the city.

FIGURE 1.48
The force of hydraulic excavation is evident in this photo taken in a typical V-shaped scarred zone of a debris avalanche. Location is near the crest of the hill in Figure 1.49. The bedrock surface is exposed.

FIGURE 1.49
Debris avalanche that covered BR 101 near Tubarao, Santa Catalina, Brazil during torrential rains in 1974. Note the minibus for scale. The debris lobe crossed the highway and continued for a distance of about 200 m and carried boulders several meters in diameter. Debris has been removed from the roadway. It is unlikely that a failure of such a magnitude could have been foreseen for this particular location.

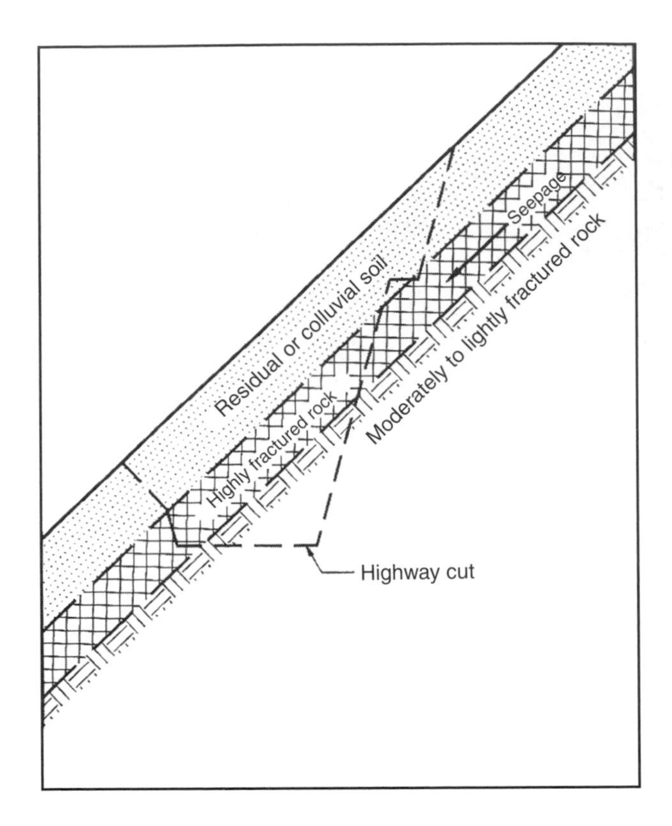

FIGURE 1.50
Typical geologic conditions in the Andes Mountains. Slopes often inclined at 35 to 40°.

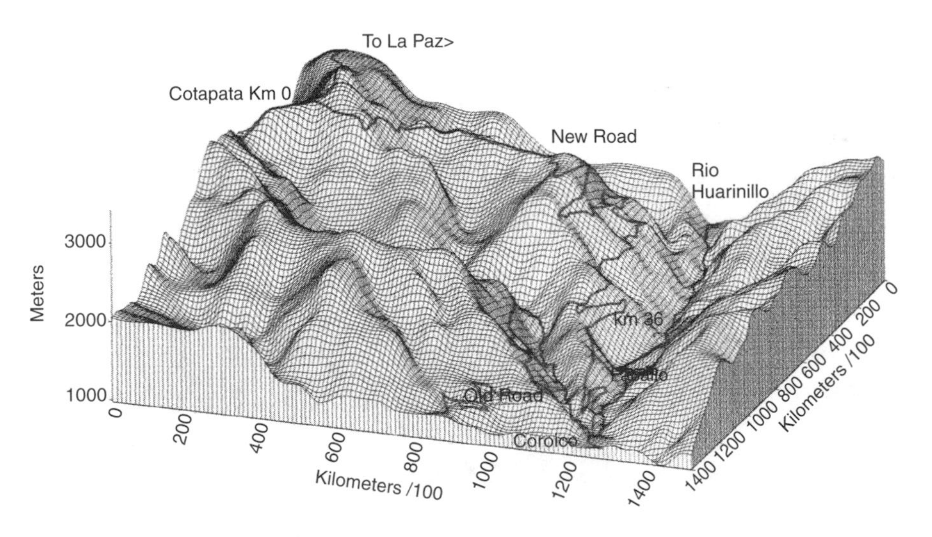

FIGURE 1.51
Three-dimensional diagram of a new roadway in the Andes of Bolivia. Steep slopes and irregular landform results in numerous cuts. See Figure 1.7 for debris avalanches above town of Pacallo, and at km 36, Figure 1.52.

Ranrahirca and Yungay, Peru

Event no. 1: One of the most disastrous debris avalanches in modern history occurred in the Andes Mountains of Peru on January 10, 1962. In a period of 7 min, 3500 lives were lost and seven towns, including Ranrahirca, were buried under a mass of ice, water, and debris (McDonald and Fletcher, 1962). The avalanche began with the collapse of Glacier 512 from

FIGURE 1.52
Debris avalanches at km 36 along the roadway shown in Figure 1.51. There were no slope failures at the beginning of construction in 1995. The photo, taken by the author, shows conditions during 2002, after several years of El Niño. At this time construction was not complete.

FIGURE 1.53
Scar of debris avalanche of February 18, 1967, which destroyed two apartment buildings and took 132 lives in the Larenjeiras section of Rio de Janeiro. The buttressed wall was constructed afterward. Debris avalanches need not be large to be destructive.

the 7300-m-high peak, Nevada Huascaran. Triggered by a thaw, 3 million tons of ice fell and flowed down a narrow canyon picking up debris and spilling out onto the fertile valley at an elevation 4000 m lower than the glacier and a distance of 15 km. The debris remaining in the towns ranged from 10 to 20 m thick.

Event No. 2: The catastrophe was almost duplicated on May 32, 1970, when the big Peruvian earthquake caused another avalanche from Nevada Huascaran that buried Yungay, adjacent to Ranrahirca, as well as Ranrahirca again, taking at least 18,000 lives (Youd, 1978). During the 1962 event, Yungay had been spared. The average velocity of the avalanche has been given as 320 km/h (200 mi/h) (Varnes, 1978), and the debris flowed upstream along the Rio Santa for a distance of approximately 2.5 km. As with the 1962 failure, the avalanche originated when a portion of a glacier on the mountain peak broke loose.

1.2.9 Debris Flows

General

Debris flows are similar to debris avalanches, except that the quantity of water in the debris-flow mass causes it to flow as a slurry; in fact, differentiation between the two forms can be difficult. The major causes are very heavy rains, high runoff, and loose surface materials.

Recognition

Occurrence is similar to debris avalanches, but debris flows are more common in steep gullies in arid climates during cloudbursts, and the failing mass can move far from its source (see Figure 1.4c).

1.2.10 Rock-Fragment Flows

General

A rock mass can suddenly break loose and move downslope at high velocities as a result of the sudden failure of a weak bed or zone on the lower slopes causing loss of support to the upper mass. Weakening can be from weathering, frost wedging, or excavation. Failure is sudden, unpredictable, and can be disastrous.

Recognition

High, steep slopes in jointed rock masses offer the most susceptible conditions. The avalanche illustrated in Figure 1.47 could also be classified as a dry rock flow because of its velocity and lack of water. In the initial stages tension cracks may develop; after the final stage a scarred surface remains over a large area, and a mass of failed debris may extend far from the toe of the slope.

Example of Major Failure

Event

The Turtle Mountain slide of the spring of 1903 destroyed part of the town of Frank, Alberta, Canada. More than 30 million m^3 of rock debris moved downslope and out onto the valley floor for a distance of over 1 km in less than 2 min.

Geology

The mountain is the limb of an anticline composed of limestone and shales as shown in the section in Figure 1.54a. Failure was sudden, apparently beginning in bedding planes in the lower shales (Krahn and Morgenstern, 1976) which are steeply inclined. The slide scar is shown in Figure 1.54b.

Cause

Terzaghi (1950) postulated that the flow was caused by joint weathering and creeping of the soft shales, accelerated by coal mining operations along the lower slopes.

1.2.11 Soil and Mud Flows

General

Soil and mud flows generally involve a saturated soil mass moving as a viscous fluid, but at times can consist of a dry mass. Major causes include earthquakes causing high pore-air pressures (loess) or high pore-water pressures; the leaching of salts from marine clays increasing their sensitivity, followed by severe weather conditions; lateral spreading followed by a sudden collapse of soil structure; and heavy rainfall on a thawing mass or the sudden drawdown of a flooded water course.

Flows occur suddenly, without warning, and can affect large areas with disastrous consequences.

Recognition

Occurrence is common in saturated or nearly saturated fine-grained soils, particularly sensitive clays, and occasional in dry loess or sands (sand runs).

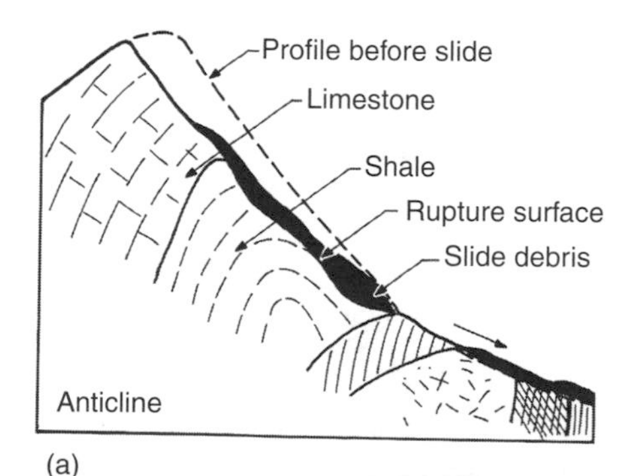

(a)

(b)

FIGURE 1.54
Turtle Mountain rock slide, Frank, Alberta in 1903. (a) Geologic section (From Krahn, J. and Morgenstern, N. R., *Proceedings of ASCE, Rock Engineering for Foundations and Slopes*, Vol. 1, 1976, pp. 309–332. With permission.) (b) Photo of scar (source unknown).

During initial stages flows may begin by slump failure followed by lateral spreading in the case of sensitive clays. During final failure, a tongue-shaped lobe of low profile extends back to a bottleneck-shaped source area with a small opening at the toe of the flow as shown in Figure 1.4f, and a distinct scarp remains at the head. Flowing masses can extend for great distances, at times measured in kilometers.

Examples of Major Failures

Achocallo Mudflow, La Paz, Bolivia

Believed triggered by an earthquake some thousands of years ago, an enormous portion of the rim of the Bolivian altiplano (elevation 4000 m), roughly 9 km across, slipped loose and flowed down the valley of the Rio Achocallo into the Rio La Paz at an elevation approximately 1500 m lower. The flow remnants extend downstream today for a distance of about 25 km, part of which are shown on the aerial oblique panorama given in Figure 1.55. The head scarp is shown on the photo (Figure 1.56).

The altiplano is the remains of an ancient lake bed, probably an extension of Lake Titicaca, underlain by a thick stratum of sand and gravel beneath which are at least several hundred meters of lacustrine clays and silts, interbedded with clays of volcanic origin. The photo (see Figure 2.36) of a large piping tunnel was taken in the bowl-shaped valley about 3 km downslope from the rim.

Province of Quebec, Canada

Event: In Saint Jean Vianney, on May 4, 1971, a mass of glaciomarine clays completely liquefied, destroying numerous homes and taking 31 lives (Tavenas et al., 1971).

Description: The flow began in the crater of a much larger 500-year-old failure (determined by carbon-14 dating). Soil stratigraphy consisted of about 100 ft of disturbed clays with sand pockets from the ancient slide debris, overlying a deep layer of undisturbed glaciomarine clay. Occurring just after the first heavy rains following the spring thaw, the flow apparently began as a series of slumps from the bank of a small creek, which formed a temporary dam. Pressure built up behind the dam, causing it to fail, and 9 million yd^3 of completely liquefied soils flowed downstream with a wavefront 60 ft in height and a velocity estimated at 16 m/h. The flow finally discharged into the valley of the Saguenay River, 2 mi from its source.

Causes: Slope failures in the marine clays of Quebec are concentrated in areas that seem to be associated with a groundwater flow regime resulting from the existence of valleys in the underlying rock surface (Tavenas et al., 1971). The valleys cause an upward flow gradient and an artesian pressure at the slope toes. The upper part of the soil profile is subjected to a downward percolation of surface water because of the existence of sand strata. The downward percolation and upward flow produce an intense leaching of the clay, resulting in a decrease of the undrained shear strength and an increase in sensitivity. The evidence tends to indicate that the leaching is a function of the gradient. Field studies have shown a close relationship between the configuration of the bedrock and the properties of the underlying clay deposit.

Norwegian "Quick" Clays

Regional geology: Approximately 40,000 km^2 of Norway has deposits of glaciomarine clays which overlie an irregular surface of granite gneiss, similar to conditions in Quebec. During postglacial times, the area has been uplifted to place the present surface about 180 m above sea level. Typical stratigraphy includes 5 to 7 m of a stiff, fissured clay overlying normally consolidated soft marine clay which extends to depths greater than 70 m in

FIGURE 1.55
Believed triggered by an earthquake some thousands of years ago, an enormous portion of the rim of the Bolivian Altiplano, roughly 9 km across (photo right), slipped loose and flowed down the valley of the Rio Achocallo into the Rio La Paz, for a total distance of over 25 km (photo left). The flow is apparent in the photo as the light-colored area in the valley.

FIGURE 1.56
The head scarp of the Achocallo mud flow, near La Paz, Bolivia (see Figure 1.55).

some locations. Rock varies from outcropping at the surface in stream valleys to over 70 m in depth. The quick clays are formed by leaching of salts, but the leaching is believed to be caused by artesian pressure in the rock fractures from below, rather than downward percolation of water (Bjerrum et al., 1969). Sensitivity values are directly related to the amount of leaching and the salt content, and are greatest where rock is relatively shallow, about 15 to 35 m.

Slope failures are common events. The natural slopes are stable at about 20° where there is a stiff clay crust. Seepage parallel to the slopes occurs in fissures in the clay, and the stiff clay acts as a cohesionless material with slopes at $i = 1/2\phi'$, and $\phi = 38°$ (see the infinite slope problem in Section 1.3.2). Stream erosion causes small slides in the weathered stiff clay; the sliding mass moves into the soft clay, which upon deformation becomes quick and flows. In one case cited by Bjerrum et al. (1969), 200,000 m^3 flowed away from the source in a few minutes.

Solution: Since stream degrading appears to be a major cause of the flows, Bjerrum et al. (1969) proposed the construction of small weirs to impede erosion in streams where failures pose hazards.

1.2.12 Seafloor Instability

General

Various forms of slope failures have been recognized offshore, including deep rotational slides (Figure 1.57) and shallow slumps, flows, and collapsed depressions (Figure 1.58). Major causes are earthquakes, storm waves inducing bottom pressures, depositional loads accumulating rapidly and differentially over weak sediments, and biochemical degradation of organic materials forming large quantities of gases *in situ* which weaken the seafloor soils.

Offshore failures can occur suddenly and unpredictably, destroying oil production platforms, undersea cables, and pipelines. Large flows, termed "turbidity currents," can move tremendous distances.

Recognition

Occurrence is most common in areas subjected to earthquakes of significant magnitude and on gently sloping seafloors with loose or weak sediments, especially in rapidly accreting deltaic zones.

After failure the seafloor is distorted and scarred with cracks, scarps, and flow lobes similar to those features which appear on land as illustrated in Figure 1.58. Active areas are explored with side-scan sonar (Figure 1.59) and high-resolution geophysical surveys (see Figure 1.57).

Examples of Major Failures

Gulf of Alaska

The major slide illustrated on the high-resolution seismic profile given in Figure 1.57 apparently occurred during an earthquake, and covers an area about 15 km in length. Movement occurred on a 1° slope and is considered to be extremely young (Molnia et al., 1977). As shown in the figure, the slide has a well-defined head scarp, disrupted bedding, and a hummocky surface.

Gulf of Mexico

Movements are continually occurring offshore of the Mississippi River Delta. During hurricane Camille in August 1969, wave-induced bottom pressures caused massive seafloor movements that destroyed two offshore platforms and caused a third to be displaced over a meter on a bottom slope that was very flat, less than 0.5% (Focht and Kraft, 1977).

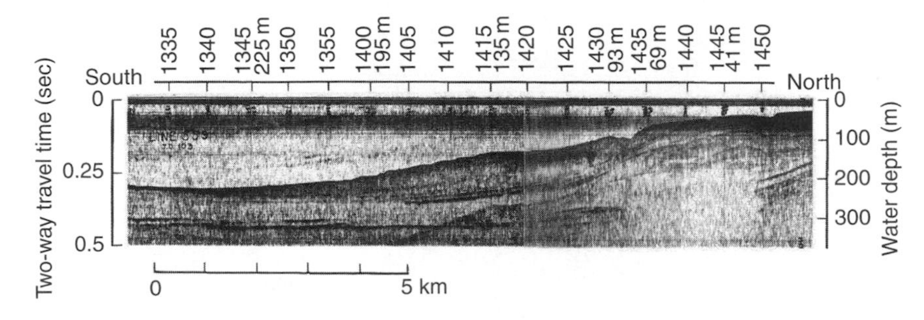

FIGURE 1.57
High-resolution seismic reflection profile showing a portion of the Kayak Trough slump slide in the Gulf of Alaska. (From Molnia, B. F. et al., in *Reviews in Engineering Geology*, Vol. VIII, *Landslides*, Coates, D.R., Ed., Geologic Society of America, 1977, pp. 137–148. With permission.)

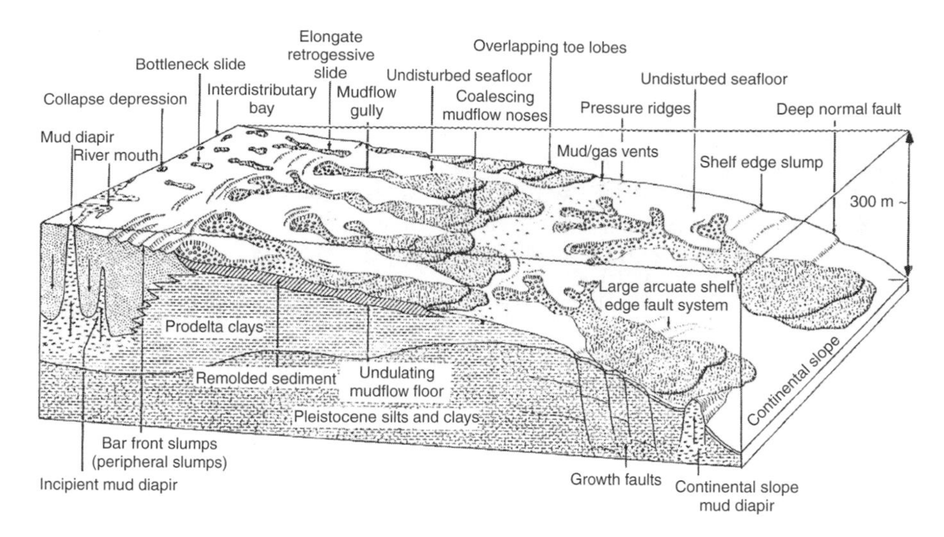

FIGURE 1.58
Schematic block diagram illustrating the various forms of slope failure in the offshore Mississippi River Delta. (From Coleman, J. M. et al., *Open File Report No. 80.01,* Bureau of Land Management, U.S. Department of Interior, 1980. With permission.)

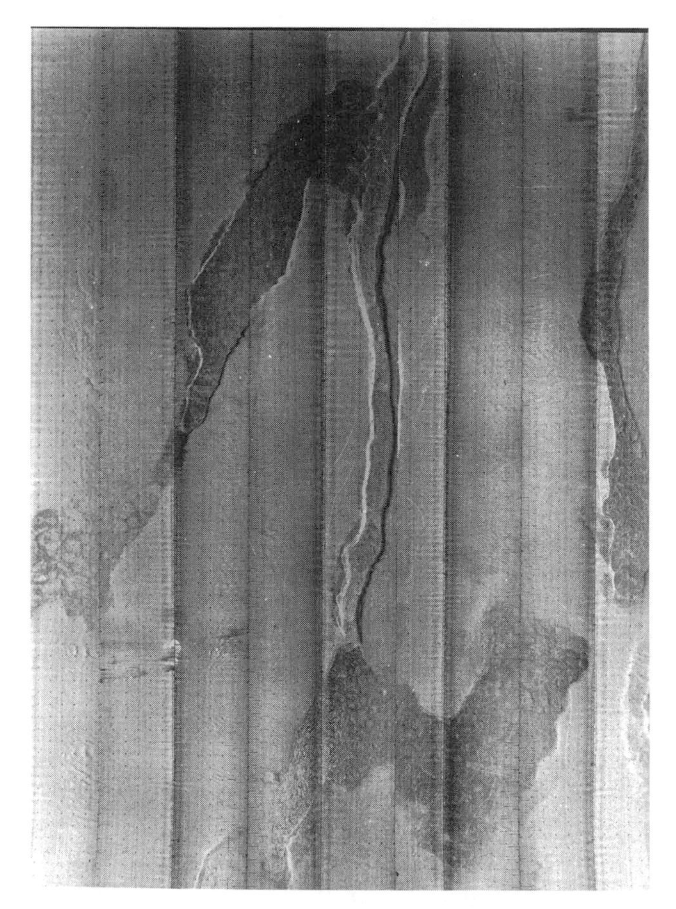

FIGURE 1.59
Side-scan sonar mosaic illustrating seafloor mudflows (offshore Gulf of Mexico). Grids are 25 m (82 ft) apart and the mosaic covers an area approximately 1.5 km in length. (Mosaic courtesy of Dr. J. M. Coleman, Coastal Studies Institute, Louisiana State University.)

Grand Banks, Newfoundland

The earthquake of November 1929 (see Section 3.3.4) caused a section of the continental shelf to break loose and subsequently mix with seawater. It moved offshore for a distance of about 925 km and broke a dozen submarine cables. Geologists called this flow a "turbidity current" (Richter, 1958).

1.3 Assessment of Slopes

1.3.1 General

Objectives

The assessment of an existing unstable or potentially unstable slope, or of a slope to be cut, provides the basis for the selection of slope treatments. Treatment selection requires forecasting the form of failure, the volume of material involved, and the degree of the hazard and risk. Assessment can be based on quantitative analysis in certain situations, but in many cases must be based on qualitative evaluation of the slope characteristics and environmental factors including weather and seismic activity.

Key Factors to Be Assessed

- History of local slope failure activity as the result of construction, weather conditions, seismic activity (see Section 3.3.4), or other factors, in terms of failure forms and magnitudes.
- Geologic conditions including related potential failure forms and their suitability for mathematical analysis, material shear strength factors (constant, variable, or subject to change or liquefaction, Section 1.3.2), and groundwater conditions.
- Slope geometry in terms of the influence of inclination, height, and shape on potential seepage forces, runoff, and failure volume.
- Surface indications of instability such as creep, scars, seepage points, and tension cracks.
- Degree of existing slope activity (see Section 1.1.2).
- Weather factors (rainfall and temperature) in terms of the relationship between recent weather history and long-term conditions (less severe, average, and more severe) in view of present slope activity, stability of existing cut slopes, groundwater levels, and slope seepage.

1.3.2 Stability Analysis: A Brief Review

General Principles

Basic Relationships

Stability analysis of slopes by mathematical procedures is applicable only to the evaluation of failure by sliding along some definable surface. Avalanches, flows, falls, and progressive failure cannot be assessed mathematically in the present state of the art.

Slide failure occurs when the shearing resistance available along some failure surface in a slope is exceeded by shearing stresses imposed on the failure surface. Static analysis of sliding requires knowledge of the location and shape of the potential failure surface, the shear strength along the failure surface, and the magnitude of the driving forces. Statically determinate failure forms may be classified as:

- Infinite slope — translation on a plane parallel to the ground surface whose length is large compared with its depth below the surface (end effects can be neglected) (Figure 1.63) (Morgenstern and Sangrey, 1978).
- Finite slope, planar surface — displacement of one or more blocks, or wedge-shaped bodies, along planar surfaces with finite lengths (Figure 1.66).
- Finite slope, curved surface — rotation along a curved surface approximated by a circular arc, log-spiral, or other definable cylindrical shape (Figure 1.76).

Failure origin: As stresses are usually highest at the toe of the slope, failure often begins there and progresses upslope, as illustrated in Figure 1.60, which shows the distribution of active and passive stresses in a slope where failure is just beginning. Failures can begin at any point along a failure surface, however, where the stresses exceed the peak strength. Because failure often is progressive, it usually occurs at some average shear strength that can be considerably less than the peak strength measured by testing techniques.

Limit Equilibrium Analysis

Most analytical methods applied to evaluate slope stability are based on limiting equilibrium, i.e., on equating the driving or shearing forces due to water and gravity to the resisting forces due to cohesion and friction.

Shearing forces result from gravity forces and internal pressures acting on a mass bounded by a failure surface. Gravity forces are a function of the weight of the materials, slope angle, depth to the failure surface, and in some cases, slope height. Pressures develop in joints in rock masses from water, freezing, swelling materials, or hydration of minerals and, in soils, from water in tension cracks and pores.

Resisting forces, provided by the shear strength along the failure surface, are decreased by an increase in pore pressures along the failure surface, by lateral strains in overconsolidated clays in clay shales, by dissolution of cementing agents and leaching, or by the development of tension cracks (which serve to reduce the length of the resisting surface).

Safety factor against rupture, given as

$$FS = \frac{\text{shearing strength available along the sliding surface}}{\text{shearing stresses tending to produce failure along the surface}} \tag{1.1}$$

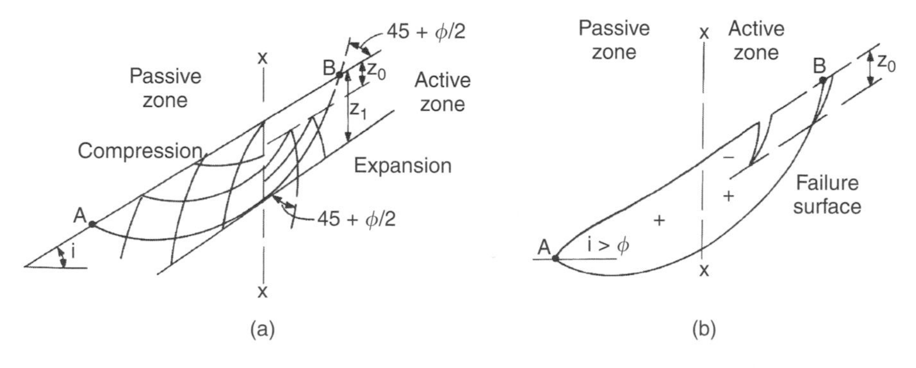

FIGURE 1.60
Active and passive stresses acting on a slope. The passive resistance at the lower portions is most significant. (a) Distribution of stresses on slope with $i > \phi$ (failure beginning). (b) A case of rupture where + indicates zones of passive stress and (−) indicates zones of active stress, subjected to tension cracks.

Shear Strength Factors

Strength Parameters

The basic strength parameters are the angle of internal friction ϕ and cohesion c. Frictional resistance based on ϕ is a function of the normal stress, and the maximum frictional shear strength is expressed as

$$S_{max}=N \tan\phi \tag{1.2}$$

Cohesion c is independent of the normal stress and acts over the area of the failure surface.

Total and Effective Stresses

In the total stress condition, the measured stress includes both pore-water pressures and stresses from grain-to-grain contact. In the effective stress condition, stresses from grain-to-grain contact are measured, which increase as pore pressures dissipate. Effective stress equals total stress minus pore pressure.

Pore-water pressures (U for total, u for unit pressures) are induced either by a load applied to a saturated specimen, or by the existence of a phreatic surface above the sliding surface. They directly reduce the normal force component N, and shearing resistance is then expressed as

$$S_{max}=(N-U)\tan\phi \tag{1.3}$$

In Figure 1.66, therefore, if pore pressures become equal to the normal component of the weight of the block, there will be no shearing resistance.

Failure Criteria

The Mohr–Coulomb criterion defines failure in terms of unit shear strength and total stresses as

$$s=c+\sigma_n\tan\phi \tag{1.4}$$

The Coulomb–Terzaghi criterion accounts for pore-water pressures by defining failure in terms of effective stresses as

$$s=c'+\sigma_n'\tan\phi' \tag{1.5}$$

where c' is the effective cohesion; σ_n' the effective normal stress ($= p - u$) with p total normal stress. In a slope, the total pressure p per unit of area at a point on the sliding surface equals $h_z\,\gamma_t/\cos^2\theta$, where h_z is the vertical distance from the point on surface of sliding to top of slope, γ_t the slope unit weight of soil plus water, and θ the inclination of surface of sliding at point with respect to horizontal; u the pore-water pressure, in a slope $u=h_w\gamma_w$, the piezometric head times the unit weight of water; and ϕ' the effective friction angle.

The strength parameters representing shearing resistance in the field are a function of the material type, slope history, drainage conditions, and time. Most soils (except purely granular materials and some normally consolidated clays) are represented by both parameters ϕ and c, but whether both will act during failure depends primarily on drainage conditions; and, the stress history of the slope.

Undrained vs. Drained Strength

Undrained conditions exist when a fully saturated slope is sheared to failure so rapidly that no drainage can occur, as when an embankment is placed rapidly over soft soils. Such conditions are rare except in relatively impervious soils such as clays. Soil behavior may then be regarded as purely cohesive and $\phi = 0$. Results are interpreted in terms of total stresses, and s_u, the undrained strength, applies. The case of sudden drawdown of an adjacent water body is an undrained condition, but analysis is based on the consolidated undrained (CU) strength of the soil before the drawdown. This strength is usually expressed in terms of the CU friction angle.

Drained or long-term conditions exist in most natural slopes, or some time after a cut is made and drainage permitted. Analysis is based on effective stresses, and the parameters ϕ' and c' will be applicable.

Peak and Residual Strength

The foregoing discussion, in general, pertains to peak strengths. When materials continue to strain beyond their peak strengths, however, resistance decreases until a minimum strength, referred to as the ultimate or residual strength, is attained. The residual strength, or some value between residual and peak strengths, normally applies to a portion of the failure surface for most soils; therefore, the peak strength is seldom developed over the entire failure surface.

Progressive failure, when anticipated, has been approximately evaluated by using the residual strength along the upper portion of the failure surface, and the peak strength at maximum normal stress along the lower zone (Conlon as reported in Peck, 1967; Barton, 1972).

Stiff fissured clays and *clay shales* seldom fail in natural slopes at peak strength, but rather at some intermediate level between peak and residual. Strength is controlled by their secondary structure. The magnitude of peak strength varies with the magnitude of normal stress, and the strain at which peak stress occurs also depends on the normal stress (Peck, 1967). Because the normal stress varies along a failure surface in the field, the peak strength cannot be mobilized simultaneously everywhere along the failure surface.

Residual strength applies in the field to the entire failure surface where movement has occurred or is occurring. Deere and Patton (1971) suggest using ϕ_r (the residual friction angle), where preexisting failure surfaces are present.

Other Strength Factors

Stress levels affect strength. Creep deformation occurs at stress levels somewhat lower than those required to produce failure by sudden rupture. A steady, constant force may cause plastic deformation of a stratum that can result in intense folding, as illustrated in Figure 1.61. Shear failure by rupture occurs at higher strain rates and stress levels and distinct failure surfaces are developed as shown in Figure 1.62. The materials are genetically the same, i.e., varved clays from the same general area.

The strength of partially saturated materials cannot be directly evaluated by effective stress analysis since both pore-air and pore-water pressures prevail. Residual soils, for example, are often partially saturated when sampled. In Brazil, effective stress analysis has sometimes been based on parameters measured from direct shear tests performed on saturated specimens to approximate the most unfavorable field conditions (Vargas and Pichler, 1957). Depending upon the degree of field saturation, the saturated strengths may be as little as 50% of the strength at field moisture. Apparent cohesion results from capillary forces in partially saturated fine-grained soils such as fine sands and silts; it constitutes a temporary strength that is lost upon saturation and, in many instances, on drying.

FIGURE 1.61
Creep deformation in varved clays (Roseton, New York).

Spontaneous liquefaction occurs and the mass becomes fluid in fine-grained, essentially cohesionless soils when the pore pressure is sufficiently high to cause a minimum of grain-to-grain contact. After failure, as the mass drains and pore pressures dissipate, the mass can achieve a strength higher than before failure. High pressures can develop in pore air or pore water.

Changes with time occur from chemical weathering, lateral strains, solution of cementing agents, or leaching of salts (see Section 1.1.4).

In Situ *Rock Strength*

Effective stress analysis normally is applicable because the permeability of the rock mass is usually high. In clay shales and slopes with preexisting failure surfaces, the residual friction angle ϕ_r is often applicable, with pore pressures corresponding to groundwater conditions.

Two aspects that require consideration regarding strength are that strength is either governed by (1) planes of weakness that divide the mass into blocks, or (2) the degree of weathering controls, and soil strength parameters apply.

Seepage or cleft-water pressures affect the frictional resistance of the rock mass in the same manner that pore pressures affect the strength of a soil mass.

FIGURE 1.62
Section of 3-in.-diameter undisturbed specimen of varved clay taken from a depth of 11m in the failure zone showing the rupture surfaces after collapse of an excavation in Haverstraw, New York.

Failure Surface Modes and Stability Relationships

General: Two Broad Modes of Failure

Infinite slope mode involves translation on a planar surface whose length is large compared with its depth. This mode is generally applicable to cohesionless sands, some colluvial and residual soil slopes underlain by a shallow rock surface, and some cases of clay shale slopes.

Finite or limited slope mode involves movement along a surface limited in extent. The movement can be along a straight line, a circular arc, a log-spiral arc, or combinations of these. There are two general forms of finite slope failures: wedges and circular failures. Wedge analysis forms are generally applicable to jointed or layered rock, intact clays on steep slopes, stratified soil deposits containing interbedded strong and weak layers, and clay shale slopes. Cylindrical failure surfaces are typical of normally consolidated to slightly overconsolidated clays and common to other cohesive materials including residual, colluvial, and glacial soils where the deposit is homogeneous.

Infinite-Slope Analysis

The infinite slope and forces acting on an element in the slope are illustrated in Figure 1.63. In the infinite-slope problem, neither the slope height nor the length of the failure surface is considered when the material is cohesionless.

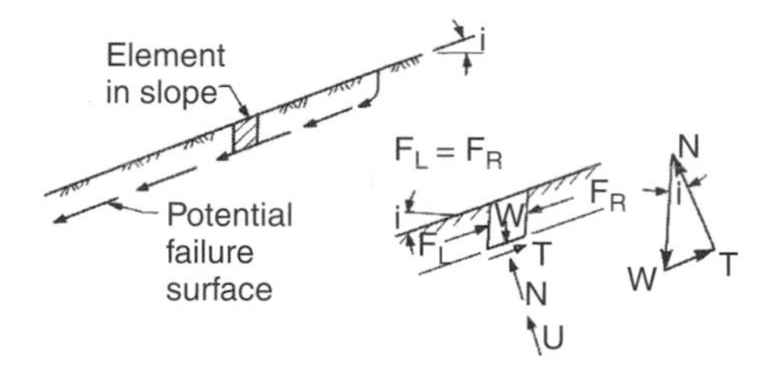

FIGURE 1.63
The infinite slope and forces on an element (total shear resistance $= T_{max} = N \tan \phi$: $N = W \cos i$).

Relationships at equilibrium between friction ϕ and the slope angle i for various conditions in a cohesionless material are given in Figure 1.64, in which T is the total shearing resistance, summarized as follows:

- Dry slope: $i = \phi$ (angle of repose for sands), $T = N \tan \phi$.
- Submerged slope: $i = \phi$, $T = N' \tan \phi'$, and

$$FS = (W \cos i) \tan \phi' / W \sin i \tag{1.6}$$

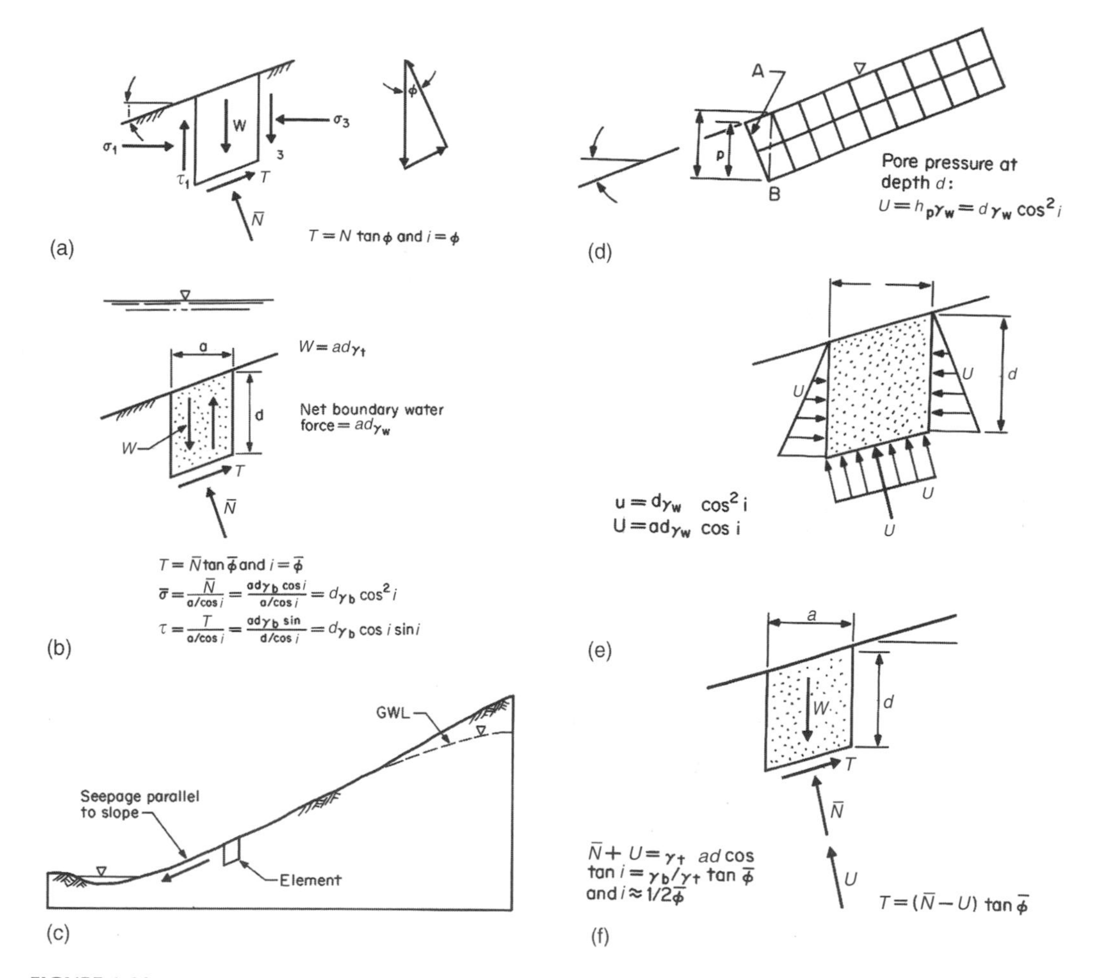

FIGURE 1.64
Equilibrium of an infinite slope in sand under dry, submerged, and slope seepage conditions: (a) slope in dry sand; (b) submerged slope in sand; (c) seepage in a natural slope; (d) flow net of seepage parallel to slope; (e) boundary pore pressures; (f) force equilibrium for element with seepage pressures. (After Lambe, T. W. and Whitman, R. V., *Soil Mechanics*, Wiley, New York, 1969. Adapted with permission of John Wiley & Sons, Inc.)

- Seepage parallel to slope with free water surface coincident with the ground surface (Figure 1.64c–f): $i=1/2\phi'$, and $T=(N'-U)\tan\phi$.
- Infinite-slope conditions can exist in soils with cohesion which serves to increase the stable slope angle i. These conditions generally occur where the thickness of the stratum and, therefore, the position of the failure surface that can develop are limited by a lower boundary of stronger material. Many colluvial and clay shale slopes are found in nature at $i=1/2\phi_r$, the case of seepage parallel to the slope with the free water surface coincident with the ground surface.

Finite Slope: Planar Failure Surface

Case 1: Single planar failure surface with location assumed, involving a single block and no water pressures (Figure 1.65). Driving force F (= $W\sin i$) is the block weight component. Resisting force $T=N\tan\phi=(W\cos i)\tan\phi$,

$$FS=(W\cos i)\tan\phi/W\sin i \tag{1.7}$$

where $i_{cr}=\phi$.

Case 2: Single block with cleft-water pressures and cohesion along the failure surface with location assumed (Figure 1.66):

$$FS=[cA+(W\cos i-U)\tan\phi]/W\sin i+V \tag{1.8}$$

where A is the block base area, V the total joint water pressure on upstream face of the block, U the total water pressure acting on the base area (boundary water pressures), c the cohesion, independent of normal stress, acting over the base area and W the total weight of block, based on γ_t.

Case 3: Simple wedge acting along one continuous failure surface with cohesion and water pressure; failure surface location known (Figure 1.67):

$$FS=[cL+(W\cos\theta-U)\tan\phi]/W\sin\theta \tag{1.9}$$

where L is the length of failure surface.

Case 4: Simple wedge with tension crack and cleft-water pressures V and U. Failure surface location known; tension crack beyond slope crest (Figure 1.68a); tension crack along

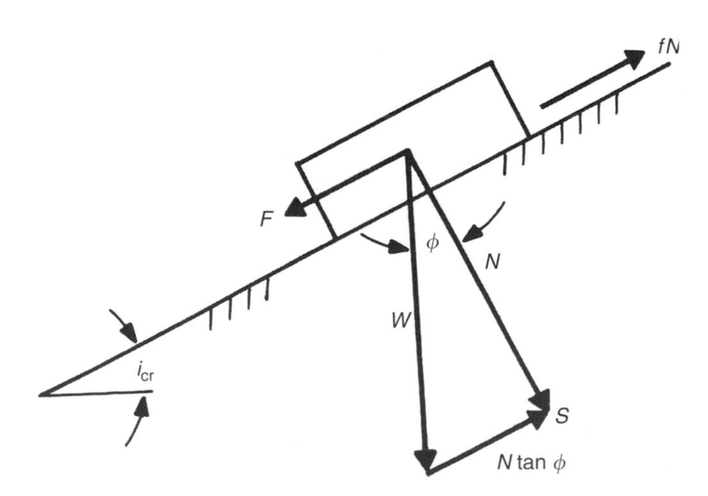

FIGURE 1.65
Simple sliding block.

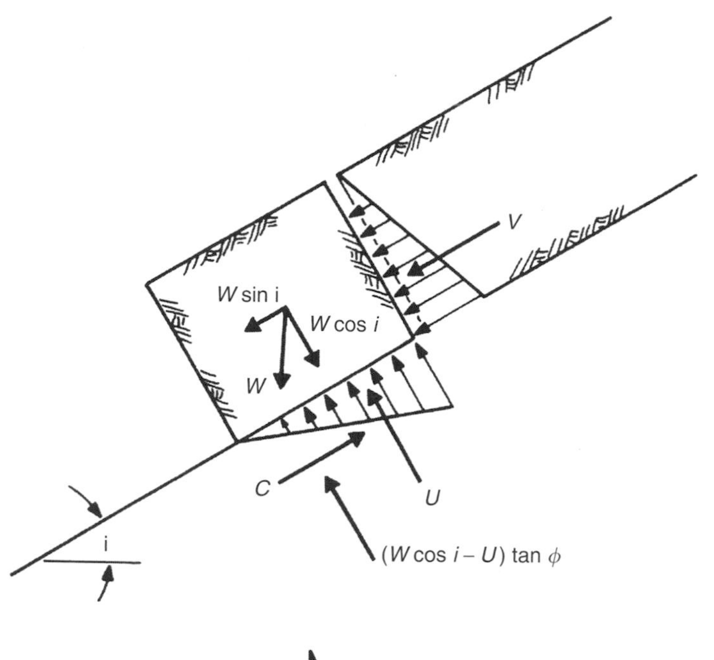

FIGURE 1.66
Block with cleft water pressures and cohesion.

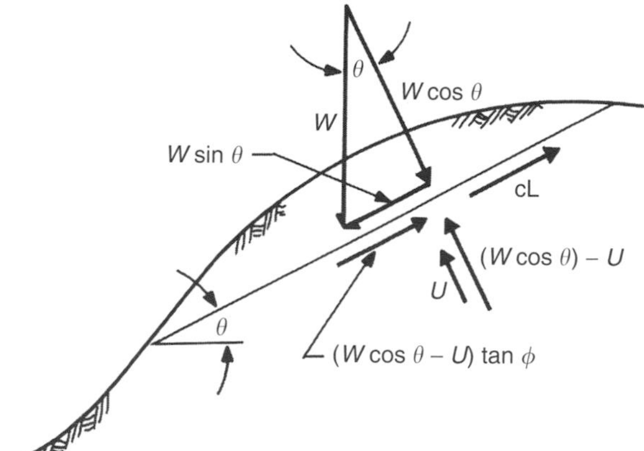

FIGURE 1.67
Simple wedge acting along continuous surface with cohesion.

slope (Figure 1.68b). Figure 1.69 gives an example of a simple wedge developing in residual soils. In Figure 1.68,

$$FS = [cL + (W \cos\theta - U - V \sin\theta) \tan\phi] / \ W \sin\theta + V \cos\theta \qquad (1.10)$$

where

$$L = (H - z) \operatorname{cosec}\theta,$$

$$U = 1/2\gamma_w \, z_w \, (H - z) \operatorname{cosec}\theta,$$

$$V = 1/2\gamma_w \, z_w^{\,2},$$

$$W = 1/2\gamma_t \, H^2 \, \{[1 - (z/H)^2]\cot\theta - \cot i\} \qquad \text{(Figure 1.68a)}$$

$$\text{or } W = 1/2\gamma_t \, H^2 \, \{[1 - z/H)^2 \cot\theta \, (\cot\theta \tan i - 1)] \qquad \text{(Figure 1.68b)}$$

Case 5: Single planar failure surface in clay: location unknown; Culmann's simple wedge (Figure 1.70). Assumptions are that the failure surface is planar and passes through the

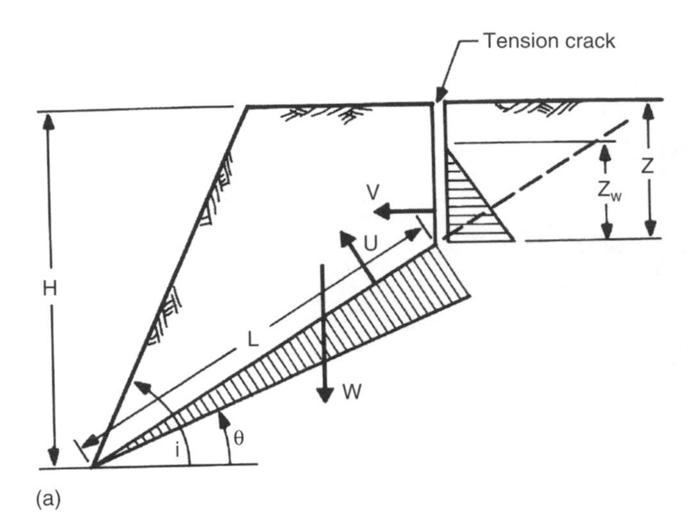

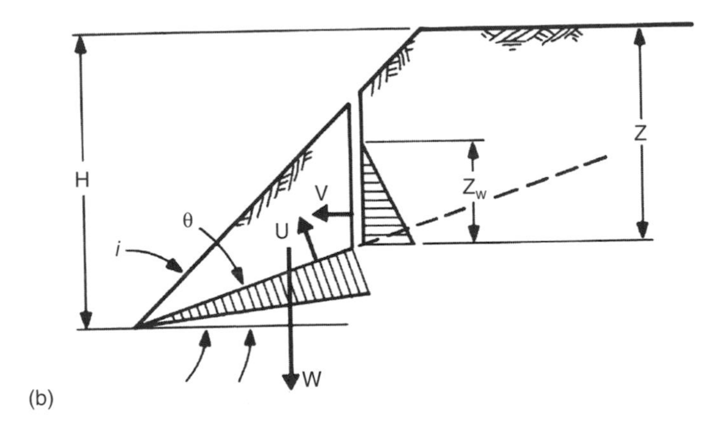

FIGURE 1.68
Plane failure analysis of a rock slope with a tension crack: (a) tension crack in upper slope surface; (b) tension crack in slope face. (From Hoek, E. and Bray, J. W., *Rock Slope Engineering*, 2nd ed., The Institute of Mining and Metallurgy, London, 1977. With permission.)

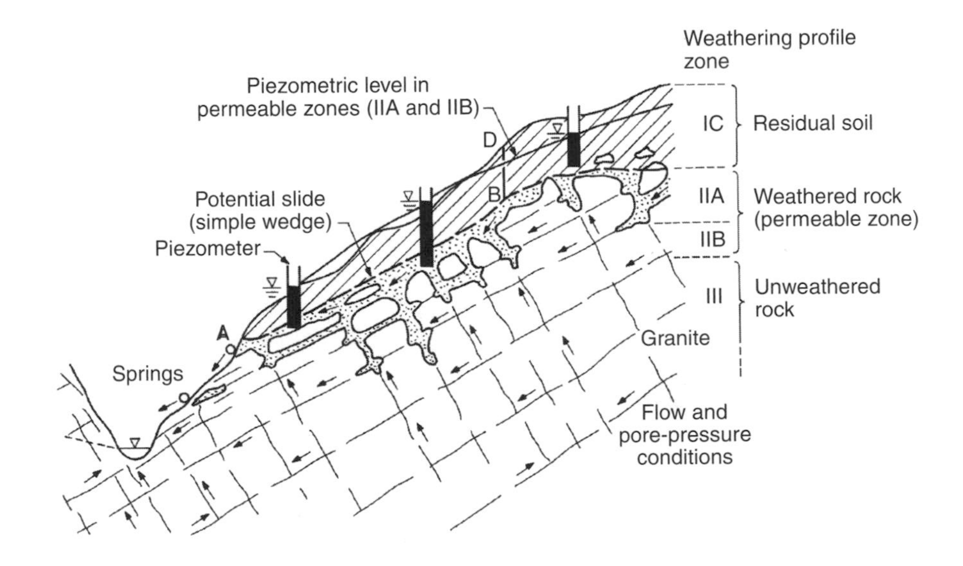

FIGURE 1.69
Development of simple wedge failure in residual soil over rock. (After Patton, F. D. and Hendron, A. J., Jr., *Proceedings of the 2nd International Congress, International Association of Engineering Geology,* São Paulo, 1974, p. V-GR 1.)

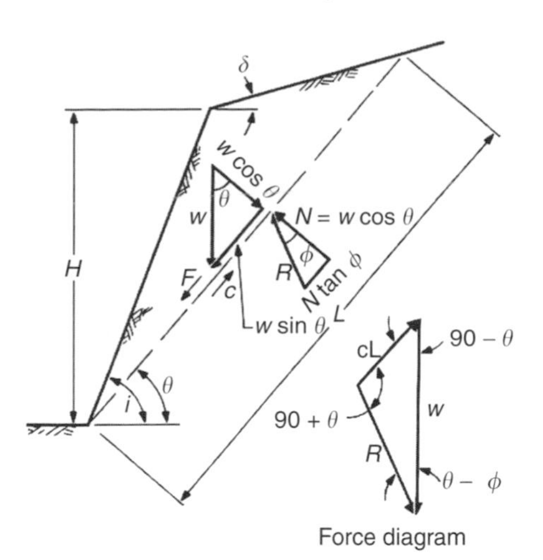

FIGURE 1.70
Culmann's simple wedge in clayey soil.

slope toe, shear strength is constant along the failure surface in a homogeneous section, and there are no seepage forces. In practice, seepage forces are applied as in Equation 1.9. The solution is generally considered to yield reasonable results in slopes that are vertical or nearly so, and is used commonly in Brazil to analyze forces to be resisted by anchored curtain walls (see Section 1.4.6). The solution requires finding the critical failure surface given by

$$\theta_{cr}=(i+\phi)/2 \tag{1.11}$$

$$FS=[cL+(W\cos\theta)\tan\phi]/\ W\sin\theta \tag{1.12}$$

where $W=1/2\gamma_t\,LH$ cosec i sin $(i-\theta)$, with

$$H_{max=}4c(\sin i)(\cos\phi)/\gamma_t\,[1-\cos(i-\phi)] \tag{1.13}$$

Case 6: Critical height and tension crack in clay (Figure 1.71). The critical height H_{cr} is defined as the maximum height at which a slope can stand before the state of tension, which develops as the slope yields, is relieved by tension cracks. Terzaghi (1943) gave the critical height in terms of total soil weight, having concluded that the tension crack would reach to one half the critical height, as

$$H'_{cr}=(2.67c/\gamma_t)\tan(45^\circ+\phi/2) \tag{1.14}$$

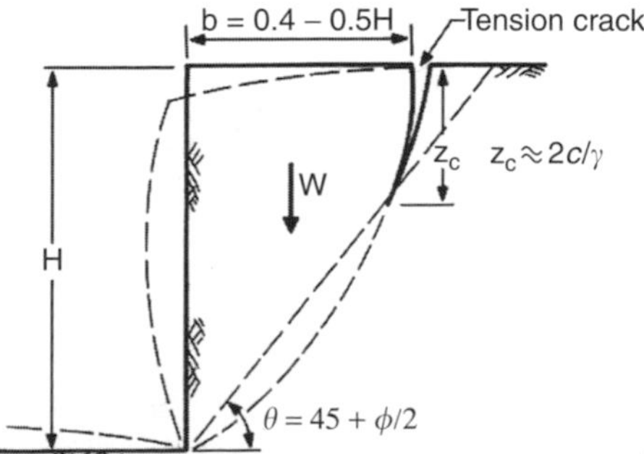

FIGURE 1.71
Critical height of a vertical slope in clay and the tension crack.

or for the case of $\phi=0$,

$$H_{cr}=2.67c/\gamma_t \tag{1.15}$$

Field observations indicate that the tension crack depth z_c ranges from $1/3H$ to $1/2H$. In practice, z_c is often taken as $1/2H_{cr}$ of an unsupported vertical cut or as

$$z_c=2c/\gamma \tag{1.16}$$

which is considered conservative (Tschebotarioff, 1973).

Case 7: Multiple planar failure surfaces are illustrated as follows, relationships not included:

- Active and passive wedge force system applicable to rock or soil and rock slopes — Figure 1.72.
- General wedge or sliding block method applicable to soil formations and earth dams — Figure 1.73.
- Intersecting joints along a common vertical plane — Figure 1.74.
- Triangular wedge failure, applicable to rock slopes — Figure 1.75.

Finite Slope: Cylindrical Failure Surface

In rotational slide failures, methods are available to analyze a circular or log-spiral failure surface, or a surface of any general shape. In all cases, the location of the critical failure surface is found by trial and error, by determining the factor of safety (FS) for various trial positions of the failure surface until the lowest value of the FS is reached. The forces acting on a free body taken from a slope are given in Figure 1.76.

Most modern analytical methods are based on dividing the potential failure mass into slices, as shown in Figure 1.77. The various methods differ slightly based on the force system assumed about each slice (Figure 1.78). Iterations of the complex equations to find the "critical circle" has led to the development of many computer programs for use with the personal computer (PC). Equation 1.17 given under the Janbu method is similar to the Simplified Bishop. In most cases, for analysis, the parameters selected for input are far more important than the method employed. The most significant parameter affecting the FS is usually the value input for shear strength; assuming even a small amount of cohesion can result in FS = 1.2 rather than 1.02, if only internal friction is assumed.

Ordinary method of slices: In 1936, Fellenius published a method of slices based on cylindrical failure surfaces which was known as the Swedish Circle or Fellenius method. Modified for effective stress analysis, it is now known as the Ordinary Method of Slices.

As illustrated in Figure 1.77, the mass above a potential failure surface is drawn to scale and divided into a number of slices with each slice having a normal force resulting from its weight. A flow net is drawn on the slope section (Figure 1.77a), or more simply, a phreatic surface is drawn. Pore pressures are determined as shown in Figure 1.77b. The equilibrium of each slice is determined and FS found by summing the resisting forces and dividing by the driving forces as shown in Figure 1.77c. The operation is repeated for other circles until the lowest safety factor is found. The method does not consider all of the forces acting on a slice (Figure 1.78a), as it omits the shear and normal stresses and pore-water pressures acting on the sides of the slice, but usually (although not always) it yields conservative results. However, the conservatism may be high.

Bishop's method of slices: This method considers the complete force system (Figure 1.78b), but is complex and requires a computer for solution. The results, however, are substantially

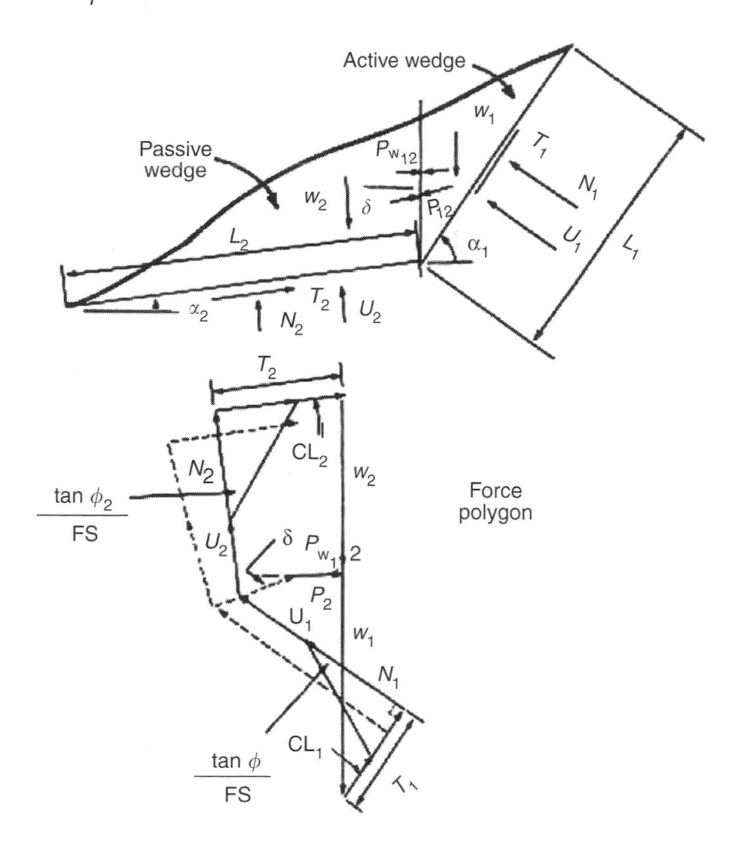

FIGURE 1.72
Forces acting on two wedges: one active, one passive. W_1, W_2 are the weights of the wedge, U_1, U_2 the resultant water pressure acting on the base of the wedge, N_1, N_2 the effective force normal to the base, T_1, T_2 the shear force acting along the base of the wedge, L_1, L_2 the length of the base, α_1, α_2 the inclination of the base to the horizontal, Pw_{12} the resultant water pressure at the interface, P_{12} the effective force at the interface, and δ the inclination of P_{12} to the horizontal. (From Morganstern, N. R. and Sangrey, D. A., *Landslides: Analysis and Control*, Schuster and Krizek, Eds., National Academy of Sciences, Washington, DC, 1978, pp. 255–272. Reprinted with permission of the National Academy of Sciences.)

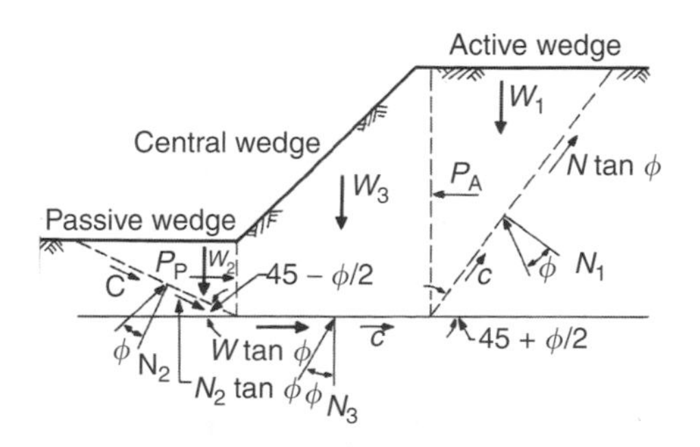

FIGURE 1.73
The general wedge or sliding block concept. (After NAVFAC, *Design Manual, Soil Mechanics, Foundation and Earth Structures*, DM-7.1, Naval Facilities Engineering Command, Alexandria, Virginia, 1982.)

more accurate than the ordinary method, and slightly more accurate than the modified Bishop's (1955) method.

Modified Bishop's method: This is a simplified Bishop's method (Janbu et al., 1956), widely used for hand calculations since it gives reasonably accurate solutions for circular failure surfaces. It is still widely used today on personal computers. The force system is given in Figure 1.78c.

Janbu's method: This is an approximate method applicable to circular as well as noncircular failure surfaces, as shown in Figure 1.79. It is sufficiently accurate for many practical

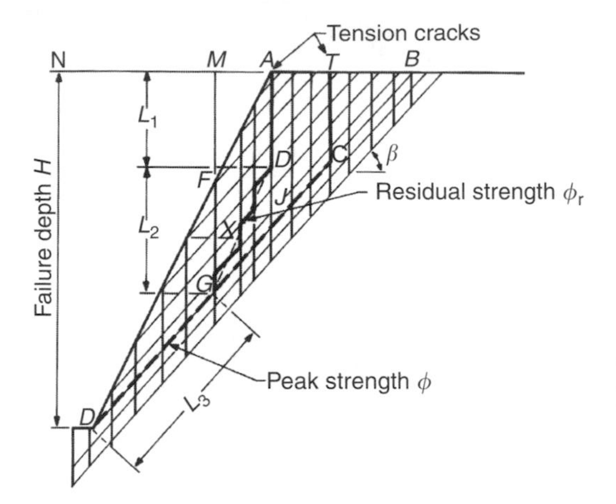

FIGURE 1.74
Geometry of two intersecting joint sets failing progressively (the multilinear failure surface). (From Barton, N., *Proceedings of ASCE, 13th Symposium on Rock Mechanics,* Urbana, Illinois [1971], 1972, pp. 139–170. With permission.)

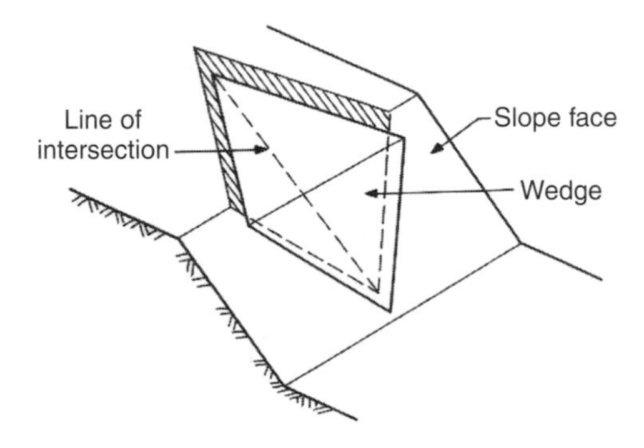

FIGURE 1.75
Geometry of a triangular wedge.

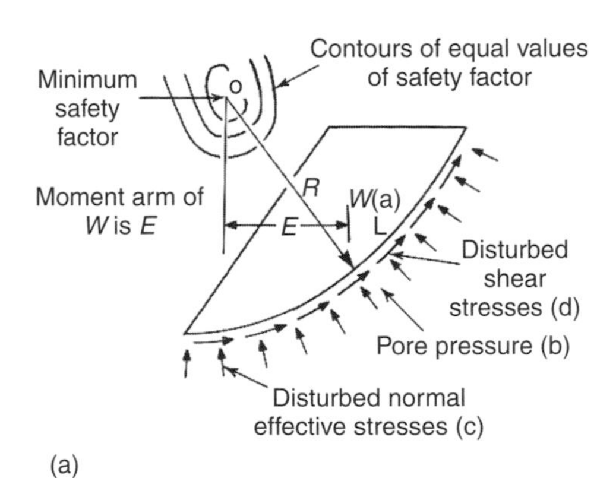

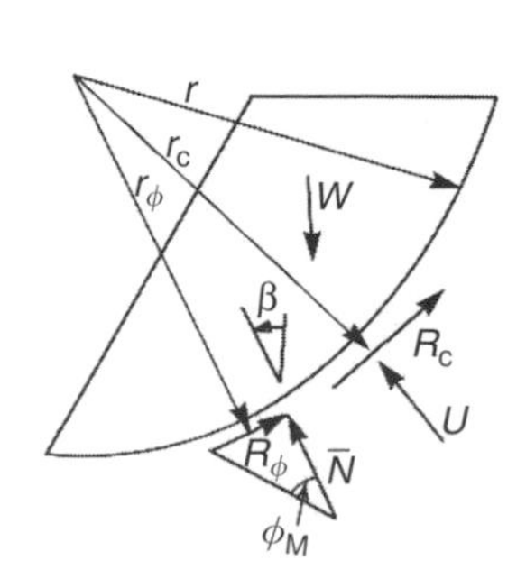

FIGURE 1.76
Forces acting on a free body with circular failure: (a) distributed stresses; (b) resultant forces.

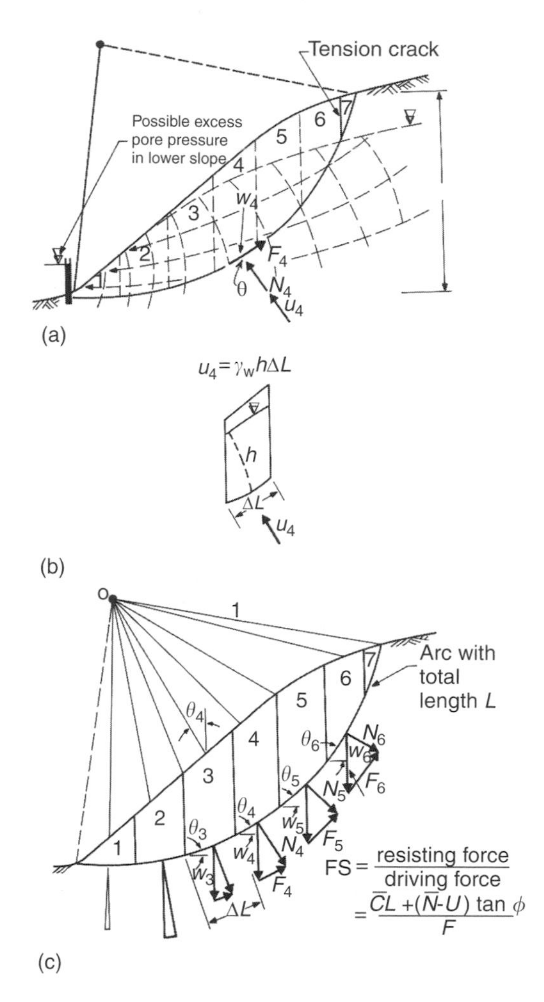

FIGURE 1.77
The ordinary method of slices. (a) Draw the slope and flow net to scale. Select failure surface and divide into equal slices of similar conditions. (b) Pore pressures for slice U_4. (c) Determine forces and safety factor:

$N_i = W_i \cos\theta - u_i L_i$ and $F = W_i \sin\theta$

$$FS = \frac{\Sigma \bar{c} \Delta L_i + \Sigma N_i \tan\phi}{\Sigma W_i \sin\theta_i}$$

Repeat for other r values to find FS_{min}.

cases (Janbu, 1973; Morganstern and Sangrey, 1978). Suitable for hand calculations it is particularly useful in slopes undergoing progressive failure on a long, noncylindrical failure surface where the location is known. It was used to analyze the failure illustrated in Figure 1.33. The equations are as follows:

$$FS = f_o \left(\Sigma\{[c'b + (W - ub)\tan\phi'] \, [1/\cos\theta \, M_i(\theta)]\}/\Sigma W \tan\theta + V\right) \quad (1.17)$$

where

$$M_i(\theta) = \cos\theta_I \, (1 + \tan\theta_i \tan\phi'/FS) \quad (1.18)$$

$$f_o = 1 + 0.50\,[d/L\text{-}1.4(d/L^2] \quad \text{for } c > 0, \phi > 0 \quad (1.19)$$

$$f_o = 1 + 0.31\,[d/L\text{-}1.4(d/L^2] \quad \text{for } c = 0$$

$$f_o = 1 + 0.66\,[d/L\text{-}1.4(d/L^2] \quad \text{for } \phi = 0$$

Data are input for each slice. Substitution of Equation 1.18 into Equation 1.17 results in FS on both sides of the equation. For solution a value for FS_1 is assumed and FS_2 calculated,

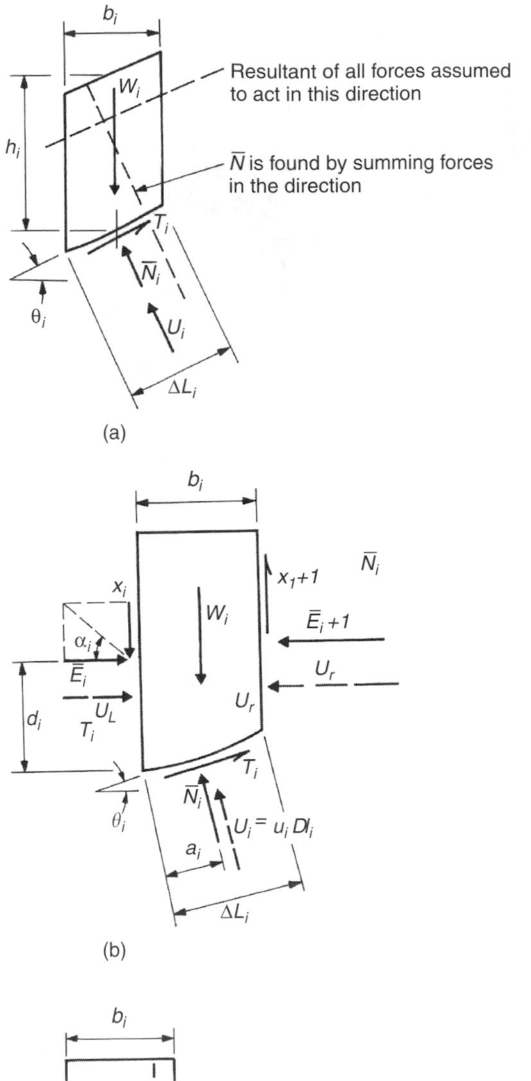

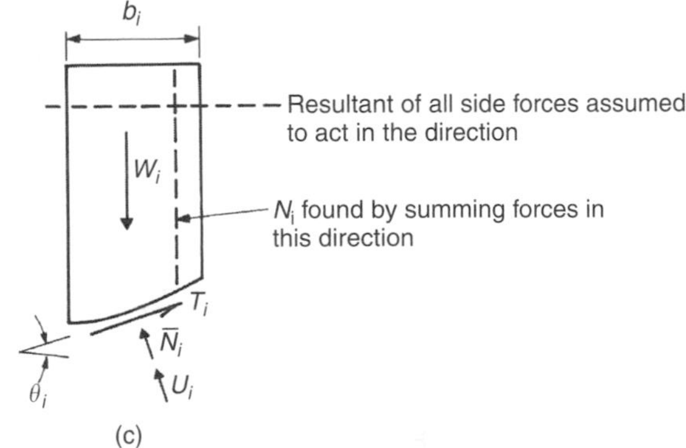

FIGURE 1.78
Force systems acting on a slice assumed by various analytical methods. (a) Ordinary method of slices. (b) Complete force system assumed by Bishop. (c) Force system for modified Bishop. (From Lambe, T. W. and Whitman, R. V., *Soil Mechanics*, Wiley, New York, 1969. Reprinted with permission of John Wiley & Sons, Inc.)

and the procedure is repeated until $FS_1 = FS_2$. Solutions converge rapidly. Parameters d and L are illustrated in Figure 1.79a. Janbu Equation 1.17 is similar to the modified Bishop, except for the parameter f_o and the denominator, which in the Bishop equation is $\Sigma W \sin\theta$.

The Morgenstern and Price method: This method (Morgenstern and Price, 1965) can be used to analyze any shape of failure surface and satisfies all equilibrium conditions. It is based on the Bishop method and requires a computer for solution. There are several

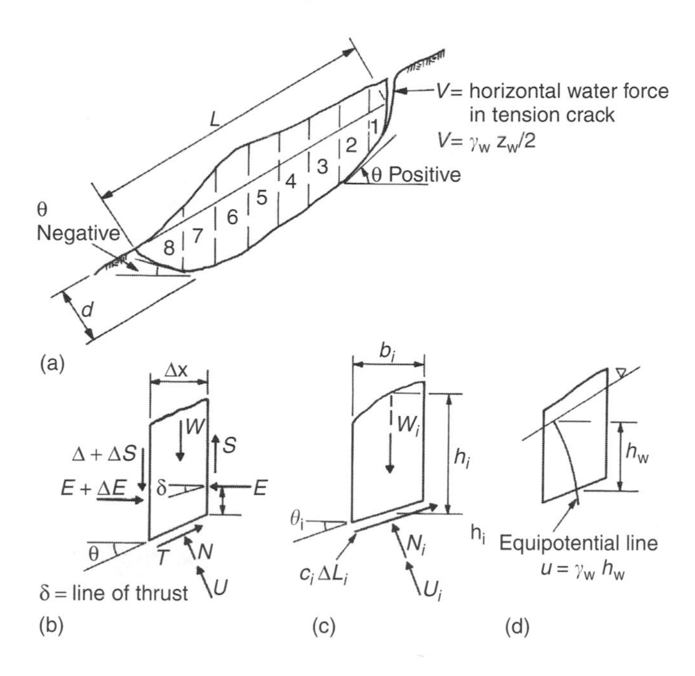

FIGURE 1.79
Nomenclature for Janbu's simplified method of analysis for noncircular failure surfaces: (a) failure mass geometry; (b) slice force system; (c) slice parameters for analysis; (d) determination of pore pressure u. Useful for cases where the failure surface is known or assumed.

theoretically possible positions for the line of action of the resultant forces between slices, and the line of action must be checked to determine if it is a possible one.

Spencer's method: This method (Spencer, 1967, 1973) is similar to the Morganstern and Price method.

Friction Circle method: See Taylor (1948).

Charts based on total stresses are used to find FS in terms of slope height and angle, and of soil parameters c, ϕ, and unit weight. The direction of the resultant normal stress for the entire free body is slightly in error because the resultant is not really tangent to the friction circle, but the analysis provides a lower bound for safety and is therefore conservative. The Taylor charts are strictly valid only for homogeneous slopes with no seepage. They consider that shear strength is mobilized simultaneously along the entire failure surface and that there is no tension crack. They are used for rough approximations and preliminary solutions of more complex cases. If the strength values vary along the failure surface they are averaged to obtain working values. This must be done with judgment and caution. For the foregoing conditions of validity, solutions using the charts are in close agreement with the method of slices described below.

Earthquake Forces

Pseudostatic methods have been the conventional approach in the past (Terzaghi, 1950). The stability of a potential sliding mass is determined for static loading conditions, and the effects of earthquake forces are accounted for by including equivalent horizontal forces acting on the mass. The horizontal force is expressed as the product of the weight of the sliding mass and a seismic coefficient that is expressed as a fraction of the acceleration due to gravity (see Section 3.3.4).

Dynamic analysis techniques provide for much more realistic results but also have limited validity. These techniques are described by Newmark (1965) and Seed (1968). The Newmark sliding block analyses are widely used for estimating the permanent displacements of slopes during earthquakes (Kramer and Smith, 1997; Wartman et al., 2003).

Summary

The applicability of mathematical analysis to various slope failure forms and the elements affecting slope failures are summarized in Table 1.4. General methods of stability analysis for sliding masses and the applicable geologic conditions are summarized in Table 1.5. Strength parameters acting at failure under various field conditions are summarized in Table 1.6.

1.3.3 Slope Characteristics

General

Qualitative assessment of slopes provides the basis for predicting the potential for failure and selecting practical methods for treatment, and for evaluating the applicability of mathematical solutions. The two major elements of qualitative assessment are slope characteristics (geology, geometry, surface conditions, and activity) and the environment (weather conditions of rainfall and temperature, and earthquake activity). The discussion in Section 1.2 presents relationships between the mode of slope failure and geologic conditions, as well as other slope characteristics, giving a basis for recognizing potential slope stability problems.

TABLE 1.4

Comparison of Elements and Classification of Geological and Engineering Failure Forms

	Elements of Slope Failures[a]						Engineering Failure Forms[b]				
Geologic Failure Forms	**Slope Inclination**	**Slope Height**	**Material Structure**	**Material Strength**	**Seepage Forces**	**Runoff**	**Infinite Slope**	**Single Planar Failure Surface (Simple Wedge and Sliding Block)**	**Multiple Planar Failure Surfaces (Oblique Surfaces Intersecting Parallel Surfaces)**	**Multiple Planar Failure Surfaces Intersecting Obliquely (Wedge)**	**Cylindrical Failure Surface**
Falls	P	N	P	P	P	N	N	N	N	P	N
Planar slides (translational block glides)	P	S	P	P	P	M	A	A	A	A	N
Rotational slides in rock	P	P	P	P	P	M	N	N	N	N	A
Rotational slides in soil	P	P	P	P	P	M	N	N	N	N	A
Lateral spreading and progressive failure	S	M	P	P	P	N	N	S	S	N	N
Debris slides	P	M	P	P	P	N	S	S	S	S	N
Debris avalanches	P	S	S	S	P	P	N	N	N	N	N
Debris flows	P	S	S	S	P	P	N	N	N	N	N
Rock fragment flow	P	S	P	P	P	N	N	N	N	N	N
Soil and mud flows	S	S	S	P	P	M	N	N	N	N	N
Submarine slides	S	S	P	P	P	N	N	N	N	N	S

[a] P — primary cause; S — secondary cause; M — minor effect; N — little or no effect.

[b] A — application; S — some application; P — poor application; N — no application.

TABLE 1.5
General Method of Stability Analysis and Applicable Geologic Condition for Slides

General Method of Analysis	Geologic Conditions
Infinite slope — (depth small compared with length of failure surface)	Cohesionless sands. Residual or colluvial soils over shallow rock. Stiff fissured clays and marine shales in the highly weathered zone
Limited slope	Sliding block
Simple wedge (single planar failure surface)	Interbedded dipping rock or soil Faulted or slickensided material Stiff to hard cohesive soil, intact, on steep slope
General wedge (multiple planar failure surfaces)	Sliding blocks in rock masses Closely jointed rock with several sets Weathered interbedded sedimentary rock Clay shales and stiff fissured clays Stratified soils Side-hill fills over colluvium
Cylindrical arc	Thick residual or colluvial soil Soft marine or clay shales Soft to firm cohesive soils

TABLE 1.6
Strength Parameters Acting at Failure under Various Field Conditions

Material	Field Conditions	Strength Parameters
(*a*) Cohesionless sands	Dry	ϕ
(*b*) Cohesionless sands	Submerged slope	$\bar{\phi}$
(*c*) Cohesionless sands	Slope seepage with top flow line coincident with and parallel to slope surface	$\bar{\phi}$
(*d*) Cohesive materials	Saturated slope, short-term or undrained conditions (ϕ=0)	s_u
(e) Cohesive materials (except for stiff fissured clays and clay shales)	Long-term stability	$\bar{\phi}$, $\bar{c}$
(*f*) Stiff fissured clays and clay shales	Part of failure surface	$\bar{\phi}_r$
(*g*) Soil or rock	Part of failure surface Existing failure surfaces	$\bar{\phi}$, $\bar{c}$ $\bar{\phi}_r$
(*h*) Clay shales or existing failure surfaces	Seepage parallel and top flow line coincident to slope surface	$\bar{\phi}_r$
(*i*) Pore-water pressures	Reduce ϕ in *e*, *f*, and *g* in accordance with seepage forces, γ applicable; or boundary water forces, γ_t applicable In (*c*) and (*h*), pore pressures reduce effectiveness by 50%	$(p-u)\tan\bar{\phi}$

Geologic Conditions

Significant Factors

- Materials forming the slope (for rock, the type and the degree of weathering; for soil, the type as classified by origin, mode of occurrence, and composition) as well as their engineering properties.
- Discontinuities in the formations, which for rock slopes include joints, shears, bedding, foliations, faults, slickensides, etc.; and, for soils include layering, slickensides, and the bedrock surface.
- Groundwater conditions: static, perched or artesian, and seepage forces.

Conditions with a High Failure Incidence

- Jointed rock masses on steep slopes can result in falls, slides, avalanches, and flows varying from a single block to many blocks.
- Weakness planes dipping down and out of the slope can result in planar failures with volumes ranging from very large to small.
- Clay shales and stiff fissured clays are frequently unstable in the natural state where they normally fail by shallow sloughing, but cuts can result in large rotational or planar slides.
- Residual soils on moderate to steep slopes in wet climates may fail progressively, generally involving small to moderate volumes, although heavy runoff can result in debris avalanches and flows, particularly where bedrock is shallow.
- Colluvium is generally unstable on any slope in wet climates and when cut can fail in large volumes, usually progressively.
- Glaciolacustrine soils normally fail as shallow sloughing during spring rains, although failures can be large and progressive.
- Glaciomarine and other fine-grained soils with significant granular components can involve large volumes in which failure may start by slumping, may spread laterally, and under certain conditions may become a flow.
- Any slope exposed to erosion at the toe, particularly by stream activity; cut too steeply; subject to unusually heavy rainfall; or experiencing deformation.

Some Examples

A general summary of typical forms of slope failures as related to geologic conditions is given in Table 1.3.

Dipping beds of sedimentary rocks in mountainous terrain are often the source of disastrous slides or avalanches (see Figure 1.18). Very large planar slides failing along a major discontinuity occur where the beds incline in the slope direction. On the opposite side of the failure in Figure 1.18 the slope is steeper and more stable because of the bedding orientation. Failures will generally be small, evolving under joint sets, although disastrous avalanches have occurred under these conditions, such as the one at Turtle Mountain, Alberta.

Orientation of joints with respect to the rock slope face controls stability and the form of failure. The near-vertical slope in the 40-year-old railroad cut illustrated in Figure 1.80 is stable in decomposed amphibolite gneiss because of the vertical jointing. The cut shown in Figure 1.81 is near that of Figure 1.80 but at a different station and on the opposite side of the tracks. Here, the slope is much flatter, approximately 1:1, but after 40 years is still experiencing failures such as that of the wedge shown in the photo that broke loose along the upper joints and slid along a slickensided surface. These examples illustrate how joint orientation controls slope stability, even in "soft" rock. The cuts were examined as part of a geologic study for 30 km of new railroad to be constructed in the same formation but some distance away.

Sea erosion undercutting jointed limestone illustrated in Figure 1.82 was causing concern over the possible loss of the roadway, which is the only link between the town of Tapaktuan, Sumatra, and its airport. A fault zone may be seen on the right-hand side of the photo. For the most part, the joints are vertical and perpendicular to the cliff face, shown as plane *a* in Figure 1.83, and the conditions are consequently stable. Where the joints are parallel to the face and inclined into it, as shown by plane *b* in the figure, a potentially unstable condition exists. This condition was judged to prevail along a short stretch of road beginning to the

FIGURE 1.80
Near-vertical slope in 40-year-old railroad cut standing stable in amphibolite gneiss because of vertical jointing. (Tres Ranchos, Goias, Brazil). Compare with Figure 1.81.

right of the photo, shown in Figure 1.84. The recommendation was to cut into the landward slope and relocate the roadway away from the sea cliff along this short stretch. The very costly alternative was to relocate the roadway inland around and over the coastal mountain.

The major cause of instability in colluvial soil slopes is illustrated in Figure 1.85. The slide debris impedes drainage at the toe and causes an increase in pore-water pressures in (b) over those in (a). The sketch also illustrates the importance of placing piezometers at different depths because of pressure variations. Conditions in (b) apply also to the case of a side-hill embankment for which a free-draining blanket beneath the fill would be necessary to provide stability.

Slope Geometry

General

The significant elements of slope geometry are inclination, height, and form. Aspects of inclination and height, as they relate to a particular point along a slope, are described in Section 1.1.4. This section is more concerned with the form and other characteristics of an entire slope as they affect seepage and runoff, which can be dispersed by the geometrical configuration of the slope or can be concentrated. The difference influences slope stability.

The examples given are intended to illustrate the *importance of considering the topography of an entire slope* during roadway planning and design, not only the immediate cut or fill area.

Topographic Expression

In both natural and cut slopes, the topographic expression has a strong influence on where failure may occur since landform provides the natural control over rainfall infiltration and runoff when geologic factors are constant. In Figure 1.86, runoff is directed away from the

FIGURE 1.81
Same cut as in Figure 1.80, but at a different station. Wedge failure along dipping joints.

convex nose form at (a) and a cut made there will be stable at a much steeper angle than at (b), where runoff is concentrated in the swale or concave form. Runoff and seepage at (c) are less severe than at (b) but still a problem to be considered. Natural slides, avalanches, and flows usually will not occur at (a), but rather at (b) and (c), with the highest incidence at (b) (see Figure 1.89 and Figure 1.90).

Location of Cut on Slope

Cuts in level ground or bisecting a ridge perpendicular to its strike will be stable at much steeper inclinations than cuts made along a slope, parallel to the strike (side-hill cuts). The side-hill cut in Figure 1.87 intercepts seepage and runoff from upslope and will be much less stable on its upslope side than on its downslope side where seepage is directed away from the cut. The treatment to provide stability, therefore, will be more extensive on the upslope side than on the opposite side.

The significance of cut locations along a steeply inclined slope in mountainous terrain in a tropical climate is illustrated in Figure 1.88. A cut made at location 1 will be much less stable than at location 3, and treatment will be far more costly because of differences in runoff and seepage quantities. River erosion protection or retention of the cut slope at 1 can be more costly than the roadway itself. Retention would not be required at 3 if a stable cut angle were selected, but might be required at 2 together with positive seepage control.

FIGURE 1.82
Sea erosion undercutting limestone and causing rockfalls. Slope failure could result in loss of roadway in photo middle (Tapaktuan, Sumatra) (see Figure 1.83 and Figure 1.84).

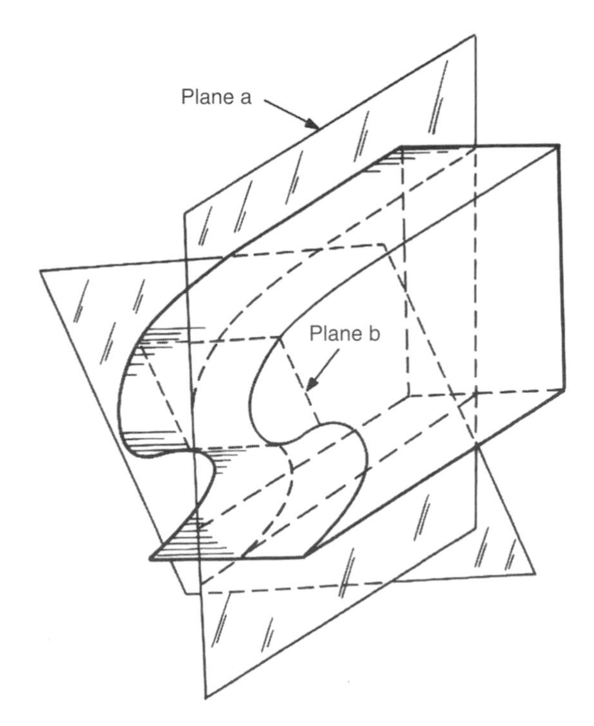

FIGURE 1.83
Orientation of fracture planes controls rock mass stability. Plane (a) represents stable conditions and plane (b) unstable.

Surface Conditions

Seepage Points

Observations of seepage points should be made in consideration of the weather conditions prevailing during the weeks preceding the visit, as well as the season of the year, and

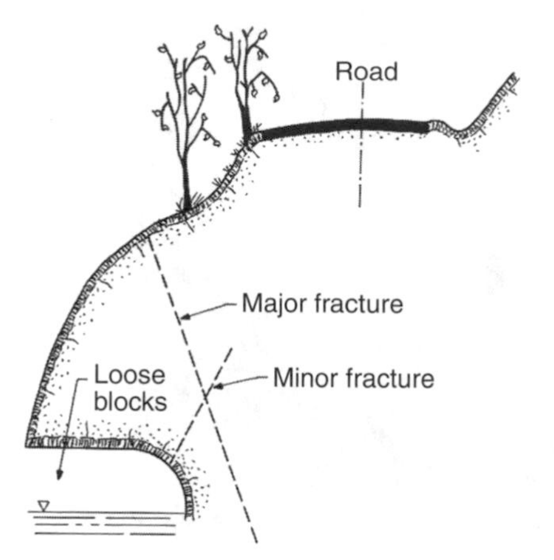

FIGURE 1.84
Possible orientation of fracture planes at km 10+750 which might lead to a very large failure (Tapaktuan, Sumatra).

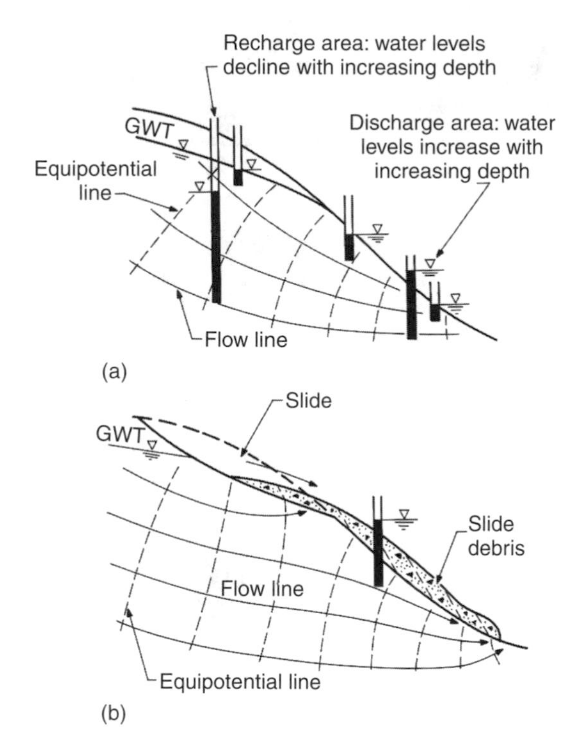

FIGURE 1.85
The effect of colluvium on groundwater flow in a slope: (a) flow in slope before slide; (b) flow in slope with mantle of slide debris. (From Patton, F. D. and Hendron, A. J., Jr., *Proceedings of the 2nd International Congress, International Association of Engineering Geology,* São Paulo, 1974, P.V–GR 1. With permission.)

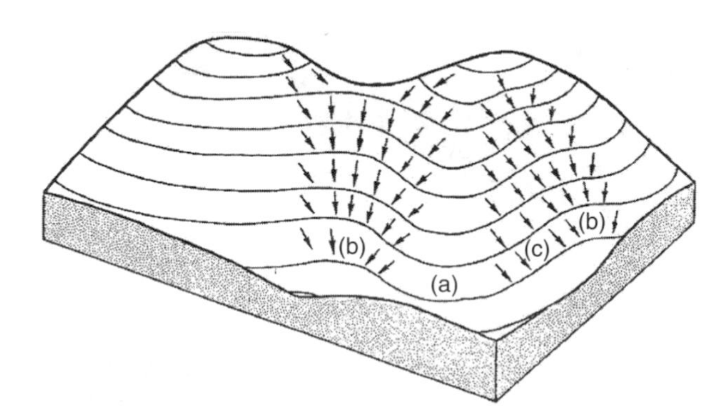

FIGURE 1.86
The influence of topography on runoff and seepage. Cuts and natural slopes at convex slopes at (a) are relatively stable compared with the concave slope at (b) or the slope at (c).

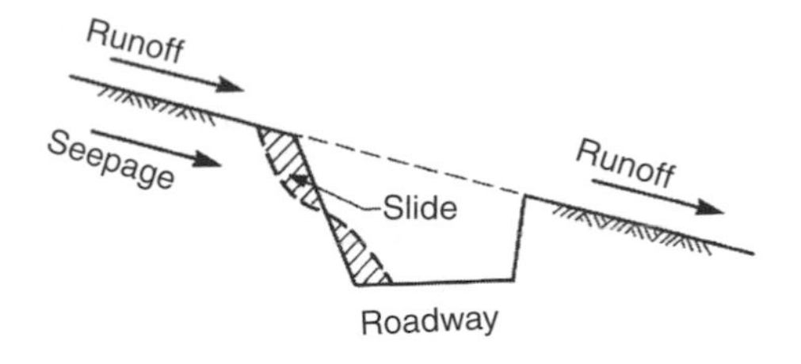

FIGURE 1.87
Upslope side of hillside cut tends to be much less stable than downslope side because of runoff and seepage.

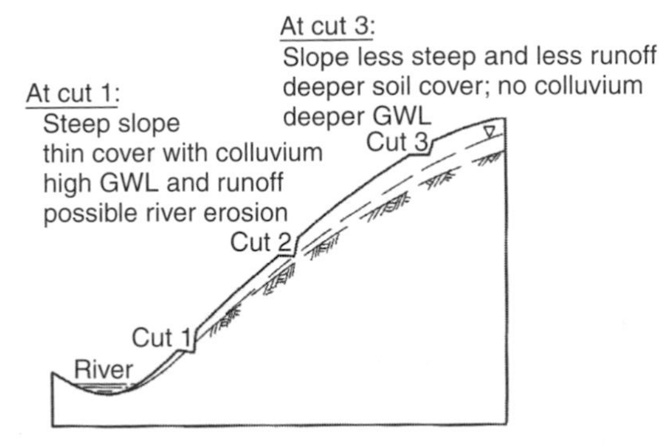

FIGURE 1.88
Stability problems may be very much related to the cut location along a steeply inclined slope in mountainous terrain in a tropical climate, and generally decrease in the upslope direction.

regional climatic history. No slope seepage during a rainy period may be considered as very favorable for stability, if there is no blockage from ice, colluvium, etc. On the other hand, seepage during a dry period signifies that a substantial increase in seepage will occur during wet periods. Toe seepage indicates a particularly dangerous condition, especially during dry periods.

Vegetation

Density of vegetation is an important factor in slope stability. Recently cleared upslope areas for logging, farming, or grazing are very likely to be locations where failures will occur in freshly made cuts, or in old cuts during severe weather conditions. Removal of vegetation permits an increase in erosion, a reduction in strength in the shallow portions of the slope from the loss of root structure, an increase in infiltration during rainy periods, and an increase in evaporation during dry spells resulting in surface desiccation and cracking.

Certain types of vegetation may be indicators of potential instability. For example, in tropical climates, such as in Brazil and Indonesia, banana plants seem to favor colluvial soil slopes, probably because colluvium has a higher moisture content than the residual soils in the same area.

Indications of Instability

Surface features indicating instability include tilted or bending tree trunks, tilted poles and fence posts, tension cracks along the slope and beyond the crest, and slump and hummocky topography, as described and illustrated in Sections 1.1.2 and 1.2.

Slope Activity

Degrees of Activity

Slopes reside at various degrees of activity, as discussed in Section 1.1.2, ranging through stable slopes with no movement, early failure stages with creep and tension cracks, intermediate failure stages with significant movement, partial total failures with substantial displacement, to complete failure with total displacement.

Evidence that a slope is unstable requires an assessment of the imminence of total collapse, and, if movement is occurring, how much time is available for treatment and stabilization. Tension cracks, in particular, serve as an early warning of impending failure and are commonly associated with the early stages of many failures. Their appearance, even together with scarps, does not necessarily mean, however, that failure is imminent. The slope shown in Figure 1.89 appears precipitous but the slide has moved very little over a 4-year period, an interval of lower than normal rainfall (see Section 1.3.4). Note the "nose" location; a massive slide has already occurred along the slope at the far left in the photo that closed the highway for a brief time. This "total failure," shown in Figure 1.90, occurred where the slope shape was concave, a few hundred meters from the slope in Figure 1.89.

Movement Velocity vs. Failure

The most significant factors indicating approaching total failure for many geologic conditions are velocities of movement and accelerations.

During surface monitoring, both vertical and horizontal movements should be measured, and evaluated in terms of velocity and acceleration. There is a lack in the literature of definitive observational data that relate velocity and acceleration to final or total failure in a manner suitable for formulating judgments as to when failure is imminent. Movement velocities before total failure and the stabilization treatment applied are summarized in Table 1.7 for a few cases from the literature. From the author's experience and literature review, it appears that, as a rule of thumb, if a slope of residual or colluvial soils is moving at a rate of the order of 2 to 5 cm/day (0.8 to 2.0 in./day) during a rainy season with the probability of storms, and if the velocity is increasing, final failure is imminent

FIGURE 1.89
Slump movement caused by cut made in residual soils for the Rio-Santos Highway (Itaorna, Brazil) has remained stable as shown for at least 4 years. (Photo taken in 1978.) The slope form is convex. The failure shown in Figure 1.90 was located around the roadway bend in the photo (left).

FIGURE 1.90
Debris avalanche scar along the Rio-Santos Highway at Praia Brava, Itaorna, Brazil. The slope form was concave; location is around the bend from Figure 1.89.

TABLE 1.7
Velocities of Slide Movements before Total Failure and Solutions[a]

Location	Material	Movement Velocity	Solution	Reference
Philippine Islands	Weathered rock (open-pit mine)	2 cm/day	Horizontal adits	Brawner (1975)
Santos, Brazil	Colluvium (cut)	2.5 cm/day	Trenches, galleries	Fox (1964)
Rio de Janeiro state, Brazil	Residuum (cut)	0.4–2.2 cm/day	None applied, no total failure in 20 years	Garga and DeCampos (1977)
Rio-Santos Highway, Brazil	Residuum (cut)	2–3.5 cm/day for first 2 weeks, 30 cm/day during 6th week, then failure	Removal of failure mass	Hunt (1978) (see Section 1.3.3)
Golden slide	Debris (cut)	2.5 cm/day	Horizontal drains, vertical wells	Noble (1973)
Pipe Organ slide	Debris (cut)	5 cm/week	Gravity drains	Noble (1973)
Vaiont, Italy	Rock-mass translation	1 cm/week, then 1 cm/day, then 20–30 cm/day and after 3 weeks, 80 cm/day and failure		Kiersch (1965)
Portuguese Bend	Lateral spreading	1956–1957, 5–12 cm/yr 1958, 15–60 cm/yr 1961–1968, 1–3m/yr 1968, dramatic increase, houses destroyed 1973, 8 cm/day during dry season 1973, 10 cm/day during wet season 1973, 15 cm/day during heavy rains No total mass failure	None	Easton (1973)

[a] From examples given in Chapter 1.

and may occur during the first heavy rain, or at some time during the rainy season. The following two case histories relate slope movements to total failure.

Case 1: A roadway cut made in residual soils in the coastal mountains of Brazil continued to show instability through creep, tension cracks, small slumps, and periodic encroachment on the roadway for a period of several years. An intermediate failure stage occurred on November 29, 1977, after a weekend of moderately heavy rain and a period during which the highway department had been removing material from the slope toe. Figure 1.91, a photo taken from a helicopter about 10 days after the failure, illustrates the general conditions. Tension cracks have opened at the base of the forward transmission tower, a large scarp has formed at midslope, and the small gabion wall at the toe has failed.

Figure 1.92 illustrates the tension crack and the distortion in the transmission tower shown in Figure 1.91. The maximum crack width was about 30 cm (12 in.) and the scarp was as high as 50 cm (20 in.). Slope movement measurements were begun immediately by optical survey and the transmission lines were quickly transferred to a newly constructed tower situated farther upslope.

For the first 2 weeks after the initial movement, a period of little rainfall, the vertical drop along the scarp was about 2 to 3.5 cm/day (0.8 to 1.4 in./day). In 5 weeks, with occasional rainy periods, the scarp had increased to 3 m (10 ft). Finally, after a weekend of heavy rains the slide failed totally in its final stage within a few hours, leaving a scarp about 30 m (100 ft) in height as shown in Figure 1.93, and partially blocking the roadway. Excavation removed the slide debris, and 2 years afterward the high scarp still remained. The tower in the photo was again relocated farther upslope. Although future failures will occur, they will be too far from the roadway to cover the pavement.

Remediating the slope by permitting failure is an example of one relatively low-cost treatment method. Cutting material from the head of the slide would be an alternate method, but there was concern that perhaps time for construction prior to total failure

FIGURE 1.91
Rotational slide at Muriqui, Rio de Janeiro, Brazil, initiated by the road cut, had been active for 4 years. Major movements began after a weekend of heavy rains, endangering the transmission tower. Tension crack and distorted transmission tower shown in the inset, Figure 1.92.

FIGURE 1.93
Total collapse of slope shown in Figure 1.91 left a high head scarp and temporarily closed the roadway. Failure occurred 6 weeks later during the rainy season after several days of heavy rains.

would be inadequate. Considering the slope activity, weather conditions, costs, and construction time, a retaining structure was considered as not practical.

Case 2: Another situation is similar to that described in Case 1 although the volume of the failing mass is greater. A 25° slope of residual soil has been moving for 20 years since it was activated by a road cut (Garga and DeCampos, 1977). Each year during the rainy season movement occurs at a velocity measured by slope inclinometer ranging from 0.4 to 2.2 cm/day (0.2 to 0.9 in./day). The movement causes slide debris to enter the roadway, from which it is removed. During the dry season movement ceases, and as of the time of the report (1977), total failure had not occurred.

1.3.4 Weather Factors

Correlations between Rainfall and Slope Failures

Significance

Ground saturation and rainfall are the major factors in slope failures and influence their incidence, form, and magnitude. Evaluating rainfall data are very important for anticipating and predicting slope failures. Three aspects are important:

1. Climatic cycles over a period of years, i.e., high annual precipitation vs. low annual precipitation.
2. Rainfall accumulation in a given year in relationship to normal accumulation.
3. Intensities of given storms.

Cumulative Precipitation vs. Mean Annual Precipitation

A study of the occurrence of landslides relative to the cumulative precipitation record up to the date of failure as a percentage of the mean annual precipitation (termed the cycle coefficient C_c) was made by Guidicini and Iwasa (1977). The study covered nine areas of the mountainous coastal region of Brazil, which has a tropical climate characterized by a wet season from January through March and a dry season, June through August.

Cumulative precipitation increases the ground saturation and a rise in the water table. A rainstorm occurring during the dry season or at the beginning of the wet season will have a lesser effect on slope stability than a storm of the same intensity occurring near the end of the wet season. A plot by month of the occurrence of failures as a function of the cycle coefficient is given in Figure 1.94. It is seen that the most catastrophic events occur toward the end of the rainy season, when the cumulative precipitation is higher than the mean annual.

Regarding rainfall intensity, Guidicini and Iwasa concluded that:

- Extremely intense rainfalls, about 12% greater than the mean annual rainfall (300 mm in 24 to 72 h) or more, can cause natural slope failures in their area, regardless of the previous rainfall history.
- Intense rainfalls, up to 12% of the mean annual, where the precipitation cycle is normal or higher, will cause failures, but if the preceding precipitation level is lower than the mean annual, failures are not likely even with intensities to 12%.
- Rainfalls of 8% or less of annual precipitation will generally not cause failures, regardless of the preceding precipitation, because the gradual increase in the saturation level never reaches a critical magnitude.
- A danger level chart (Figure 1.95) was prepared by Guidicini and Iwasa for each study area, intended to serve communities as a guide for assessing failure hazard in terms of the mean cumulative precipitation for a given year.

Such data can be used for temporarily closing mountain roadways, subject to slope failures, when large storms are expected during the rainy season. At the least warnings signs can be placed.

Evaluating Existing Cut Slope Stability

General

It is often necessary to evaluate a cut slope that appears stable to formulate judgments as to whether it will remain so. If a cut slope has been subjected during its lifetime to conditions

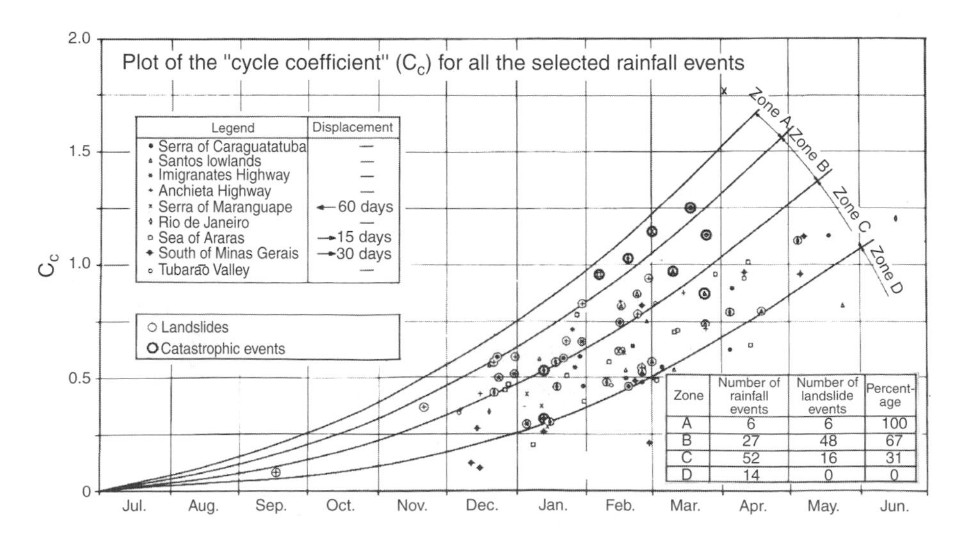

FIGURE 1.94
Comparison of landslide events and rainfall cycle coefficient C_c for coastal mountains of Brazil. (From Guidicini, G. and Iwasa, O. Y., *Bull. Int. Assoc. Eng. Geol.*, 16, 13–20, 1977. With permission.)

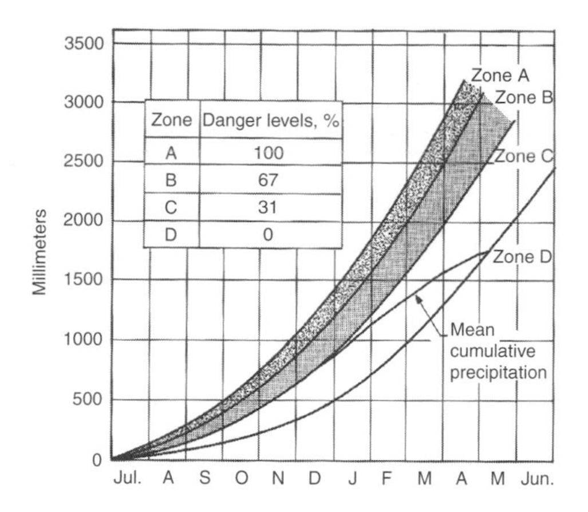

FIGURE 1.95
Correlation between rainfall and landslides (Serra de Caraguatatuba, São Paulo, Brazil). Chart based on rainfall record station No. E2-65-DAEE, installed in May 6, 1928, in the town of Caraguatatuba. Mean annual precipitation of 1905 mm based on 46 year record. Approximate station elevation of 10 m. (From Guidicini, G. and Iwasa, O. Y., *Bull. Int. Assoc. Eng. Geol.*, 16, 13–20, 1977. With permission.)

drier than normal and there have been no major storms, it can be stated that the slope has *not* been tested under severe weather conditions. It may be concluded that it is not necessarily a potentially stable slope. If a cut is failing under conditions of normal rainfall, it can be concluded that it will certainly undergo total failure at some future date during more severe conditions.

Case 1: In a study of slope failures on the island of Sumatra, examination was made of several high, steep slopes cut in colluvium which were subjected to debris and slump slides during construction. Failure occurred during a normal rainy season of 500 mm (20 in.) for the month of occurrence. The cuts were reshaped with some benching and flatter inclinations and have remained stable for a year of near-normal rainfall of about 2500 mm (98 in.) with monthly variations from 80 to 673 mm (3 to 26 in.).

Rainfall records were available during the study for only a 5-year period, but during the year before the cuts were made 1685 mm (66 in.) were reported for the month of December during monsoon storms. The cuts cannot be considered stable until subjected to a rainfall of this magnitude, unless there is an error in the data or the storm was a very unusual occurrence. Neither condition appears to be the case for the geographic location.

Case 2: A number of examples have been given of slope failures along the Rio Santos Highway that passes through the coastal mountains of Brazil. Numerous cuts were made in the years 1974 and 1975 without retention, and a large number of relatively small slides and other failures have occurred. The solution to the problem adapted by the highway department, in most cases, is to allow the failures and subsequently clean up the roadway. As of 1980, except for short periods, during the slide illustrated in Figure 1.93, the road has remained in service.

A review of the rainfall records for the region during the past 40 years (from 1984) revealed that the last decade had been a relatively dry period with rainfall averaging about 1500 to 2000 mm (59 to 79 in.). During the previous 30 years, however, the annual rainfall averaged 2500 to 3500 mm (98 to 138 in.). One storm in the period dropped 678 mm (27 in.) in 3 days (see Section 1.2.8). In view of the already unstable conditions along the roadway, if the weather cycle changes from the currently dry epoch to the wetter cycle of the previous epoch, a marked increase in incidence and magnitude of slope failures can be anticipated.

Temperatures

Freezing temperatures and the occurrence of frost in soil or rock slopes are highly significant. Ground frost can wedge loose rock blocks and cause falls, or in the spring months can block normal seepage, resulting in high water pressures which cause falls, debris avalanches, slides, and flows. A relationship among the number of rock falls, mean monthly temperature, and mean monthly precipitation is given in Figure 1.96. It is seen that the highest incidence for rock falls is from November through March in the Fraser Canyon of British Columbia.

1.3.5 Hazard Maps and Risk Assessment

Purpose

Degrees of slope-failure hazards along a proposed or existing roadway or other development can be illustrated on slope hazard maps. Such maps provide the basis not only for establishing the form of treatment required, but also for establishing the degree of urgency for such treatment in the case of existing works, or the programming of treatment for future works. They represent the product of a regional assessment.

Hazard Rating Systems

In recent years, various organizations have developed hazard rating systems. In the United States, the system apparently most commonly used is the "Rockfall Hazard Rating System" (RHRS) developed for the Federal Highway Administration by the Oregon State Highway Department (Pierson et al., 1990). Highway departments use the RHRS to inventory and classify rock slopes according to their potential hazard to motorists, and to identify those slopes that present the greatest degree of hazard and formulate cost estimates for treatments (McKown, 1999).

Some states, such as West Virginia (Lessing et al., 1994), have prepared landslide hazard maps. The Japanese have studied the relationships between earthquake magnitude and epicenter distance to slope failures in Japan and several other countries and have proposed procedures for zoning the hazard (Orense, 2003).

Example

The Problem

A 7-km stretch of existing mountain roadway with a 20-year history of slope failures, including rotational slides, debris slides, avalanches, and rock falls, was mapped in detail with respect to slope stability to provide the basis for the selection of treatments and the establishment of treatment priorities. A panoramic photo of the slopes in the higher elevations along the roadway is given in Figure 1.97.

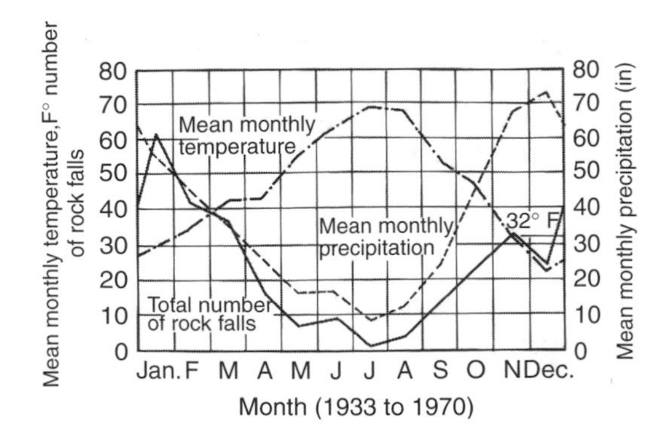

FIGURE 1.96
Number of rock falls, mean monthly temperature, and mean monthly precipitation in the Fraser Canyon of British Columbia for 1933–1970. (Peckover, 1975; from Piteau, D. R., *Reviews in Engineering Geology*, Vol. III, Coates, D. R., Ed., Geologic Society of America, Boulder, Colorado, 1977, pp. 85–112. With permission.)

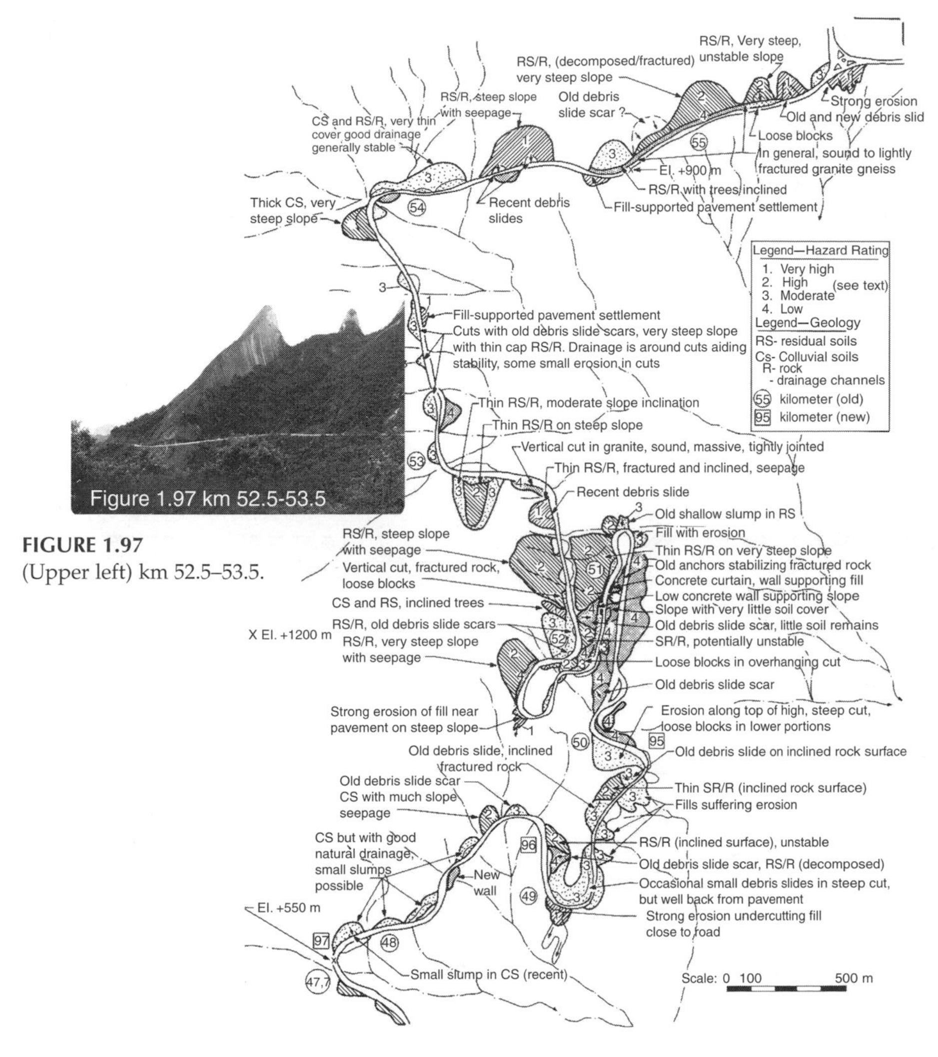

FIGURE 1.97
(Upper left) km 52.5–53.5.

FIGURE 1.98
Slope failure hazard map for an existing roadway subject to numerous failures (BR-116, Rio de Janeiro, Brazil). A panorama of the road between km 52 and 53.5 is shown in the photo (Figure 1.97). A debris avalanche that occurred at km 56 is shown in Figure 1.10.

The Slope Failure Hazard Map

The map (Figure 1.98), prepared by the author during a 1979 study, illustrates the location of cuts and fills, drainage, and the degree of hazard. Maps accompanying the report gave geologic conditions and proposed solutions. The maps were prepared by enlarging relatively recent aerial photographs to a scale of 1:10,000 to serve as a base map for plotting, since more accurate maps illustrating the topography and locations of cuts and fills were not available.

Five degrees of hazard were used to describe slope conditions:

1. *Very high*: Relatively large failures will close the roadway. Slopes are very steep with a thin cover of residual or colluvial soils over rock, and substantial water penetrates the mass. Fills are unstable and have suffered failures.

2. *High*: Relatively large failures probably will close the road. Failures have occurred already in residual or colluvial soils and in soils over rock on moderately steep slopes. Fills are unstable.
3. *Moderate*: In general, failures will not close the road completely. Relatively small failures have occurred in residual soils on steep slopes, in colluvial soils on moderate slopes with seepage, in vertical slopes with loose rock blocks, and in slopes with severe erosion.
4. *Low*: Low cuts, cuts in strong soils or stable rock slopes, and fills. Some erosion is to be expected, but in general slopes are without serious problems.
5. *No Hazard*: Level ground, or sound rock, or low cuts in strong soils.

Note: Several years after the study, during heavy rains, a debris avalanche occurred at km 52.9 (a high-hazard location) carrying a minibus downslope over the side of the roadway resulting in a number of deaths

1.4 Treatment of Slopes

1.4.1 General Concepts

Selection Basis

Basic Factors

The first factor to consider in the selection of a slope treatment is its purpose, which can be placed in one of two broad categories:

- Preventive treatments which are applied to stable, but potentially unstable, natural slopes or to slopes to be cut or to side-hill fills to be placed
- Remedial or corrective treatments which are applied to existing unstable, moving slopes or to failed slopes

Assessment is then made of other factors, including the degree of the failure hazard and risk (see Section 1.1.3) and the slope condition, which can be considered in four general groupings as discussed below.

Slope Conditions

Potentially unstable natural slopes range from those subjects to falls or slides where development along the slope or at its base can be protected with reasonable treatments, to those where failures may be unpreventable and will have disastrous consequences. The potential for the latter failures may be recognizable, but since it cannot be known when the necessary conditions may occur, the failures are essentially unpredictable. Some examples of slope failures that could not have been prevented with any reasonable cost include those at Nevada Huascaran in the Andes that destroyed several towns (see Section 1.2.8), the Achocallo mudflow near La Paz (see Section 1.2.11), and the thousands of debris avalanches and flows that have occurred in a given area during heavy storms in mountainous regions in tropical climates (see Section 1.2.8).

Unstable natural slopes undergoing failure may or may not require treatment depending upon the degree of hazard and risk, and in some instances, such as Portuguese Bend (see Section 1.2.6), stabilization may not be practical because of costs.

Unstable cut slopes in the process of failure need treatment, but stabilization may not be economically practical.

New slopes formed by cutting or filling may require treatment by some form of stabilization.

Initial Assessment

An initial assessment is made of the slope conditions, the degree of the hazard, and the risk. There are then three possible options to consider for slope treatment:

- avoid the high-risk hazard, or
- accept the failure hazard, or
- stabilize the slope to eliminate or reduce the hazard.

Treatment Options

Avoid the High-Risk Hazard

Conditions: Where failure is essentially not predictable or preventable by reasonable means and the consequences are potentially disastrous, as in mountainous terrain subject to massive planar slides or avalanches, or slopes in tropical climates subject to debris avalanches, or slopes subject to liquefaction and flows, the hazard should be avoided.

Solutions: Avoid development along the slope or near its base and relocate roadways or railroads to areas of lower hazard where stabilization is feasible, or avoid the hazard by tunneling.

Accept the Failure Hazard

Conditions: Low to moderate hazards, such as partial temporary closure of a roadway, or a failure in an open-pit mine where failure is predictable but prevention is considered uneconomical, may be accepted.

Open-pit mines: Economics dictates excavating the steepest slope possible to minimize quantities to be removed, and most forms of treatment are not feasible; therefore, the hazard is accepted. Slope movements are monitored to provide for early warning and evacuation of personnel and equipment. In some instances, measures may be used to reduce the hazard where large masses are involved, but normally failures are simply removed with the equipment available.

Roadways: Three options exist besides avoiding the hazard, i.e., accept the hazard, reduce the hazard, or eliminate it. Acceptance is based on an evaluation of the degree of hazard and the economics of prevention. In many cases involving relatively small volumes failure is self-correcting and most, if not all, of the unstable material is removed from the slope by the failure; it only remains to clean up the roadway. These nuisance failures commonly occur during or shortly after construction when the first adverse weather arrives. The true economics of this approach, however, depends on a knowledgeable assessment of the form and magnitude of the potential failure, and assurance that the risk is low to moderate. Conditions may be such that small failures will evolve into very large ones, or that a continuous and costly maintenance program may be required. Public opinion regarding small but frequent failures of the nuisance type also must be considered.

Eliminate or Reduce the Hazard

Where failure is essentially predictable and preventable, or is occurring or has occurred and is suitable for treatment, slope stabilization methods are applied. For low- to moderate-risk conditions, the approach can be either to eliminate or to reduce the hazard, depending on comparative economics. For high-risk conditions the hazard should be eliminated.

Slope Stabilization

Methods

Slope stabilization methods may be placed in five general categories:

1. Change slope geometry to decrease the driving forces or increase the resisting forces.
2. Control surface water infiltration to reduce seepage forces.
3. Control internal seepage to reduce the driving forces and increase material strengths.
4. Provide retention to increase the resisting forces.
5. Increase soil strength with injections. In a number of instances the injection of quicklime slurry into predrilled holes has arrested slope movements as a result of the strength increase from chemical reaction with clays (Handy and Williams, 1967; Broms and Bowman, 1979). Strength increase in saltwater clays, however, was found to be low.

Stabilization methods are illustrated generally in Figure 1.99 and summarized in Table 1.8 with respect to conditions and general purpose. "General Purpose" indicates whether the aim is to prevent failure or to treat the slope by some remedial measure.

Selection

In the selection of the stabilization method or methods, consideration is given to a number of factors including:

- Material types composing the slope and intensity and orientation of the discontinuities.
- Slope activity.
- Proposed construction, whether cut or side-hill .
- Form and magnitude of potential or recurring failure (summary of preventive and remedial measures for the various failure forms is given in Table 1.9).

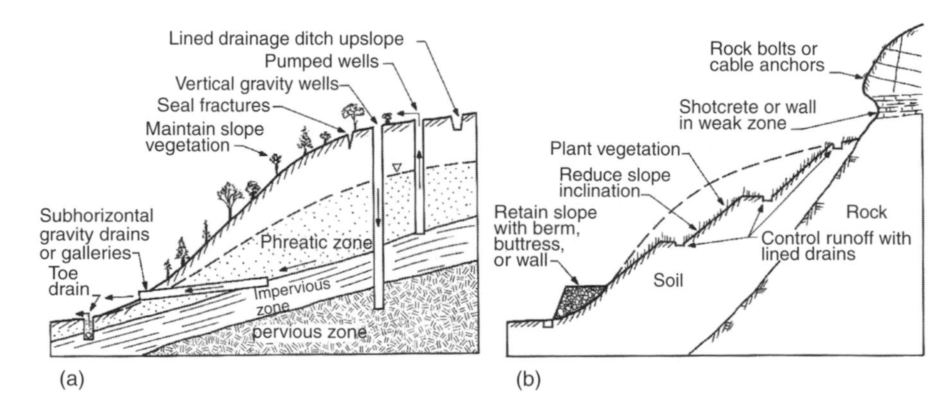

FIGURE 1.99
The general methods of slope stabilization: (a) control of seepage forces; (b) reducing the driving forces and increasing the resisting forces.

TABLE 1.8
Summary of Slope Treatment Methods for Stabilization

Treatment	Conditions	General Purpose (Preventive or Remedial)
	Change Slope Geometry	
Reduce height	Rotational slides	Prevent/treat during early stages
Reduce Inclination	All soil/rock	Prevent/treat during early stages
Add weight to toe	Soils	Treat during early stages
	Control Surface Water	
Vegetation	Soils	Prevent
Seal cracks	Soil/rock	Prevent/treat during early stages
Drainage system	Soil/decomposing rock	Prevent/treat during early stages
	Control Internal Seepage	
Deep wells	Rock masses	Temporary treatment
Vertical gravity drains	Soil/rock	Prevent/treat during early stages
Subhorizontal drains	Soil/rock	Prevent/treat early to Intermediate stages
Galleries	Rock/strong soils	Prevent/treat during early stages
Relief wells or toe trenches	Soils	Treat during early stages
Interceptor trench drains	Soils (cuts/fills)	Prevent/treat during early stages
Blanket drains	Soils (fills)	Prevent
Electroosmosis[a]	Soils (silts)	Prevent/treat during early stages: temporarily
Chemical[a]	Soils (clays)	Prevent/treat during early stages
	Retention	
Concrete pedestals	Rock overhang	Prevent
Rock bolts	Jointed or sheared rock	Prevent/treat sliding slabs
Concrete straps and bolts	Heavily jointed or soft rock	Prevent
Cable anchors	Dipping rock beds	Prevent/treat early stages
Wire meshes	Steep rock slopes	Contain falls
Concrete impact wells	Moderate slopes	Contain sliding or rolling blocks
Shotcrete	Soft or jointed rock	Prevent
Rock-filled buttress	Strong soils/soft rock	Prevent/treat during early stages
Gabion wall	Strong soils/soft rock	Prevent/treat during early stages
Crib wall	Moderately strong soils	Prevent
Reinforced earth wall	Soils/decomposing rock	Prevent
Concrete gravity walls	Soils to rock	Prevent
Anchored concrete curtain walls	Soils/decomposing rock	Prevent/treat — early to intermediate stages
Bored or root piles	Soils/decomposing rock	Prevent/treat — early stages

[a] Provides strength increase.

- Time available for remedial work on failed slopes, judged on the basis of slope activity, movement velocity and acceleration, and existing and near-future weather conditions.
- Degree of hazard and risk.
- Necessity to reduce or eliminate the hazard.

Hazard Elimination

The hazard can be eliminated by sufficient reduction of the slope height and inclination combined with an adequate surface drainage system or by retention.

Retention of rock slopes is accomplished with pedestals, rock bolts, bolts and straps, or cable anchors; retention of soil slopes is accomplished with the addition of adequate material at the toe of the slope or with properly designed and constructed walls.

TABLE 1.9
Failure Forms: Typical Preventive and Remedial Measures

Failure Form	Prevention during Construction	Remedial Measures
Rockfall	Base erosion protection Controlled blasting excavation Rock bolts and straps, or cables Concrete supports, large masses Remove loose blocks Shotcrete weak strata	Permit fall, clean roadway Rock bolts and straps Concrete supports Remove loose blocks Impact walls
Soil fall	Base erosion protection	Retention
Planar rock slide	Small volume: remove or bolt Moderate volume: provide stable inclination or bolt to retain Large volume: install internal drainage or relocate to avoid	Permit slide, clean roadway Remove to stable inclination or bolt Install internal drainage or relocate to avoid
Rotational rock slide	Provide stable inclination and surface drainage system Install internal drainage	Remove to stable inclination Provide surface drainage Install internal drains
Planar (debris) slides	Provide stable inclination and surface drainage control Retention for small to moderate volumes Large volumes: relocate	Allow failure and clean roadway Use preventive measures
Rotational soil slides	Provide stable inclination and surface drainage control, or retain	Permit failure, clean roadway Remove to stable inclination, provide surface drainage, or retain Subhorizontal drains for large volumes
Failure by lateral spreading	Small scale: retain Large scale: avoid and relocate, prevention difficult	Small scale: retain Large scale: avoid
Debris avalanche	Prediction and prevention difficult Treat as debris slide Avoid high-hazard areas	Permit failure, clean roadway: eventually self-correcting Otherwise relocate Small scale: retain or remove
Flows	Prediction and prevention difficult Avoid susceptible areas	Small scale: remove Large scale: relocate

Hazard Reduction

The hazard can be decreased by partially reducing the height and inclination or adding material at the toe; by planting vegetation, sealing cracks, installing surface drains, and shotcreting rock slopes; and by controlling internal seepage. In the last case, one can never be certain that drains will not clog, break off during movements, or be overwhelmed by extreme weather conditions.

Time Factor

Where slopes are in the process of failing, the time factor must be considered. Time may not be available for carrying out measures that will eliminate the hazard; therefore, the hazard should be reduced and perhaps eliminated at a later date. The objective is to arrest the immediate movement. To the extent possible, treatments should be performed during the dry season when movements will not close trenches, break off drains, or result in even larger failures when cuttings are made.

In general, the time required for various treatment measures are as follows:

- Sealing surface cracks and constructing interceptor ditches upslope are performed within several days, at most.
- Excavation at the head of a slide or the removal of loose blocks may require 1 to 2 weeks.
- Relief of internal water pressures may require 1 to 4 weeks for toe drains and trenches and 1 to 2 months for the installation of horizontal or vertical drains.
- Counterberms and buttresses at the toe require space, but can be constructed within 1 to 2 weeks.
- Retention with concrete walls can require 6 months or longer.

1.4.2 Changing Slope Geometry

Natural Slope Inclinations

Significance

In many cases, the natural slope represents the maximum long-term inclination, but in other cases the slope is not stable. The inclination of existing slopes should be noted during field reconnaissance, since an increase in inclination by cutting may result in failure.

Some examples of natural slope inclinations

- *Hard, massive rocks*: Maximum slope angle and height is controlled by the concentration and orientation of joints and by seepage. The critical angle for high slopes of hard, massive rock with random joint patterns and no seepage acting along the joints is about 70° (Terzaghi, 1962).
- *Interbedded sedimentary rocks*: Extremely variable, depending upon rock type, climate, and bedding thickness as well as joint orientations and seepage conditions. Along the river valleys, natural excavation may have reduced stresses sufficiently to permit lateral movement along bedding planes and produce bedding-plane mylonite shear zones. On major projects such shears should be assumed to exist until proven otherwise.
- *Clay shales*: 8 to 15°, but often unstable. When interbedded with sandstones, 20 to 45°.
- *Residual soils*: 30 to 40°, depending upon parent rock type and seepage.
- *Colluvium:* 10 to 20°, and often unstable.
- *Loess*: Often stands vertical to substantial heights.
- *Sands:* dry and "clean," are stable at the angle of repose ($i = \phi$).
- *Clays*: Depends upon consistency, whether intact or fissured, and the slope height.
- *Sand–clay mixtures*: Often stable at angles greater than repose as long as seepage forces are not excessive.

Cut Slopes in Rock

Excavation

The objective of any cut slope is to form a stable inclination without retention. Careful blasting procedures are required to avoid excessive rock breakage resulting in numerous blocks. Line drilling and presplitting during blasting operations minimize disturbance of the rock face.

Typical Cut Inclinations

Hard masses of igneous or metamorphic rocks, widely jointed, are commonly cut to 1H:4V (76°) as shown in Figure 1.100. Hard rock masses with joints, shears, or bedding representing major discontinuities and dipping downslope are excavated along the dip of the discontinuity as shown in Figure 1.101, although all material should be removed until the original slope is intercepted. If the dip is too shallow for economical excavation, slabs can be retained with rock bolts (see Section 1.4.6).

Hard sedimentary rocks with bedding dipping vertically and perpendicularly to the face as in Figure 1.102 or dipping into the face; or horizontally interbedded hard sandstone

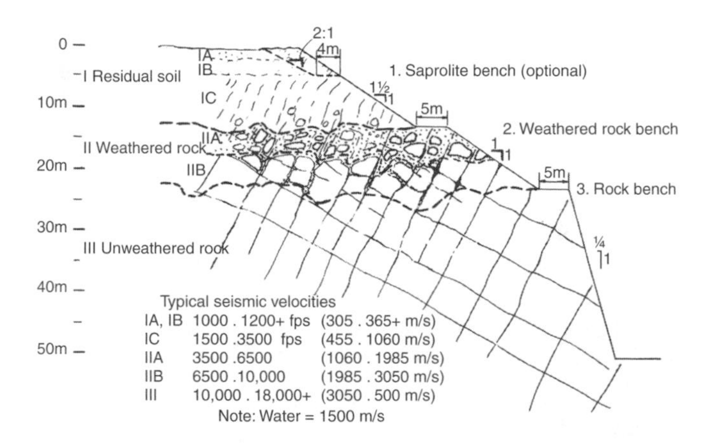

FIGURE 1.100
Typical cut slope angles for various rock and soil conditions. (From Deere, D. U. and Patton, F. D., *Proceedings of ASCE, 4th Pan American Conference on Soil Mechanics and Foundation Engineering,* San Juan, Puerto Rico, 1971, pp. 87–170. With permission.)

FIGURE 1.101
The face of the major joint surface in siltstone is stable but the overhanging blocks will fail along the same surface unless removed or retained (Sidikalang, Sumatra).

FIGURE 1.102
Folded, vertical beds in highly fractured, hard arenaceous shale are stable in near-vertical cuts (Sidikalang, Sumatra). Decreasing the slope inclination would increase susceptibility to erosion.

and shale are often cut to 1H:4V. The shale beds should be protected from weathering with shotcrete or gunite if they have expansive properties or are subjected to intense fracturing and erosion from weathering processes. Note that the steeper the cut slope the more resistant it is to erosion from rainfall.

Clay shale, unless interbedded with sandstone, is often excavated to 6H:1V (1.5°).

Weathered or closely jointed masses (except clay shale and dipping major discontinuities) require a reduction in inclination to between 1H:2V 1H:1V (63–45°) depending on conditions, or require some form of retention.

Benching

Benching is common practice in high cuts in rock slopes but there is disagreement among practitioners as to its value. Some consider benches as undesirable because they provide takeoff points for falling blocks (Chassie and Goughnor, 1975). To provide for storage they must be of adequate width. Block storage space should always be provided at the slope toe to protect the roadway from falls and topples.

Cut Slopes in Soils

Typical Inclinations

Thin soil cover over rock (Figure 1.103): The soil should be removed or retained as the condition is unstable.

Soil–rock transition (strong residual soils to weathered rock) such as in Figures 1.100 and 1.104 are often excavated to between 1H:1V to 1H:2V (45 to 63°), although potential failure along relict discontinuities must be considered. Saprolite is usually cut to 1.5H:1V (Figure 1.100).

Most soil formations are commonly cut to an average inclination of 2H:1V (26°) (Figure 1.100), but consideration must be given to seepage forces and other physical and environmental factors to determine if retention is required. Slopes between benches are usually steeper.

Benching and Surface Drainage

Soil cuts are normally designed with benches, especially for cuts over 25 to 30 ft (8 to 10 m) high. Benches reduce the amount of excavation necessary to achieve lower inclinations because the slope angle between benches may be increased.

FIGURE 1.103
Cuts in colluvium over inclined rock are potentially highly unstable and require either removal of soil or retention (Rio de Janeiro, Brazil).

FIGURE 1.104
Cut slopes at the beginning of a 300-m-deep excavation in highly decomposed igneous and metamorphic rocks for a uranium mine. Bench width is 20 m, height is 16 m, and inclination is 1H:1.5V (57°). Small wedge failure at right occurred along kaolinite-filled vertical relict joint. Part of similar failure shows in lower left.

Drains are installed as standard practice along the slopes and the benches to control runoff as illustrated in Figure 1.105 to Figure 1.107.

Failing Slopes

If a slope is failing and undergoing substantial movement, the removal of material from the head to reduce the driving forces can be the quickest method of arresting movement of relatively small failures. Placing material at the toe to form a counterberm increases the resisting forces. Benching may be effective in the early stages, but it did not fully stabilize the slope illustrated in Figure 1.29, even though a large amount of material was removed. An alternative is to permit movement to occur and remove debris from the toe; eventually the mass may naturally attain a stable inclination.

Changing slope geometry to achieve stability once failure has begun usually requires either the removal of very large volumes or the implementation of other methods. Space is seldom available in critical situations to permit placement of material at the toe, since very large volumes normally are required.

1.4.3 Surface Water Control

Purpose

Surface water is controlled to eliminate or reduce infiltration and to provide erosion protection. External measures are generally effective, however, only if the slope is stable and there is no internal source of water to cause excessive seepage forces.

Infiltration Protection

Planting the slope with thick, fast-growing native vegetation not only strengthens the shallow soils with root systems, but discourages desiccation which causes fissuring. Not all vegetation works equally well, and selection requires experience. In the Los Angeles area of California, for example, Algerian ivy has been found to be quite effective in stabilizing steep slopes (Sunset, 1978). Newly cut slopes should be immediately planted and seeded.

Sealing cracks and fissures with asphalt or soil cement will reduce infiltration but will not stabilize a moving slope since the cracks will continue to open. Grading a moving area results in filling cracks with soil, which helps to reduce infiltration.

Surface Drainage Systems

Cut slopes should be protected with interceptor drains installed along the crest of the cut, along benches, and along the toe (Figure 1.105). On long cuts the interceptors are connected to downslope collectors (Figure 1.106). All drains should be lined with nonerodible materials, free of cracks or other openings, and designed to direct all concentrated runoff to discharge offslope.

With failing slopes, installation of an interceptor along the crest beyond the head of the slide area will reduce runoff into the slide. But the interceptor is a temporary expedient, since in time it may break up and cease to function as the slide disturbance progresses upslope.

Roadway storm water drains should be located so as to not discharge on steep slopes immediate adjacent to the roadway. The failure shown in Figure 1.6 was caused by storm water discharge through a drainpipe connecting the catch basin on the upslope side of the roadway with a pipe beneath the roadway which exited on the slope.

1.4.4 Internal Seepage Control

General

Purpose

Internal drainage systems are installed to lower the piezometric level below the potential or existing sliding surface.

System Selection

Selection of the drainage method is based on consideration of the geologic materials, structure, and groundwater conditions (static, perched, or artesian), and the location of the phreatic surface.

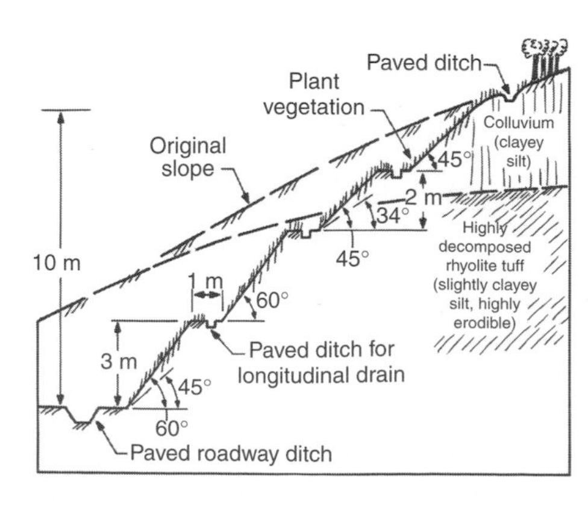

FIGURE 1.105
Benching scheme for cut in highly erodible soils in a tropical climate. Low benches permit maximum inclination to reduce the effect of runoff erosion.

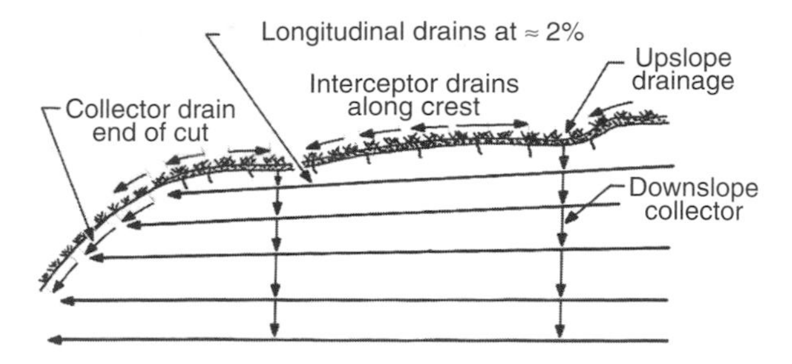

FIGURE 1.106
Sketch of slope face in Figure 1.105 showing system of longitudinal and downslope drains to control erosion.

Monitoring

As the drains are installed, the piezometric head is monitored by piezometers and the efficiency of the drains is evaluated. The season of the year and the potential for increased flow during wet seasons must be considered, and if piezometric levels are observed to rise to dangerous values (as determined by stability analysis or from monitoring slope movements), the installation of additional drains is required.

Cut Slopes

Systems to relieve seepage forces in cut slopes are seldom installed in practice, but they should be considered more frequently, since there are many conditions where they would aid significantly in maintaining stability.

Failing Slopes

The relief of seepage pressures is often the most expedient means of stabilizing a moving mass. The primary problem is that, as mass movement continues, the drains may be cut off and cease to function; therefore, it is often necessary to install the drains in stages over a period of time. Installation must be planned and performed with care, since the use of water during drilling could possibly trigger a total failure.

Methods (see Figure 1.99a)

Deep wells have been used to stabilize many deep-seated slide masses, but they are costly since continuous or frequent pumping is required. Check valves normally are installed so that when the water level rises, pumping begins. Deep wells are most effective if installed in relatively free-draining material below the failing mass.

Vertical, cylindrical gravity drains are useful in perched water-table conditions, where an impervious stratum overlies an open, free-draining stratum with a lower piezometric level. The drains permit seepage by gravity through the confining stratum and thus relieve hydrostatic pressures (see Section 1.2.7, discussion of the Pipe Organ Slide). Clay strata over granular soils, or clays or shales over open-jointed rock, offer favorable conditions for gravity drains where a perched water table exists.

Subhorizontal drains is one of the most effective methods to improve stability of a cut slope or to stabilize a failing slope. Installed at a slight angle upslope to penetrate the phreatic zone and permit gravity flow, they usually consist of perforated pipe, of 2 in. diameter or larger, forced into a predrilled hole of slightly larger diameter than the pipe. Subhorizontal drains have been installed to lengths of more than 300 ft (100 m). Spacing depends on the type of material being drained; fine-grained soils may require spacing as close as 10 to 30 ft (3 to 8 m), whereas, for more permeable materials, 30 to 50 ft (8 to 15 m) may suffice.

Santi et al. (2003) report on recent installations of subhorizontal *wick drains* to stabilize slopes. Composed of geotextiles (polypropylene) they have the important advantages of stretching and not rupturing during deformation, and are resistant to clogging. Installation proceeds with a disposal plate attached to the end of a length of wick drain that is inserted into a drive pipe. The pipe, which can be a wire line drill rod, is pushed into the slope with a bulldozer or backhoe. Additional lengths of wicks and pipe are attached and driven into the slope. When the final length is installed, the drive pipe is extracted.

Drainage galleries are very effective for draining large moving masses but their installation is difficult and costly. They are used mostly in rock masses where roof support is less of a problem than in soils. Installed below the failure zone to be effective, they are often backfilled with stone. Vertical holes drilled into the galleries from above provide for drainage from the failure zone into the galleries.

Interceptor trench drains or slots are installed along a slope to intercept seepage in a cut or sliding mass, but they must be sufficiently deep. As shown in Figure 1.107, slotted pipe is laid in the trench bottom, embedded in sand, and covered with free-draining material, then sealed at the surface. The drain bottom should be sloped to provide for gravity drainage to a discharge point. Interceptor trench drains are generally not practical on steep, heavily vegetated slopes because installation of the drains and access roads requires stripping the vegetation, which will further decrease stability.

Relief trenches or slots relieve pore pressures at the slope toe. They are relatively simple to install. Excavation should be made in sections and quickly backfilled with stone so as not to reduce the slope stability and possibly cause a total failure. Generally, relief trenches are most effective for slump slides (Figure 1.25) where high toe seepage forces are the major cause of instability.

Electro-osmosis has been used occasionally to stabilize silts and clayey silts, but the method is relatively costly, and not a permanent solution unless operation is maintained.

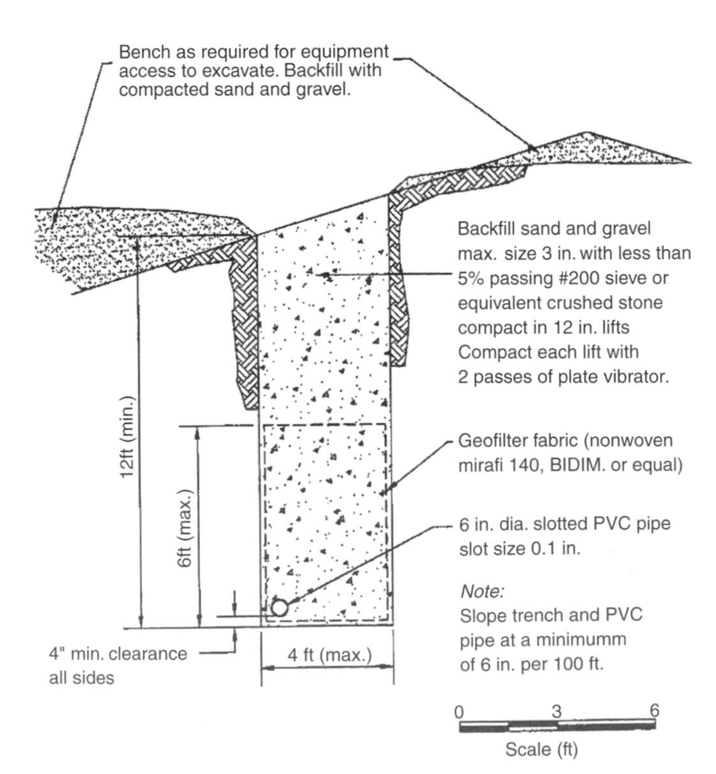

FIGURE 1.107
Typical slope trench drain.

Examples

Case 1: Open-Pit Mines (*Brawner, 1975*)

General: Problems encountered in open-pit mines in soft rock (coal, uranium, copper, and asbestos) during mining operations include both bottom heave of deep excavations (of the order of several hundred meters in depth) and slides, often involving millions of tons.

Solutions: Deep vertical wells that have relieved artesian pressures below mine floors where heave was occurring have arrested both the heave and the associated slope instability. Horizontal drainage in the form of galleries and boreholes as long as 150 m installed in the toe zone of slowly moving masses arrested movement even when large failures were occurring. In some cases, vacuum pumps were installed to place the galleries under negative pressures. Horizontal drains, consisting of slotted pipe installed in boreholes, relieve cleft-water pressures in jointed rock masses.

Case 2: Failure of a Cut Slope (*Fox, 1964*)

Geological conditions: The slope in Figure 1.108 consists of colluvial soils of boulders and clay overlying schist interbedded with gneiss. Between the colluvium and the relatively sound rock is a zone of highly decomposed rock.

Slide history: An excavation was made to a depth of 40 m into a slope with an inclination of about 28°. Upon its completion cracks opened, movement began, and springs appeared on the surface. The excavation was backfilled and the ground surface was graded to a uniform slope and covered with pitch. Monuments were installed to permit observations of movements. Even after the remedial measures were invoked, movement continued to endanger nearby structures. The greatest movement was about 2.5 cm/day. Failure had

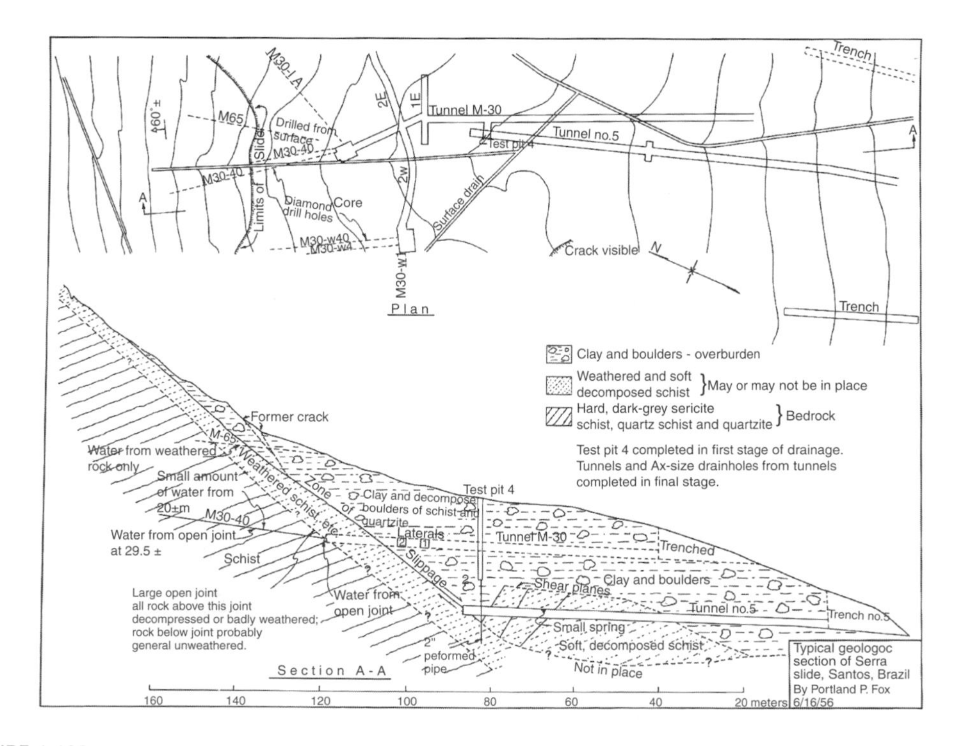

FIGURE 1.108
Stabilization of a failure in a colluvial soil slope using lateral drains and galleries. (From Fox, P.P., *Engineering Geology Case Histories Numbers 1–5,* The Geological Society of America, Engineering Geology Division, 1964, pp. 17–24. With permission.)

reduced the preexisting strengths to the extent that the original slope inclination was unstable. Piezometers installed as part of an investigation revealed that the highest pore pressures were in the fractured rock zone, under the colluvium.

Remedial measures: To correct the slide, Dr. Karl Terzaghi had a number of horizontal drill holes and galleries extended into the fractured rock as shown in Figure 1.108. The holes drained at rates of 10 to 100 L/min, and the water level in the piezometers continued to fall as work progressed. The slide was arrested and subsequent movements were reported to be minor.

Case 3: Construction of a Large Cut Slope (*D'Appolonia et al., 1967*)

Geologic conditions: As illustrated on the section, (Figure 1.109), conditions were characterized by colluvium, overlying sandstones and shales, and granular alluvium. Explorations were thorough and included test pits which revealed the overburden to be slickensided, indicating relict failure surfaces and a high potential for instability.

Treatment: Construction plans required a cut varying from 6 to 18 m in height in the colluvium along the slope toe. To prevent any movement, a system of trenches, drains, and galleries was installed. A cutoff trench, vertical drain, and gallery were constructed upslope, where the colluvium was relatively thin, to intercept surface water and water entering the colluvium from a pervious siltstone layer. A 2-m-diameter drainage gallery was excavated in the colluvium at about midslope to intercept flow from a pervious sandstone stratum and to drain the colluvium. Sand drains were installed downslope, near the proposed excavation, to enable the colluvium to drain by gravity into the underlying sand and gravel lying above the static water level, thereby reducing pore pressures in the colluvium. An anchored sheet-pile wall was constructed to retain the cut face; the other systems were installed to maintain the stability of the entire slope and reduce pressures on the wall.

1.4.5 Side-Hill Fills

Failures

Construction of a side-hill embankment using slow-draining materials can be expected to block natural drainage and evaporation. As seepage pressures increase, particularly at the toe as shown in Figure 1.110a, the embankment strains and concentric tension cracks form.

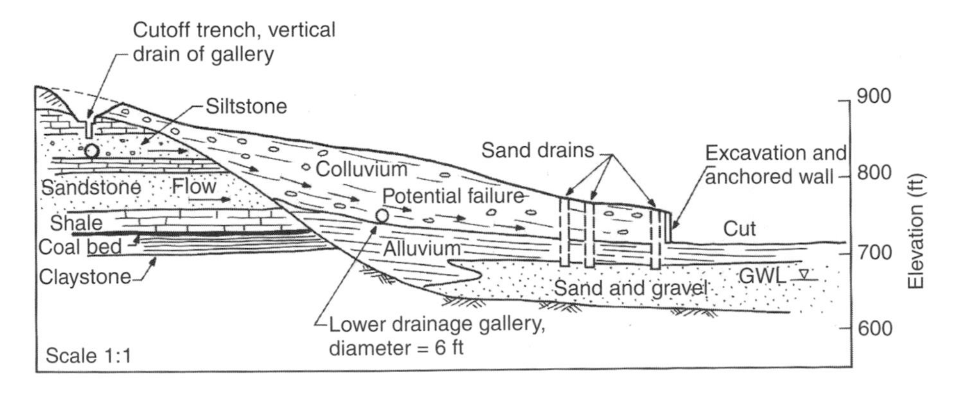

FIGURE 1.109
Stabilization of a colluvial soil slope (Weirton, West Virginia) with vertical drains and galleries. (From D'Appolonia, E. D. et al., 1967.)

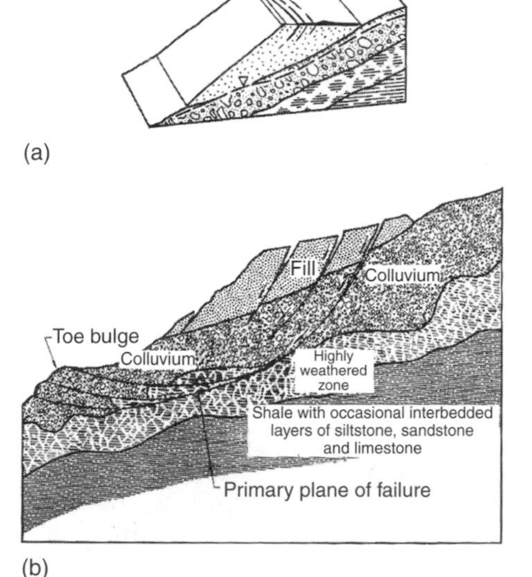

FIGURE 1.110
Development of rotational failure in side-hill fill with inadequate subsurface drainage. (a) Early failure stage: concentric cracks show in pavement. (b) Rotational failure of side-hill fill over thick colluvium. (Part b from Royster, D. L., *Bull. Assoc. Eng. Geol.*, X, 1973. With permission.)

The movements develop finally into a rotational failure as shown in Figure 1.110b, a case of deep colluvium. Figure 1.111 illustrates a case of shallow residual soils.

Side-hill fills placed on moderately steep to steep slopes of residual or colluvial soils, in particular, are prone to be unstable unless seepage is properly controlled, or the embankment is supported by a retaining structure.

Stabilization

Preventive

Interceptor trench drains should be installed along the upslope side of all side-hill fills as standard practice to intercept flow as shown in Figure 1.112. Perforated pipe is laid in the trench bottom, embedded in sand, covered by free-draining materials, and then sealed at the surface. Surface flow is collected in open drains and all discharge, including that from the trench drains, is directed away from the fill area.

A *free-draining blanket* should be installed between the fill and the natural slope materials to relieve seepage pressures from shallow groundwater conditions wherever either the fill or the natural soils are slow-draining, as shown in Figure 1.112. It is prudent to strip

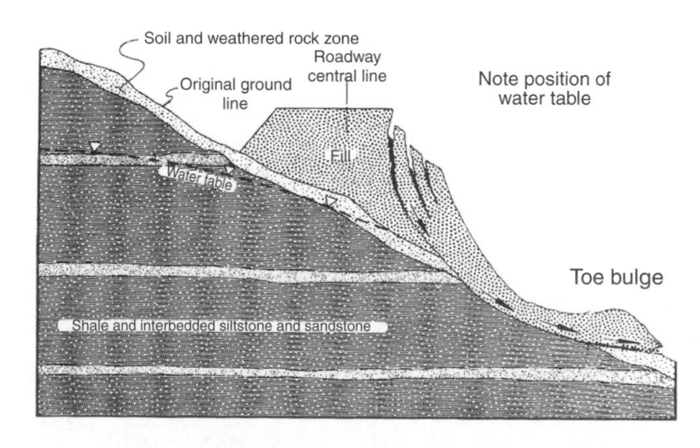

FIGURE 1.111
Development of rotational failure in side-hill fill underlain by thin formation of residual soils and inadequate subsurface drainage. (From Royster, D. L., *Bull. Assoc. Eng. Geol.*, X, 1973. With permission.)

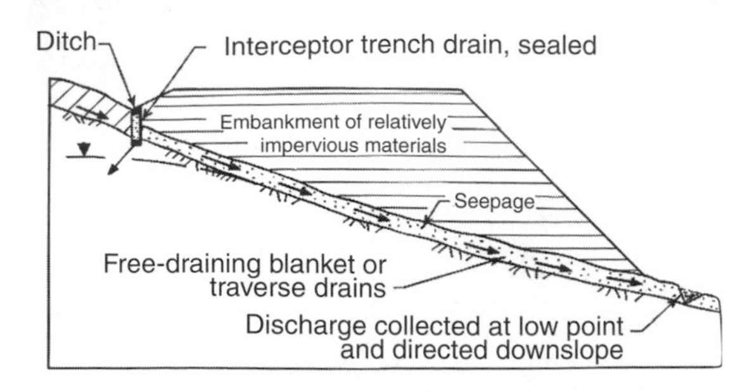

FIGURE 1.112
Proper drainage provisions for a side-hill fill.

potentially unstable upper soils, which are often creeping on moderately steep to steep slopes, to a depth where stronger soils are encountered, and to place the free-draining blanket over the entire area to be covered by the embankment. Discharge should be collected at the low point of the fill and drained downslope in a manner that will provide erosion protection.

Transverse drains extending downslope and connecting with the interceptor ditches upslope, parallel to the roadway, may provide adequate subfill drainage where anticipated flows are low to moderate.

Retaining structures may be economical on steep slopes that continue for some distance beyond the fill if stability is uncertain (see Figure 1.125).

Corrective

After the initial failure stage, subhorizontal drains may be adequate to stabilize the embankment if closely spaced, but they should be installed during the dry season since the use of water to drill holes during the wet season may accelerate total failure. An alternative is to retain the fill with an anchored curtain wall (see Figure 1.128).

After total failure, the most practical solutions are either reconstruction of the embankment with proper drainage, or retention with a wall.

1.4.6 Retention

Rock Slopes

Methods Summarized

The various methods of retaining hard rock slopes are illustrated in Figure 1.113 and described briefly below.

- Concrete pedestals are used to support overhangs, where their removal is not practical because of danger to existing construction downslope, as illustrated in Figure 1.114.
- Rock bolts are used to reinforce jointed rock masses or slabs on a sloping surface.
- Concrete straps and rock bolts are used to support loose or soft rock zones or to reduce the number of bolts as shown in Figure 1.115 and Figure 1.120.
- Cable anchors are used to reinforce thick rock masses.
- Wire meshes, hung on a slope, restrict falling blocks to movement along the face (Figure 1.116).
- Concrete impact walls are constructed along lower slopes to contain falling or sliding blocks or deflect them away from structures (see Figure 1.16).
- Shotcrete (Figure 1.117) is used to reinforce loose fractured rock, or to prevent weathering or slaking of shales or other soft rocks, especially where interbedded

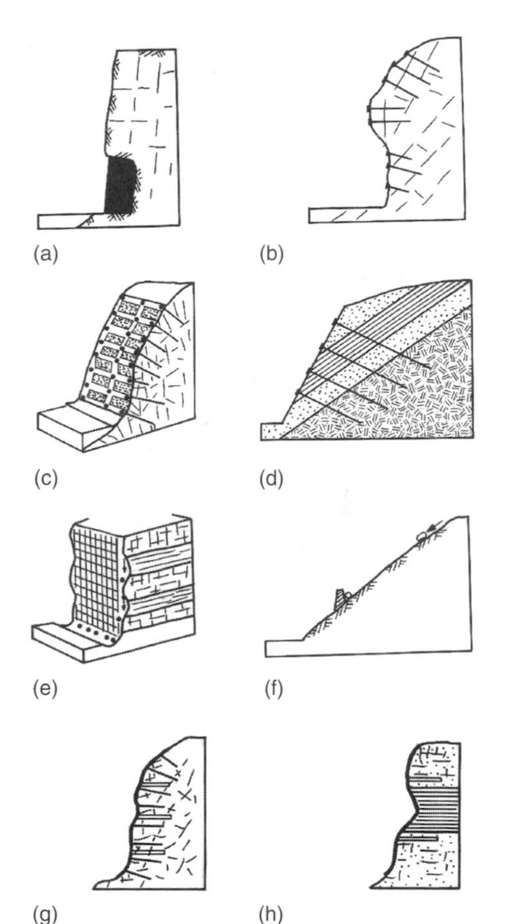

FIGURE 1.113
Various methods of retaining hard rock slopes: (a) concrete pedestals for overhangs; (b) rock bolts for jointed masses; (c) bolts and concrete straps for intensely jointed masses: (d) cable anchors to increase support depth; (e) wire mesh to constrain falls; (f) impact walls to deflect or contain rolling blocks; (g) shotcrete to reinforce loose rock, with bolts and drains; (h) shotcrete to retard weathering and slaking of shales.

with more resistant rocks. Shotcrete is normally used with wire mesh and dowels, bolts or nails as discussed below.

- Gunite is similar to shotcrete except that the aggregate is smaller.

Reinforcing Rock Slopes

Rock anchors are tensile units, fixed at one end, used to place large blocks in compression, and should be installed as near to perpendicular to a joint as practical. The ordinary types consist of rods installed in drill holes either by driving and wedging, driving and expanding, or by grouting with mortar or resins as illustrated in Figure 1.118. Bolt heads are then attached to the rod and torqued against a metal plate to impose the compressive force on the mass. Weathering of rock around the bolt head may cause a loss in tension; therefore, heads are usually protected with concrete or other means, or used in conjunction with concrete straps in high-risk conditions.

Fully grouted rock bolts, illustrated in Figure 1.119, provide a more permanent anchor than those shown in Figure 1.118. The ordinary anchor is subject to loss in tension with time from several possible sources including corrosion from attack by aggressive water, anchorage slip or rock spalling around and under the bearing plate, and block movement along joints pinching the shaft. Care is required during grouting to minimize grout spread, which results in decreasing mass drainage, especially where bolts are closely spaced. Drain holes may be required.

(a)

(b)

FIGURE 1.114
Support of granite overhang with pedestals (Rio de Janeiro). Ancient slide mass of colluvium appears on lower slopes. (a) Side view and (b) face view.

A major installation of bolts and straps is illustrated in Figure 1.120, part of a 60-m-high rock slope (Figure 1.121), at the base of which is to be constructed a steel mill. The consultant selected the support system rather than shaving and blasting loose large blocks for fear of leaving a weaker slope in a high-risk situation.

Rock Dowels are fully grouted rock bolts, usually consisting of a ribbed reinforcing bar, installed in a drill hole and bonded to the rock over its full length (Franklin and Dusseult, 1989). Rock movement results in the dowels being self-tensioned. Grouting with resins is becoming more and more common because of easy installation and the rapid attainment of capacity within minutes of installation. Sausage-shaped resin packages are installed in the drill hole and a ribbed bar inserted and rotated to open the packages which contain resin and a catalyst (Figure 1.119c).

Shotcrete, when applied to rock slopes, usually consists of a wet-mix mortar with aggregate as large as 2 cm (3/4 in.) which is projected by air jet directly onto the slope face. The

FIGURE 1.115
Stabilization of exfoliating granite with rock bolts and concrete straps (Rio de Janeiro).

force of the jet compacts the mortar in place, bonding it to the rock, which first must be cleaned of loose particles and loose blocks. Application is in 8 to 20 cm (3 to 4 in.) layers, each of which is permitted to set before application of subsequent layers. Originally, weep holes were installed to relieve seepage pressures behind the face, but modern installations include geocomposite drainage strips placed behind the shotcrete. Since shotcrete acts as reinforcing and not as support, it is used often in conjunction with rock bolts. The tensile strength can be increased significantly by adding 25-mm-long wire fibers to the concrete mix. A typical installation is illustrated in Figure 1.124.

Soil Layer over Rock Slopes

As shown in Figure 1.50, cuts in mountainous terrain are inherently unstable where a relatively thin layer of soil overlies rock. The upper portion of the underlying rock normally is fractured and a conduit for seepage. Investigations made during the dry season may not encounter seepage in the rock, but flow during the wet season often is common and must be considered during evaluations.

Some typical solutions are given in Figure 1.122. In (a), design provides for inclining the cut in the rock and the soil; in (b), the soil is cut to a stable inclination and the rock cut made steeper by retention with shotcrete and rock nails; and in (c), the soil is retained with a top down wall (Figure 1.123) and the rock with shotcrete and nails (Figure 1.124).

FIGURE 1.116
Wire mesh installed to prevent blocks of gneiss from falling on the Harlem River Drive, New York City.

Soil Slopes

Purpose

Walls are used to retain earth slopes where space is not available for a flat enough slope or excessive volumes of excavation are required, or to obtain more positive stability under certain conditions. Except for anchored concrete curtain walls, other types of walls that require cutting into the slope for construction are seldom suitable for retention of a failing slope.

Classes

The various types of walls are illustrated in Figure 1.125. They may be divided into four general classes, with some wall types included in more than one class: gravity walls, nongravity walls, rigid walls, and flexible walls.

FIGURE 1.117
Shotcrete applied to retain loose blocks of granite gneiss in cut. Untreated rock exposed in lower right (Rio de Janeiro).

Gravity walls provide slope retention by either their weight alone, or their weight combined with the weight of a soil mass acting on a portion of their base or by the weight of a composite system. They are free to move at the top thereby mobilizing active earth pressure. Included are rock-filled buttresses, gabion walls, crib walls, reinforced earth walls, concrete gravity walls, cantilever walls, and counterfort walls. A series of reinforced earth walls with a combined height of 100 ft is shown in Figure 1.126. Concrete walls are becoming relatively uncommon due to costs, construction time, and the fact that the slope is unsupported during construction.

Gebney and McKittrick (1975) report on a complex system of gravity walls installed to correct a debris slide along Highway 39, Los Angeles County, California. The reconstructed roadway was supported on a reinforced earth wall in turn supported by an embankment and a buttress at the toe of the 360-ft-high slope. Horizontal and longitudinal drains were installed to relieve hydrostatic pressures in that part of the slide debris which was not totally removed.

Nongravity walls are restrained at the top and not free to move. They include basement walls, some bridge abutments, and anchored concrete curtain walls. *Anchored concrete curtain walls*, such as the one illustrated in Figure 1.127, can be constructed to substantial heights and have a very high retention capacity. They are constructed from the top down by excavation of a series of benches into the slope and formation of a section of wall, retained by anchors, in each bench along the slope. Since the slope is thus retained completely during the wall construction, the system is particularly suited to potentially unstable or unstable slopes. An example of an anchored curtain wall retaining a side-hill fill is shown in Figure 1.128. A variation of the anchored curtain wall consists of anchored premolded concrete panels. The advantage of the system is that the wall conforms readily to the slope configuration, as shown in Figure 1.129.

An alternate "top-down" procedure, common to the United States, is to install anchored (tiebacks) soldier piles (Figure 1.123). As the excavation proceeds, breasting boards are installed in the soldier pile flanges to support the slope.

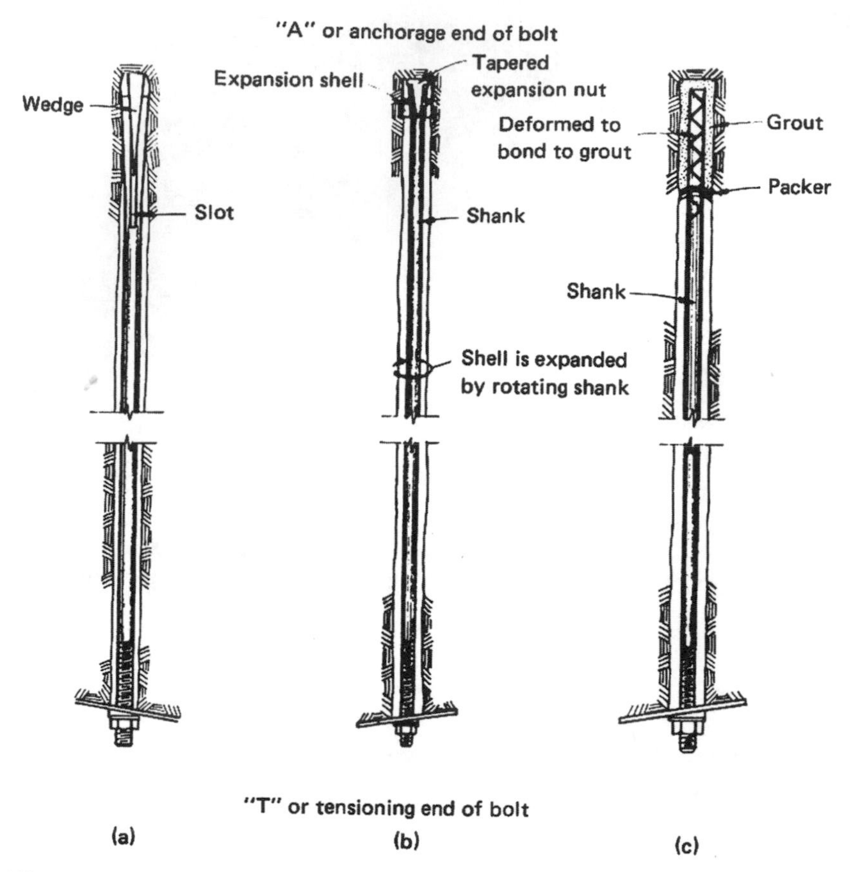

FIGURE 1.118
Types of ordinary rock bolts (anchors): (a) drive-set or slot and wedge bolt; (b) torque-set or expansion bolt; (c) grouted bolt. (From Lang, T. A., *Bull. Assoc. Eng. Geol.*, 9, 215–239, 1972. With permission.)

Rigid walls include concrete walls: gravity and semigravity walls, cantilever walls, and counterfort walls. Anchored concrete curtain walls are considered as semirigid.

Flexible walls include rock-filled buttresses, gabion walls, crib walls, reinforced earth walls, and anchored sheet-pile walls.

Soil nailing is an *in situ* soil reinforcement technique that is finding increasing application. Long rods (nails) are installed to retain excavations or stabilize existing slopes. Nails are driven for temporary installations or drilled and grouted for permanent installations similar to the procedures described for shotcreting rock masses. Cohesive soils with LL > 50 and PI > 20 require careful assessment for creep susceptibility. Soil nailing is discussed in detail in Elias and Juran (1991).

Wall Characteristics

The general characteristics of retaining walls are summarized in Table 1.10. Also included are bored piles and root piles, not shown in Figure 1.125.

Wall Selection and Design Elements

The wall type is tentatively selected on the basis of an evaluation of the cut height, materials to be supported, wall purpose, and a preliminary economic study.

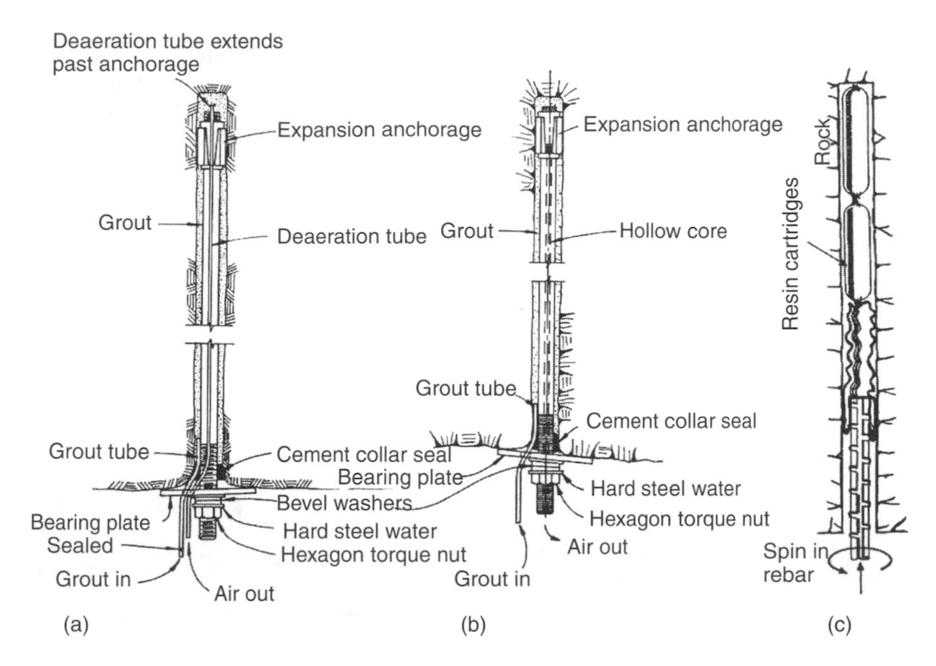

FIGURE 1.119
Fully grouted rock bolts or anchors: (a) grouted solid expansion anchorage bolt; (b) hollow-core grouted rock bolt. (a and b from Lang, T. A., *Bull. Assoc. Eng. Geol.*, 9, 215–239, 1972. With permission.) (c) Rebar and resin cartridges.

FIGURE 1.120
Loose granite gneiss blocks retained with rock bolts, and blocks of gneiss and a soft zone of schist retained with bolts and concrete straps on natural slope (Joao Monlevade, M. G., Brazil). (Courtesy of Tecnosolo. S. A.)

FIGURE 1.121
The 60-m-high slope of Figure 1.116 before treatment. The lower portions have been excavated into relatively sound gneiss. The workers (in circle) give the scale. Several large blocks weighing many tons broke loose during early phases when some scaling of loose blocks was undertaken.

Earth pressures are determined (magnitude, location, and direction), as influenced by the slope inclination and height; location and magnitude of surcharge loads; wall type, configuration and dimensions; depth of embedment; magnitude and direction of wall movement; soil parameters for natural materials and borrowed backfill; and seepage forces.

Stability of gravity walls is evaluated with respect to adequacy against overturning, sliding along the base, foundation bearing failure, and settlement. The slope must be evaluated with respect to formation of a possible failure surface beneath the wall.

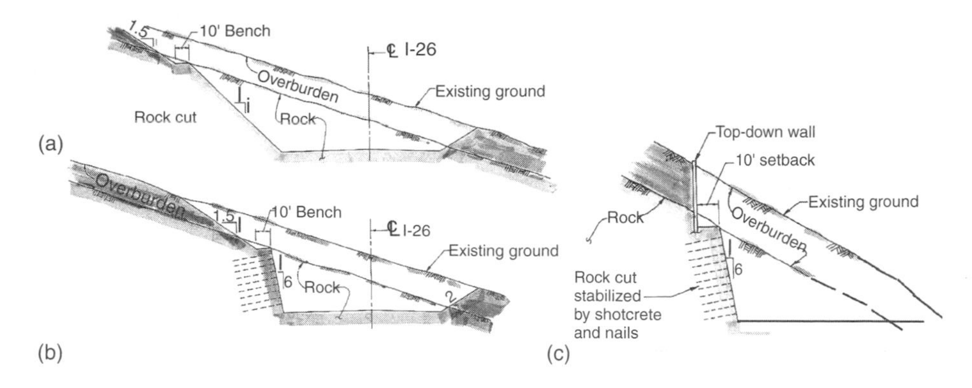

FIGURE 1.122
Various solutions for roadway cuts in soil over fractured rock. (a) Typical roadway rock cut; (b) roadway rock cut with nails; (c) roadway rock cut with top-down wall.

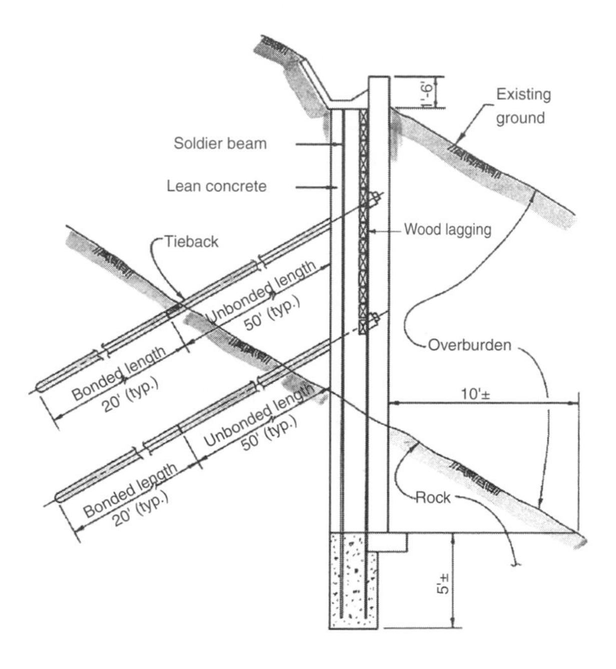

FIGURE 1.123
Typical top-down wall with anchored soldier beam and lagging.

Structural design proceeds when all of the forces acting on the wall have been determined. Beyond the foregoing discussion, the design of retaining walls is not within the scope of this volume.

1.5 Investigation: A Review

1.5.1 General

Study Scopes and Objectives

Regional Planning

Regional studies are performed to provide the basis for planning urban expansion, transportation networks, large area developments, etc. The objectives are to identify areas

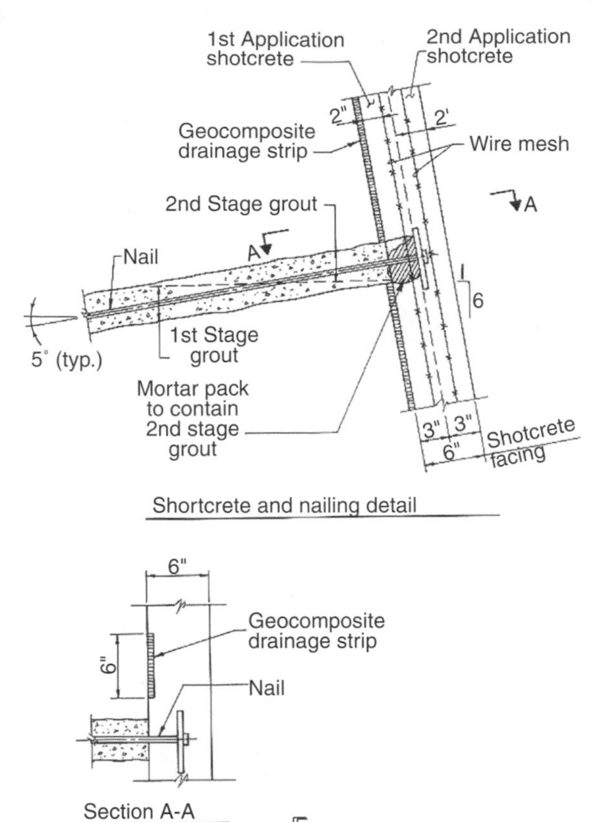

FIGURE 1.124
Retention with shotcrete and nails or bolts in soil or rock.

prone to slope failures, and the type, magnitude, and probability of occurrence. Hazard maps illustrate the findings.

Individual Slopes

Individual slopes are studied when signs of instability are noted and development is endangered, or when new cuts and fills are required for development. Studies should be performed in two phases: Phase 1, to establish the overall stability, is a study of the entire slope from toe to crest to identify potential or existing failure forms and their failure surfaces, and Phase 2 is a detailed study of the immediate area affected by the proposed cut or fill.

Considerations

Failure Forms and Hazard Degrees

Engineers and geologists must be aware of which natural slope conditions are hazardous, which can be analyzed mathematically with some degree of confidence, which are very sensitive to human activities on a potentially catastrophic scale, which can be feasibly controlled, and which are to be avoided. They should also be aware that in the present state of the art there are many limitations in our abilities to predict, analyze, prevent, and contain slope failures.

Rotational slides are the forms most commonly anticipated, whereas the occurrence of other forms is often neglected during slope studies. They are generally the least catastrophic of all forms, normally involve a relatively small area, give substantial warning in the form of surface cracking, and usually result in gradual downslope movement during the initial development stages. Several potential failure forms can exist in a given slope, however.

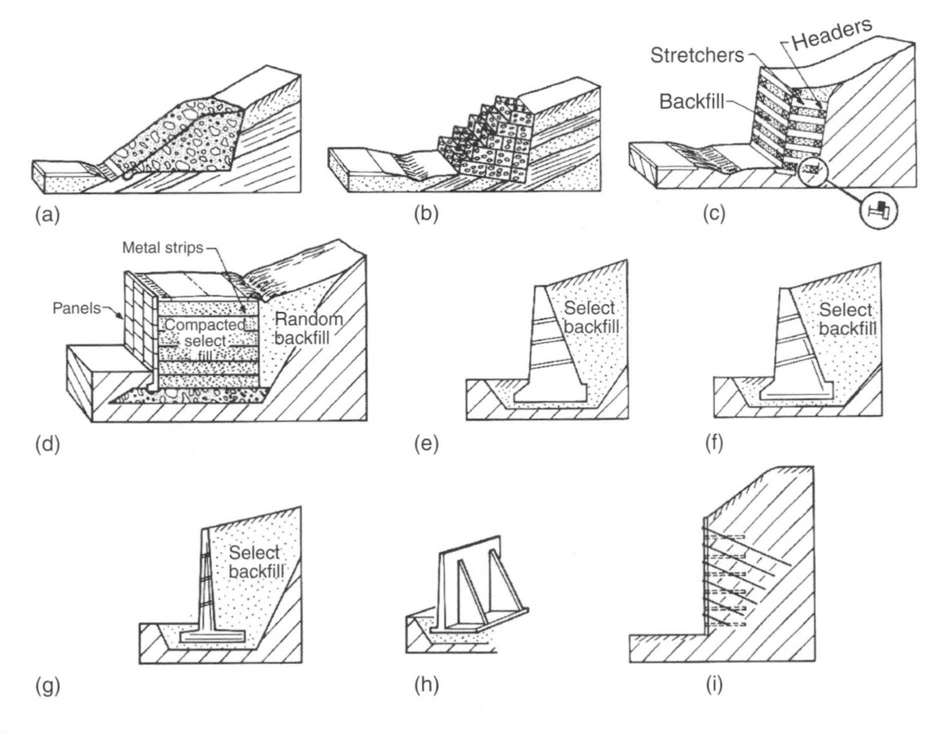

FIGURE 1.125
Various types of retaining walls: (a) rock-filled buttress; (b) gabion wall; (c) crib wall; (d) reinforced earth wall; (e) concrete gravity wall; (f) concrete-reinforced semigravity wall; (g) cantilever wall; (h) counterfort wall; (i) anchored curtain wall.

FIGURE 1.126
Reinforced earth walls over 100 ft in combined height support roadway fill, I–26, Sams Gap, Tennessee.

FIGURE 1.127
Anchored curtain wall being constructed to a height of 25 m and length of 150 m, completed to 15 m height (Joao Monlevade, M. G., Brazil). The wall, of maximum thickness of 50 cm, is constructed in sections 1.5 m high from the top down, with each section anchored to provide continuous support. Geologic Section is shown in the inset. (From Hunt, R. E. and Costa Nunes, A. J. da, *Civil Engineering, ASCE,* 1978, pp. 73–75. With permission.)

FIGURE 1.128
Anchored concete curtain wall supports fill placed over colluvium. Wall is pile-supported to rock. Slope movements are occurring downslope to the right but the wall is stable (Highway BR 277, Parana, Brazil).

FIGURE 1.129
Stabilization of a slump slide with anchored premolded concrete panels which conform readily to the shape of the slope (Itaorna, R. J., Brazil).

Planar slides in mountainous terrain, which usually give warning and develop slowly, can undergo sudden total failure, involving huge volumes and high velocities with disastrous consequences.

Falls, avalanches, and flows often occur suddenly without warning, move with great velocities, and can have disastrous consequences.

Stability Factors

Slope geometry and geology, weather conditions, and seismic activity are the factors influencing slope stability, but conditions are frequently transient. Erosion, increased seepage forces, strength deterioration, seismic forces, tectonic activity, as well as human activity, all undergo changes with time and work to decrease slope stability.

Selection of Slope Treatments

Slope treatments are selected primarily on the basis of judgment and experience, and normally a combination of methods is chosen.

For *active slides of large dimensions*, consideration should be given chiefly to external and internal drainage; retaining structures are seldom feasible.

Active slides of small dimensions can be stabilized by changing their geometry, improving drainage, and when a permanent solution is desired, containing them by walls. An alternative, which is often economically attractive, is to permit the slide to occur and to remove material continuously from the toe until a stable slope has been achieved naturally. The risk of total failure, however, must be recognized.

Cut slopes are first approached by determining the maximum stable slope angle; if too much excavation is required or if space does not permit a large cut, alternative methods employing retention are considered. It must be noted that side-hill cuts are potentially far

TABLE 1.10

Retaining Wall Characteristics

Wall Type	Description	Comments
Rock-filled buttress	Constructed of nondegradable, equidimensional rock fragments with at least 50% between 30 and 100 cm and not more than 10% passing 2-in. sleeve (Royster, (1979).	Gradation is important to maintain free-draining characteristics amid high friction angle, which combined with weight provides retention. Capacity limited by ϕ of approximately=40° and space available for construction.
Gabion wall	Wire baskets, about 50 cm each side, are filled with broken stone about 10 to 15 cm across. Baskets are then stacked in rows.	Free-draining. Retention is obtained from the stone weight and its interlocking and frictional strength. Typical wall heights are about 5 to 6 m, but capacity is limited by ϕ.
Crib wall	Constructed by forming interconnected boxes from timber, precast concrete, or metal members and then filling the boxes with crushed stone or other coarse granular material. Members are usually 2 m in length.	Free-draining. Height of single wall is limited to an amount twice the member length. Heights are increased by doubling box sections in depth. High walls are very sensitive to transverse differential settlements, and the weakness of cross members precludes support of high surcharge loads.
Reinforced earth walls	A compacted backfill of select fill is placed as metal strips, called ties, are embedded in the fill to resist tensile forces. The strips are attached to a thin outer skin of precast concrete panels to retain the face.	Free-draining and tolerant of different settlements, they can have high capacity and have been constructed to heights of at least 18 m. Relatively large space is required for the wall.
Concrete gravity wall	A mass of plain concrete.	Requires weep holes, free-draining backfill, large excavation. Can take no tensile stresses and is uneconomical for high walls.
Semigravity concrete wall	Small amount of reinforcing steel is used to reduce concrete volume and provide capacity for greater heights.	Requires weep holes, free-draining backfill, and large excavation. Has been constructed to heights of 32 m (Kulhawy, 1974).
Cantilever wall	Reinforced concrete with stem connected to the base. The weight of earth acting on the heel is added to the weight of the concrete to provide resistance.	Requires weep holes and free-draining backfill; smaller excavation than gravity walls but limited to heights of about 8 m because of inherent weakness of the stem-base connection.
Counterfort wall	A cantilever wall strengthened by the addition of counterforts.	Used for wall heights over 6 to 8 m.
Buttress wall	Similar to counterfort walls except that the vertical braces are placed on the face of the wall rather than on the backfill side.	As per cantilever and counterfort walls.
Anchored reinforced-concrete curtain wall	A thin wall of reinforced concrete is tied back with anchors to cause the slope and wall to act as a retaining system. A variation by Tecnosolo S. A. uses precast panels as shown in Figure 1.129.	Constructed in the slope from the top down in sections to provide continuous retention of the slope during construction. (All other walls require an excavation which remains open while the wall is erected.) Retention capacity is high and they have been used to support cuts in residual soils over 25 m in height. Drains are installed through the wall into the slope. *See also* Figure 1.127 and Figure 1.128.
Anchored steel sheet-pile wall	Sheet piles driven or placed in an excavated slope and tied back with anchors to form a flexible wall.	Seldom used to retain slopes because of its tendency to deflect and corrode and its costs, although it has been used successfully to retain a slope toe in conjunction with other stabilization methods (*see* Figure 1.109).
Bored piles	Bored piles have been used on occasion to stabilize failed slopes during initial stages and cut slopes.	Height is limited by pile capacity in bending. Site access required for large drill rig unless holes are hand-excavated.
Root piles	Three-dimensional lattice of small-diameter, cast-in-place, reinforced-concrete piles, closely spaced to reinforce the earth mass.	Trade name "Fondedile." A retaining structure installed without excavation. Site access for large equipment required.

more dangerous than cuts made in level ground, even with the same cut inclination, depth, and geologic conditions. The significant difference is likely to be seepage conditions.

Side-hill fills must always be provided with proper drainage, and on steep slopes retention usually is prudent.

Slope Activity Monitoring

Where potentially dangerous conditions exist, monitoring of slope activity with instrumentation is necessary to provide early warning of impending failures.

Hazard Zoning

In cities and areas where potentially dangerous conditions exist and failures would result in disastrous consequences, such as on or near high, steep slopes or on sensitive soils near water bodies or courses, development should be prohibited by zoning regulations. Pertinent in this respect is a recent slide in Goteburg, Sweden (ENR, 1977), a country with a long history of slope failures in glaciolacustrine and glaciomarine deposits. Shortly after heavy rains in early December 1977, a slide occurred taking at least eight lives and carrying 67 single-family and row houses into a shallow ravine. Damage was over $7 million. The concluding statement in the article: "Last week's slide is expected to spark tighter controls of construction in questionable areas."

1.5.2 Regional and Total Slope Studies

Preliminary Phases

Objectives and Scope

The objectives of the preliminary phases of investigation, for either regional studies or for the study of a particular area, are to anticipate forms, magnitudes, and incidences of slope failures.

The study scope includes collection of existing data, generation of new data through terrain analysis, field reconnaissance, and evaluation.

Existing Data Collection

Regional data to be collected include: slope failure histories, climatic conditions of precipitation and temperature, seismicity, topography (scales of 1:50,000 and 1:10,000), and remote-sensing imagery (scales 1:250,000 to 1:50,000).

At the project location, data to be collected include topography (scales of 1:10,000 to 1:2,000, depending upon the area to be covered by the project, and contour intervals of 2 to 4 m, or 5 to 10 ft), and remote-sensing imagery (scale of 1:20,000 to 1:6,000). Slope sections are prepared at a 1:1 scale showing the proposed cut or fill in its position relative to the entire slope.

Landform Analysis

On a regional basis, landform analysis is performed to identify unstable and potentially unstable areas, and to establish preliminary conclusions regarding possible failure forms, magnitudes, and incidence of occurrence. A preliminary map is prepared, showing topography, drainage, active and ancient failures, and geology. The preliminary map is developed into a hazard map after field reconnaissance. At the project location, more detailed maps are prepared illustrating the items given above, and including points of slope seepage.

Field Reconnaissance

The region or site location is visited and notations are made regarding seepage points, vegetation, creep indications, tension cracks, failure scars, hummocky ground, natural slope inclinations, and exposed geology. The data collected during landform analysis provide a guide as to the more significant areas to be examined.

Preliminary Evaluations

From the data collected, preliminary evaluations are made regarding slope conditions in the region or project study area, the preliminary engineering geology and hazard maps are modified, and an exploration program is planned for areas of particular interest.

Explorations

Geophysical Surveys

Seismic refraction profiling is performed to determine the depth to sound rock and the probable groundwater table, and is most useful in differentiating between colluvial or residual soils and the fractured-rock zone. Typical seismic velocities from the weathering profile that develops in igneous and metamorphic rocks in warm, humid climates are given in Figure 1.100. Surveys are made both longitudinal and transverse to the slope. They are particularly valuable on steep slopes with a deep weathering profile where test borings are time-consuming and costly.

Resistivity profiling is performed to determine the depth to groundwater and to rock. Profiling is generally only applicable to depths of about 15 to 30 ft (5 to 10 m), but very useful in areas of difficult access. In the soft, sensitive clays of Sweden, the failure surface or potential failure surface is often located by resistivity measurements since the salt content, and therefore the resistivity, often changes suddenly at the slip surface (Broms, 1975).

Test Boring Program

Test borings are made to confirm the stratigraphy determined by the geophysical explorations, to recover samples of the various materials, and to provide holes for the installation of instrumentation. The depth and number of borings depend on the stratigraphy and uniformity of conditions, but where the slope consists of colluvial or residual soils, borings should penetrate to rock. In other conditions, the borings should extend below the depth of any potential failure surface, and always below the depth of cut for an adequate distance.

Sampling should be continuous through the potential or existing rupture zone, and in residual soils and rock masses care should be taken to identify slickensided surfaces. Groundwater conditions must be defined carefully, although the conditions existing at the time of investigation are not likely to be those during failure.

In Situ *Measurements*

Piezometers yield particularly useful information if in place during the wet season. In clayey residual profiles, confined water-table conditions can be expected in the weathered or fractured rock zone near the interface with the residual soils, or beneath colluvium. A piezometer set into fractured rock under these conditions may disclose artesian pressures exceeding the hydraulic head given by piezometers set into the overlying soils, even when they are saturated (Figure 1.130).

Instrumentation is installed to monitor surface deformations, to measure movement rates, and to detect the rupture zone if the slope is considered to be potentially unstable or is undergoing movement.

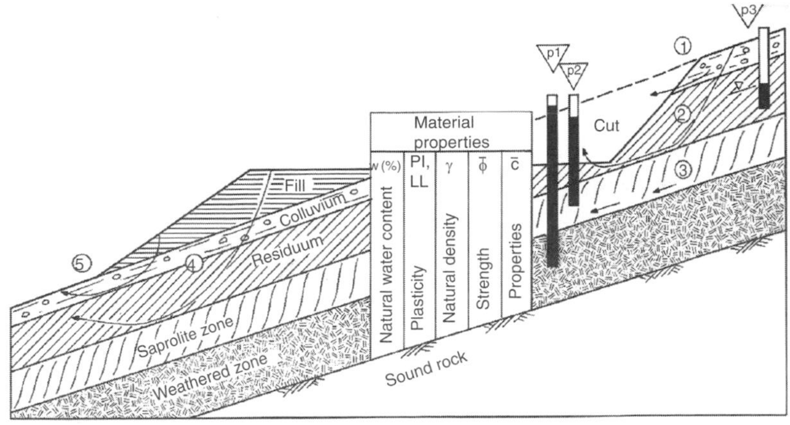

Legend:

(1) Shallow slump in colluvium; (2) total failure of cut in residuum caused by very high-seepage pressures in saprolite; (3) large-scale failure of slope caused by very high-seepage pressures in the fractured rock zone; (4) large rotational failure of fill through residuum; (5) slump failure at fill toe through the colluvium.

FIGURE 1.130
Schematic of typical section prepared at scale of 1:1 showing tentative cut and fill imposed on 30° slope in residual soil profile as basis for analysis. Piezometers P1 and P2 show excess water pressures in saprolite and fractured rock compared with saturated zone in residuum at P3. Several possible failure conditions requiring evaluation are shown.

Nuclear probes lowered into boreholes measure density and water content, and have been used to locate a failure surface by monitoring changes in these properties resulting from material rupture. In a relatively uniform material, the moisture and density logs will show an abrupt change in the failure zone from the average values (Cotecchia, 1978).

Dating Relict Slide Movements

Radiometric dating of secondary minerals in a ruptured zone or on slickensided surfaces, or of organic strata buried beneath colluvium, provides a basis for estimating the age of previous major movements.

Growth ring counts in trees that are inclined in their lower portion and vertical above also provide data for estimating the age of previous major slope movements. The date of the last major movement can be inferred from the younger, vertical-growing segments (Cotecchia, 1978). Slope failures cause stresses in the tree wood which result in particular tissues (reaction or compressed wood), which are darker and more opaque than normal unstressed wood. On the side toward which the tree leans there is an abrupt change from the growth rings of normal wood to those of compression wood. By taking small cores from the tree trunk it is possible to count the rings and estimate when the growth changes occurred and, thus, to date approximately the last major slope movement.

Slope Assessment

Data Presentation

A plan of the slope area is prepared showing contours, drainage paths, seepage emerging from the slope, outcrops, tension cracks and other failure scars, and other significant information. Sections are prepared at a 1:1 scale illustrating the stratigraphy and groundwater conditions as determined from the explorations, as well as any relict failure surfaces.

Evaluations and Analyses

Possible failure forms are predicted and existing failures are delineated as falls, slides, avalanches, or flows, and the degree of the hazard is judged. Depending upon the degree of risk, the decision is made to avoid the hazard or to eliminate or reduce it. For the cases of falls, avalanches, flows, and failures by lateral spreading, the decision is based on experience and judgment. Slides may be evaluated by mathematical analysis, but in recognition that movements may develop progressively.

Preliminary analysis of existing or potential failures by sliding includes the selection of potential failure surfaces by geometry in the case of planar slides, or analytically in the case of rotational slides, or by observation in the case of an existing slide. An evaluation is made of the safety factor against total failure on the basis of existing topographic conditions, then under conditions of the imposed cut or fill. For preliminary studies, shear strengths may be estimated from published data, or measured by laboratory or *in situ* testing. In the selection of the strength parameters, consideration is given to field conditions (Table 1.6) as well as to changes that may occur with time (reduction from weathering, leaching, solution). Other transient conditions also require consideration, especially if the safety factor for the entire slope is low and could go below unity with some environmental change.

1.5.3 Detailed Study of Cut, Fill, or Failure Area

General

Detailed study of the area of the proposed cut or fill, or of the failure, is undertaken after the stability of the entire slope is assessed. The entire slope is often erroneously neglected in studies of cuts and side-hill fills, and is particularly important in mountainous terrain.

Explorations

Seismic refraction surveys are most useful if rock is anticipated within the cut, and there are boulders in the soils that make the delineation of bedrock difficult with test and core borings.

Test and core borings, and *test pits* are made to recover samples, including undisturbed samples, for laboratory testing. In colluvium, residuum, and saprolite, the best samples are often recovered from test pits, but these are usually limited to depths of 10 to 15 ft (3 to 5 m) because of practical excavation considerations.

In situ testing is performed in materials from which undisturbed samples are difficult or impossible to procure.

Laboratory Testing

Laboratory strength testing should duplicate the field conditions of pore-water pressures, drainage, load duration, and strain rate that are likely to exist as a consequence of construction operations, and samples should usually be tested in a saturated condition. It must be considered that conditions during and at the end of construction (short-term) will be different than long-term stability conditions. In this regard, the natural ability of the slope to drain during cutting plays a significant role.

Evaluation and Analysis

Sections illustrating the proposed cut, fill, or failure imposed on the slope are prepared at a 1:1 scale. The selection of cut slope inclination is based on the engineer's judgment of stability and is shown on the section together with the stratigraphy, groundwater conditions measured, and the soil properties as shown in Figure 1.130.

Analyses are performed to evaluate stable cut angles and sidehill fill stability, and the necessity for drainage and retention. Consideration must be given to the possibility of a number of failure forms and locations as shown in Figure 1.130, as well as to changing groundwater and other environmental conditions.

1.5.4 Case Study

Background

A roadway was constructed during the early 1990s beginning on the western coastal plain of Ecuador, crossing over the Andes Mountains, and terminating after 110 km at the city of Cuenca. Landslides began at numerous locations where the roadway climbed the steep mountain slopes, usually 35° or steeper. The general landform along a portion of the roadway is illustrated on the 3D diagram in Figure 1.131. Slope failures increased significantly during the El Niño years of 1997 and 1998.

Investigation

Initially, investigation included a number of trips along the roadway during which the slope failures were photographed and cataloged. Pairs of aerial photos were examined stereoscopically. Eventually a helicopter fly-over was made and the roadway continuously photographed. Debris avalanches, occurring on the upslope side of the roadway, were the most common form of slope failure (Figure 1.5 and Figure 1.132). More than 125 failure locations were identified. The landslide debris was bulldozed from the roadway onto the downslope side (Figure 1.132) further destabilizing the slopes and contributing to erosion and choking of streams downslope.

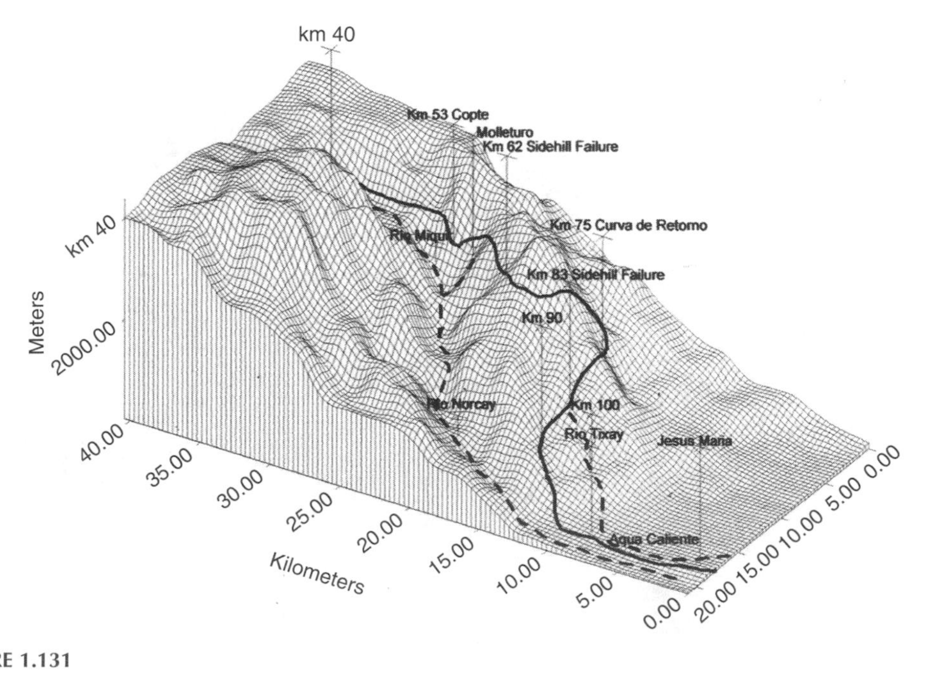

FIGURE 1.131
Three-dimensional diagram of a portion of roadway over the Andes Mountains in Ecuador. Side-hill failure at km 62 is shown in Figure 1.6. Debris avalanche at km 88 shown in Figure 1.132. Large slump slide shown in Figure 1.133.

FIGURE 1.132
Debris avalanches, km 83 along roadway in Figure 1.131.

Other slope failures included a few large "slump" slides (Figure 1.133) and numerous failures on the downslope of the roadway resulting from erosion from discharge of roadway storm drains (Figure 1.6).

FIGURE 1.133
Large slump slide at km 75 located in Figure 1.131. Large tension cracks appeared upslope and the two roadways in the photo center have dropped several meters.

Budget limitations required that detailed investigations using seismic refraction surveys and test borings be limited to four of the more critical locations.

Evaluations and Treatments

Debris Avalanches

The debris avalanches were occurring along 80 km of roadway beginning near the lowlands, where the dominant conditions are relatively soft volcanic rocks with a moderately thick cover of residual soils, and in places, colluvium. The natural slopes usually were inclined at about 35° or steeper. Initially, vegetation was removed and roadway cuts were inclined at 53°, much steeper than the original slopes. Subsequently rainfall and seepage resulted in the residual soils sliding along the fractured rock surface.

Unsupported cuts were made because the large number of cuts would cause support with retaining walls to make roadway construction prohibitive. A more stable alternate slope design could have included 10-m-high 45° cuts with benches to result in an overall cut slope inclination of 38°. Construction costs would be increased but roadway slope failures and maintenance costs should be decreased.

In many cases where failure has occurred the residual soil has been removed and the slope is now self-stabilized, although in some cases, there remains potentially unstable material upslope of the failure scar. Future failures will be removed from the roadway and disposed of in designated spoil areas, rather than dumping over the sides of the roadway.

Downslope Roadway Drains

It was recommended that all roadway drains be relocated so that discharge downslope is where erosion will not endanger the roadway. Drainage channels should be lined to prevent erosion. The failure shown in Figure 1.6 resulted from storm water discharge eroding the slope below a shotcreted gabion wall that was supporting the roadway. Storm water entered the catch basin shown on the upslope side of the roadway and discharged through a pipe exiting on the downslope side. The failure in Figure 1.6 was corrected with the construction of an anchored wall and relocating the storm water discharge point.

Large Slump Slide

At km 75 the roadway makes an abrupt switch-back as shown in Figure 1.133. At this location is a large slump slide evidenced by tension cracks upslope and roadway movement. When first visited during 1999 the portion of the roadway in the middle of the photo (Figure 1.133) had dropped about 2 m; when visited the following year, the roadway had dropped an additional 2 m and was almost impassable. This was considered a priority site for remediation.

Explorations with seismic refraction surveys and test borings determined geologic conditions to include about 15 m of colluvium overlying 5 to 20 m of fractured rock grading to hard rock.

Treatments recommended for stabilization included:

1. Surface drains constructed upslope to collect runoff and to discharge away from the failing area.
2. Subhorizontal drains installed along the toes of the cut slopes areas shown on the photo.
3. Upslope roadway cut slope to be reshaped with benches and covered with shotcrete.

4. Shotcrete placed to cover the shallow rock slope between the upper and lower roadways.
5. Low anchored concrete walls to be installed along the toe of the upslope and downslope side of the roadways.
6. The roadway grade is not to be raised as this would add load to the unstable mass.

1.5.5 Instrumentation and Monitoring

Purpose

Instrumentation is required to monitor changing conditions that may lead to total failure where slope movement is occurring and safety factors against sliding are low, or where a major work would become endangered by a slope failure.

Slope-stability analysis is often far from precise, regardless of the adequacy of the data available, and sometimes the provision for an absolutely safe slope is prohibitively costly. In this case, the engineer may wish to have contingency plans available such as the installation of internal drainage systems or the removal of material from upslope, etc., if the slope shows signs of becoming unstable.

In unstable or moving slopes, instrumentation is installed to locate the failure surface and determine pore-water pressures for analysis, and to measure surface and subsurface movements, velocities, and accelerations which provide indications of impending failure. In cut slopes, instrumentation monitors movements and changing stress conditions to provide early warning and permit invoking remedial measures when low safety factors are accepted in design.

Instrumentation Methods Summarized

Surface movements are monitored by survey nets, tiltmeters (on benches), convergence meters, surface extensometers, and terrestrial photography. Accuracy ranges from 0.5 to 1.0 mm for extensometers, to 30 mm for the geodimeter, and to 300 mm for the theodolite (Blackwell et al., 1975). GPS systems are showing promise for continuously monitoring and recording slope movements.

Subsurface deformations are monitored with inclinometers, deflectometers, shear-strip indicators, steel wire and weights in boreholes, and the acoustical emissions device. Accuracy for extensometers and inclinometers usually ranges from 0.5 to 1.0 mm, but the accuracy depends considerably on the deformation pattern and in many instances cannot be considered better than 5 to 10 mm.

Pore-water pressures are monitored with piezometers. All instruments should be monitored periodically and the data plotted as it is obtained to show changing conditions. Movement accelerations are most significant.

GPS Installations

Mission Peak Landslide, Fremont, California

On the Internet during 2003, the U.S. Geological Survey (USGS) reported on a global positioning system (GPS) installation to monitor the Mission Peak Landslide in Fremont, California. Installed in January 2000, the system included a field station with a GPS antenna, receiver, controller card, and radio modem that sent data to the base station which included a radio modem, personal computer connected to a phone line or the

Internet for graphical output. The massive block at the head of the landslide was found initially to be moving at less than 1cm/week, then accelerating to 2 cm/week apparently in response to rainfall. At the cessation of seasonal rains it remained moving at the rate of 1 mm/week for a 4-month period from February to June 2000. GPS measurements were reported to typically show repeatability ±1 cm horizontally and ±2 cm vertically.

Lishan Slope, Xian, China

Orense (2003) describes the landslide hazard threatening the Huaqing Palace, in Xian, China. Built during the Tang dynasty (618–907), the Palace is located at the foot of the Lishan slope that shows visible deformation. The potential failure mass is a large-scale rock slide. Although in an area of earthquake activity, it is believed that subsidence in the valley from extensive groundwater withdrawal has resulted in activating slope movements. Geologic conditions generally consist of a layer of loess overlying gneiss bedrock. A site plan is given in Figure 1.134. The potential failure mass has been divided into three possible blocks. The dashed line represents the limits of a thick loess deposit that has already slid.

Studies were begun in 1991 by the Disaster Prevention Institute of Kyoto University, Japan, and the Xian Municipal Government. An extensive automated monitoring system was installed as shown in Figure 1.135. Included in the system were short- and long-span extensometers (lines A and B, Figure 1.134), total station surveying, GPS survey, borehole inclinometers, and ground motion seismographs. Data are transmitted periodically to Kyoto University via satellite.

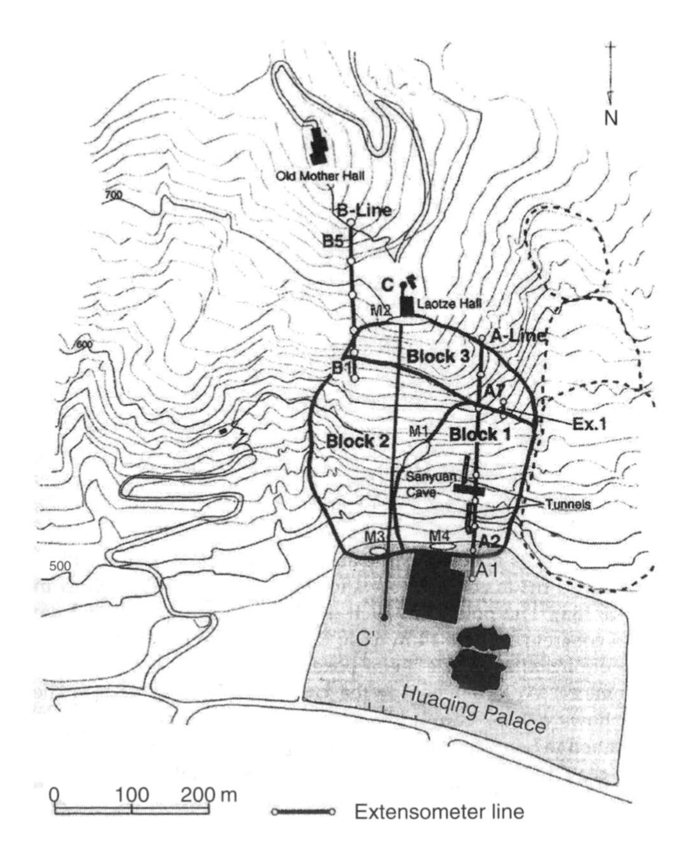

FIGURE 1.134
Plan map of the Lishan slope in China with potential landslide blocks and extensometer lines (From Orense, R. P., *Geotechnical Hazards: Nature, Assessment and Mitigation,* University of the Philippines Press, Quezon City, 2003. With permission. After Sassa, K. et al., *Proceedings of the International Symposium on Landslide Hazard Assessment,* 1–24, Xian, China, 1997.)

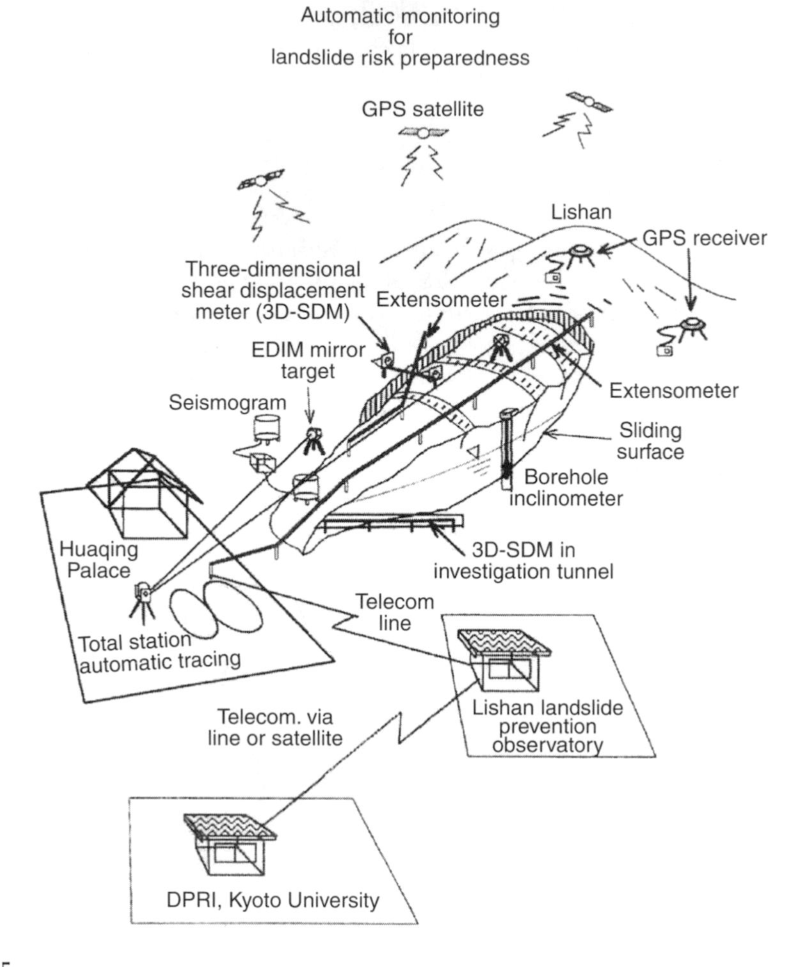

FIGURE 1.135
Monitoring system for the Lishan slope prepared for landslide risk assessment. (From Orense, R. P., *Geotechnical Hazards: Nature, Assessment and Mitigation,* University of the Philippines Press, Quezon City, 2003. With permission. After Sassa, K. et al., *Proceedings of the International Symposium on Landslide Hazard Assessment,* 1–24, Xian, China, 1997.)

References

Alden, W. C., Landslide and Flood at Gros Ventre, Wyoming, *Focus on Environmental Geology,* Tank R., Ed., Oxford University Press, New York (1973), 1928, pp. 146–153.

Banks, D. C., Study of Clay Shale Slopes, *Stability of Rock Slopes, Proceedings of the ASCE, 13th Symposium on Rock Mechanics,* Urbana, IL, (1971), 1972, pp. 303–328.

Banks, D. C. and Strohn, W. E., Calculation of Rock Slide Velocities, *Proceedings of the 3rd International Congress for Rock Mechanics, International Society for Rock Mechanics,* Denver, Vol. 118, 1974, pp. 839–847.

Barton, N., Progressive Failure of Excavated Rock Slopes, *Stability of Rock Slopes, Proceedings of the ASCE 13th Symposium on Rock Mechanics,* Urbana, IL, 1971, 1972, pp. 139–170.

Binger, W. V., Analytical Studies of Panama Canal Slides," *Proceedings of the 2nd International Conference on Soil Mechanics and Foundation Engineering,* Rotterdam, Vol. 2, 1948, pp. 54–60.

Bishop, A. W., The use of the slip circle in the stability analysis of earth slopes, *Geotechnique,* 5, 7–17, 1955.

Bjerrum, L., Progressive Failures in Slopes of Overconsolidated Plastic Clay and Clay Shales, *Terzaghi Lectures 1963–1972, ASCE,* (1974), 1966, pp. 139-189.

Bjerrum, L., Loken, T., Heiberg, S. and Foster, H., A Field Study of Factors Responsible for Quick Clay Slides, *Proceedings of the 7th International Conference on Soil Mechanics and Foundation Engineering*, Mexico City, 1969, pp. 531–540, Vol. 2.

Blackwell, G., Pow, D., and Klast, L., Slope Monitoring at Brenda Mine, *Proceedings of the 10th Canadian Rock Mechanics Symposium,* Kingston, Ontario, September 1975, pp. 45–79.

Brawner, C. O., Case examples of instability of rock slopes, *J. Assoc. Prof. Eng. Brit. Columb.*, 26, 1975.

Broms, B. B., Landslides, *Foundation Engineering Handbook*, Winterkorn, H. F. and Fang, H.-Y., eds., Van Nostrand Reinhold Co., NY, 1975, pp. 373–401, Chap. 11.

Broms, B. B. and Boman, P., Lime Columns — A New Foundation Method, *Proc. ASCE J. Geotech. Eng. Div.*, 105, 539–556, 1979.

Chassie, R. G. and Goughnor, R. D., States Intensifying Efforts to Reduce Highway Landslides, Civil Engineering, ASCE, New York, April, 1976, p. 65.

Coleman, J. M., Prior, D. B., and Garrison, L. E., Subaqueous Sediment Instabilities in the Offshore Mississippi River Delta, Open File Report No. 80.01, Bureau of Land Management, U.S. Department of Interior, 1980.

Cotecchia, V., Systematic Reconnaissance Mapping and Registration of Slope Movements, *Int. Assoc. of Eng. Geol., Bull. No. 17*, 5–37, 1978.

D'Appolonia, E. D., Alperstein, R. A. and D'Appolonia, D. J., Behavior of a colluvial soil slope, *Proc. ASCE J. Soil Mech. Found. Eng. Div.*, 93, 447–473, 1967.

Deere, D. U. and Patton, F. D., Slope Stability in Residual Soils, *Proceedings of the ASCE 4th Pan American Conference on Soil Mechanics Foundation Engineering,* San Juan, P R, 1971, pp. 87–170.

Easton. W. H., Earthquakes, Rain and Tides of Portuguese Bend Landslide, California, *Bull. Assoc. Eng. Geol.*, 8, 1971.

Eden, W. J., Fletcher, E. B., and Mitchell, R. J., South Nation River Landslide, 16 May 1971, *Can. Geotech. J.*, 8, 1971.

Elias, V. and Juran, I., Soil Nailing for Stabilization of Highway Slopes and Excavations, Report No. FHWA-RD-89-198, NTIS Springfield, VA, 1991.

ENR, Landslide kills eight, destroys 76 homes, *Engineering News Record,* Dec. 8, 1977, p.11.

Focht, Jr. J. A. and Kraft, Jr. L. M., Progress in marine geotechnical engineering, *Proc. ASCE J. Geotech. Eng. Div.*, 103, 1097–1118, 1977.

Fox, P. P., Geology, Exploration, and Drainage of the Serra Slide, Santos, Brasil, Engineering Geology Case Histories 1–5, The Geological Society of America, Engineering Geological Divsion, 1964 pp. 17–24.

Franklin, J. A. and Dusseault, M. B., *Rock Engineering,* McGraw-Hill Book Co., New York, 1989, pp. 507–536.

Garga, V. K. and De Campos, T. M. P., A Study of Two Slope Failures in Residual Soils, *Proceedings of the 5th Southeast Asian Conference on Soil Engineering,* Bangkok, 1977, pp. 189–200.

Gebney, D. S. and McKittrick, D. P., Reinforced Earth: A New Alternative for Earth-Retention Structures, Civil Engineering, ASCE, 1975.

Guidicini, G. and Iwasa, O. Y., Tentative correlation between rainfall and landslides in a humid tropical environment, *Bull. Int. Assoc. Eng. Geol.*, 13–20, 1977.

Habib, P., Production of gaseous pore pressure during rock slides, *J. Int. Soc. Rock Mech.*, 7, 193, 1975.

Hamel, J. V., Geology and slope stability in Western Pennsylvania, *Bull. AEG*, XVII, 1–26, 1980.

Hamel, J. V., The slide at Brilliant Cut, in *Stability of Rock Slopes, Proceedings of the ASCE, 13th Symposium on Rock Mechanics,* Urbana, IL, (1971), 1972, pp. 487–572.

Handy, R. L. and Williams, W. W., Chemical Stabilization of an Active Landslide, *Civil Engineering,* ASCE, 1967, pp. 62–65.

Heinz, R. A. *In situ* Soils Measuring Devices, Civil Engineering, ASCE, October 1975, pp. 62–65.

Hoek, E. and Bray, J. W., *Rock Slope Engineering*, 2nd ed., The Institute of Mining and Metallurgy, London, 1977.

Hunt, R.E., Miller, M., and Bump, V., The Forest City Landslide, *Proceedings of the 3rd International Conference, Case Histories in Geotechnical Engineering*, St. Louis, MO, 1993.

Hunt, R. E. and Costa Nunes, A. J. da, Retaining Walls: Taking It from the Top, Civil Engineering ASCE, May 1978, pp. 73–75.
Hunt, R. E. and Santiago, W. B., A fundo critica do engenheiro geologo em estudos de implantaclo de ferrovias, *Proceedings of the 1st Congress*, Brasileiro de Geologia de Engenharia, Rio de Janeiro, Aug., Vol. 1, 1976, pp. 79–98.
Jahns, R. H. and Vonder Linden, C., Space–Time Relationships of Landsliding on the Southerly Side of Palos Verdes Hills, California, *Geology, Seismicity and Environmental Impact,* Spec. Pub. Assoc. Engrg. Geol., Los Angeles, 1973, pp. 123–138.
Janbu, N., Slope Stability Computations, in *Embankment Dam Engineering*, Hirschfield, R. C. and Poulos, S. J., eds., J. Wiley, New York, 1973, pp. 47–86.
Janbu, N., Bjerrum, L., and Kjaernsli, B., *Soil Mechanics Applied to Some Engineering Problems*, Norwegian Geotechnical Institute, Oslo, Publ. 16, 1956, pp. 5–26.
Jones, F. O., Landslides of Rio de Janeiro and the Serra das Araras Escarpment, Brasil, U.S. Geological Survey Paper No. 697, U.S. Govt. Printing Office, Washington, DC, 1973.
Kiersch, G. A., The Vaiont Reservoir Disaster, *Focus on Environmental Geology*, Tank, R., Ed., The Oxford University Press (1973), Art. 17, 1965, pp. 153–164.
Krahn, J. and Morgenstern, N. R., Mechanics of the Frank Slide, *Proc. ASCE Rock Eng. Found. Slopes*, 309–332, 1976.
Kramer, S.L. and Smith, M.W., Modified newmark model for seismic displacements of complaint slopes, *J. Geotec. Geoenviron. Eng.*, ASCE, 123, 635–644, 1993.
Kulhawy, F. H., Analysis of a High Gravity Retaining Wall, *Proceedings of the ASCE Conference on Analysis and Design in Geotechnical Engineering*, Univ. of Texas, Austin, I, 1974, pp. 159–171.
Lambe, T. W. and Whitman, R. V., *Soil Mechanics*, J. Wiley, New York, 1969.
Lang, T. A., Rock reinforcement, *Bull. Assoc. Eng. Geol.*, IX, 215–239, 1972.
Lessing, P., Dean, S.L., and Kulander, B.R., Geological Evaluation of West Virginia Landslides, *Bull. Assoc. Eng. Geolog.*, XXXI, 191–202, 1994.
McDonald, B. and Fletcher, J. E., Avalanche — 3500 Peruvians Perish in Seven Minutes, *National Geographic Magazine,* June, 1962.
McKown, A.F., *Early Warning System, Civil Engineering,* ASCE, May, pp. 56–59, 1999.
Molnia, B. F., Carlson, P. R., and Bruns, T. R., Large Submarine Slide in Kayak Trough, Gulf of Alaska, *Reviews in Engineering Geology*, VII, Landslides, Coates, D. R., Ed., Geologic Society of America, 1977, pp.137–148.
Mollard. J. D., Regional landslide types in Canada, *Reviews in Engineering Geology, Vol. III, Landslides,* Geological Society of America, 1977, pp. 29–56.
Morgenstern, N. R. and Price, V. E., The analysis of the stability of general slip surfaces, in *Geotechnique*, 15, 79–93, 1965.
Morgenstern, N. R. and Sangrey, D. A., Methods of Stability Analysis, in *Landslides: Analysis and Control,* Schuster, R. L. and Krizek, R. J., eds., Special. Report 176, National Academy of Sciences, Washington, DC, 1978, pp. 255–172, Chap. 7.
NAVFAC, *Design Manual, Soil Mechanics, Foundations and Earth Structures*, DM-7.1, Naval Facilities Engineering Command, Alexandria, VA, March, 1982.
Newmark, N. M., Effects of earthquakes on dams and embankments, *Geotechnique,* 15, London, 1965.
Noble, H. L., Residual strengths and landslides in clay and shale, *Proc. ASCE J. Soil Mechs. Found. Eng. Div.*, 99, 1973.
Orense, R.P., *Geotechnical Hazards: Nature, Assessment and Mitigation,* University of the Philippines Press, Quezon City, 2003.
Palladino, D. J. and Peck, R. B., Slope failures in an over-consolidated clay, Seattle, WA, *Geotechnique*, 22, 563–595, 1972.
Patton, F. D. and Hendron, Jr., A. J., General Report on Mass Movements, *Proceedings of the 2nd International Cong. Intl. Assoc. Engrg. Geol., Sao Paulo*, 1974, p. V-GR 1.
Peck, R. B., Stability of natural slopes, *Proc. ASCE J. Soil Mech. Found. Engrg. Div.*, 93, 1967.
Pierson, L.A., Davis, S.A., and Van Vickle, R., *Rockfall Hazard Rating System Implementation Manual*, FHWA Report No. OR-EG-90-01, Oregon DOT, Highway Div., Salem, OR, 1990.
Piteau, D. R., Regional Slope-Stability Controls and Engineering Geology of the Fraser Canyon, British Columbia, *Reviews in Engineering Geology, Vol. III*, Coates, D. R., Ed., Geologic Society of America, Boulder, CO, 1977, pp. 85–112.

Richter, C. F., *Elementary Seismology*, W. H. Freeman and Co., San Francisco, 1958, p. 125.
Royster, D. L., Highway landslide problems along the Cumberland Plateau, Tennessee, *Bull. Assoc. Eng. Geol.*, X, 1973.
Royster, D. L., Landslide remedial measures, *Bull. Assoc.*, XVI, 301–352, 1979.
Santi, P.M., Crenshaw, B.A., and Elifrits, C.D., Demonstration Projects Using Wick Drains to Stabilize Landslides, *Environ. Eng. Sci.*, IX, 339–350, Joint Pub. AEG-GSA, 2003.
Sassa, K., Fukuoka, H., Yang, Q.J., and Wang, F.W., Landslide Hazard Assessment in Cultural Heritage, Lishan, Xian, *Proceedings International Symposium on Landslide Hazard Assessment*, 1–24, Xian, China, 1997.
Seed, H. B., A method for earthquake resistant design of earth dams, *Proc. ASCE J. Soil Mech. Found. Engrg. Div.*, 92, 13–41, 1966.
Seed, H. B. and Wilson, S. D., The Turnagain Heights landslide, Anchorage, Alaska, *Proc. ASCE J. Soil Mech. Found. Engrg. Div.*, 93, 325–353, 1967.
Spencer, E., A method of analysis of the stability of embankments assuming parallel inter-slice forces, *Geotechnique*, 17, 11–26, 1967.
Spencer, E., Thrust line stability in embankment stability analysis, *Geotechnique*, 23, 85-100, 1973.
Sunset, If hillside slides threaten in Southern California, November is planting action month, *Sunset Mag.*, November, 1978, pp. 122–126.
Tavenas, F., Chagnon, J. Y., and La Rochelle, P., The Saint-Jean-Vianney landslide: observations and eyewitness accounts, *Can. Geotech. J.*, 8, 1971.
Taylor, D. W., *Fundamentals of Soil Mechanics*, Wiley, New York, 1948.
Terzaghi, K., *Theoretical Soil Mechanics*, Wiley, New York, 1943.
Terzaghi, K., Mechanism of landslides, in *Engineering Geology (Berkey Volume)*, The Geologic Society of America, pp. 83–123, 1950.
Terzaghi, K., Stability of steep slopes on hard, unweathered rock, *Geotechnique*, 12, 251–270, 1962.
Tschebotarioff, G. P., *Foundations, Retaining and Earth Structures*, McGraw-Hill Book Co., New York, 1973.
Vargas, M. and Pichler, E., Residual Soils and Rock Slides in Santos, Brazil, *Proceedings of the 4th International Conference Soil Mechanics Foundation Engineering*, Vol. 2, 1957, pp. 394–398.
Varnes, D. J., Landslide Types and Processes, *Landslides and Engineering Practice*, Eckel, E. B., Ed., Highway Research Board Spec. Report No. 29, Washington, DC, 1958.
Varnes, D. J., Slope Movement Types and Processes, Landslides: analysis and Control, Special Report No. 176, National Academy of Sciences, Washington. DC, 1978, pp. 11–35, Chap. 2.
Wartman, J., Bray, J.D., and Seed, R.B., Inclined plane studies of the newmark sliding block procedure, *J. Geotec. Geoenviron. Eng.*, ASCE, 129, 673–684, 2003.
Youd, T. L., Major Cause of Earthquake Damage is Ground Failure, *Civil Engineering*, ASCE, pp. 47–51, 1978.

Further Reading

Barata, F. E., Landslides in the Tropical Region of Rio de Janeiro, *Proceedings of the 7th International Conference on Soil Mechanics and Foundation Engineering*, Mexico City, Vol. No. 2, 1969, pp. 507–516.
Coates, D. R., Ed., *Landslides, Reviews in Engineering Geology*, Vol. III, The Geologic Society of America, Boulder, CO, 1977.
Costa Nunes, A. J. da, Slope Stabilization—Improvements in the Prestressed Anchorages in Rocks and Soils, *Proceedings of the 1st International Conference on Rock Mechanics*, Lisbon, 1966.
Costa Nunes, A. J. da, Landslides in Soils of Decomposed Rock Due to Intense Rainstorms, *Proceedings of the 7th International Conference on Soil Mechanics & Foundation Engineering*, Mexico City, 1969.
Costa Nunes, A. J. da, Fonseca, A.M.C.C., and Hunt, R. E., A broad view of landslides in Brasil, in *Geology and Mechanics of Rockslides and Avalanches*, Voight, B., Ed., Elsevier Scientific Pub. Co., Amsterdam, 1980.

Dunn, J. R. and Banino, G. M., Problems with Lake Albany Clays, *Reviews in Engineering Geology*, Vol. III, Landslides, the Geologic Society of America, Boulder, CO, 1977, pp. 133–136.

ENR, Italian pile system supports slide in its first major U.S. application, *Engineering News-Record*, Nov. 24, 1977b, p. 16.

Guidicini, G. and Nieble, C. M., *Estabilidade de Taludes Naturais e de Escavacao*, Editora Edgard Blucher Ltda., Sao Paulo, 1976, 170.

Hunt, Roy E., *Geotechnical Engineering Investigation Manual*, McGraw-Hill Book Co., New York, 1984.

Hunt, R. E. and Deschamps, R. J., Stability of Slopes, in *The Civil Engineering Handbook*, 2nd ed., Chen, W. F. and Liew, J. Y. R., Eds., CRC Press, Boca Raton, FL, 2003.

Hunt, R. E. and Shea, G. P., Avoiding Landslides on Highways in Mountainous Terrain, *Proceedings of the International Road Federation, IV IRF African Highway Conference*, Nairobi, January, 1980.

Hutchinson, J. N., Assessment of the effectiveness of corrective measures in relation to geological conditions and types of slope movement, *Bull. Int. Assoc. Eng. Geol.*, Landslides and Other Mass Movements, 131–155, 1977.

Kalkani, E. C. and Piteau, D. R., Finite element analysis of toppling failure at Hell's Gate Bluffs, British Columbia, *Bull. Assoc. Eng. Geol.*, XIII, 1976.

Kennedy, B. A., Methods of Monitoring Open Pit Slopes, *Stability of Rock Slopes, Proceedings of the ASCE 13th Symposiumon on Rock Mechanics*, Urbana, IL, (1971), 1972, pp. 537–572.

Kenney, T. C., Pazin, M., and Choi, W. S., Design of horizontal drains for soil slopes, *Proc., ASCE J. Geotech. Eng. Div.*, 103, GT11, 1311–1323, 1977.

Koerner, R. M., Lord, Jr., A. E., and McCabe, W. M., Acoustic emission monitoring soil stability, *Proc., ASCE J. Geotech. Eng.* Div., 104, 571–582, 1978.

La Rochelle, P., Regional geology and landslides in the marine clay deposits of eastern Canada, *Can. Geotech. J.*, 8, 1970.

Lo, K. Y. and Lee, C. F., Analysis of Progressive Failure in Clay Slopes, *Proceedings of the 8th International Conference, Soil Mechanics, Foundation Engineering*, Moscow, Vol. 1, 1973, pp. 251–258.

Pinckney, C. J., Streiff, D., and Artim, E., The influence of bedding-plane faults in sedimentary formations on landslide occurrence western San Diego County, California, *Bull. Assoc. Eng. Geol.*, Spring, XVI, 289–300, 1979.

Piteau, D. R. and Peckover, F. L., Engineering of Rock Slopes, in *Landslides: Analysis and Control, Schuster and Krizek*, Eds., Special Report No. 176, National Academy of Sciences, Washington, DC, 1978, pp. 192–227.

Quigley, R. M. et al., Swelling clay in two slope failures at Toronto, Canada, *Can. Geotech. J.*, 8, 1971.

Skempton, A. W. and Hutchinson, J., *Stability of Natural Slopes and Embankment Foundations*, State-of-the-Art Paper, *Proceedings of the 7th International Conference on Soil Mechanics and Foundation Engineering*, Mexico City, 1969, pp. 291–340.

Taylor, D. W., Stability of Earth Slopes, Contributions to Soil Mechanics 1925 to 1940, Boston Soc. Civil Engrs., 1940.

Terzaghi, K., Varieties of Submarine Slope Failures, *Proceedings of the 8th Texas Conference on Soil Mechanics and Foundation Engineering*, 1956, Chap. 3.

Terzaghi. K., Selected Professional Reports — Concerning a Landslide on a Slope Adjacent to a Power Plant in South America, reprinted in *From Theory to Practice in Soil Mechanics*, Wiley, New York, 1960.

Terzaghi, K. and Peck, R. B., *Soil Mechanics in Engineering Practice*, Wiley, New York, 1967.

Vanmarcke, E. H., Reliability of Earth Slopes, *Proc. ASCE J. Geotech. Eng. Div.*, 103, pp. 1247–1265, 1977.

Zruba, Q. and Mencl, V., *Landslides and Their Control*, Elsevier Scientific Pub. Co., Amsterdam, 1969.

2

Ground Subsidence, Collapse, and Heave

2.1 Introduction

2.1.1 General

Origins

The hazardous vertical ground movements of subsidence, collapse, and heave, for the most part, are the results of human activities that change an environmental condition. Natural occurrences, such as earthquakes and tectonic movements, also affect the surface from time to time.

Significance

Subsidence, collapse, and heave are less disastrous than either slope failures or earthquakes in terms of lives lost, but the total property damage that results each year probably exceeds that of the other hazards. A positive prediction of their occurrence is usually very difficult, and uncertainties always exist, although the conditions favorable to their development are readily recognizable.

2.1.2 The Hazards

A summary of hazardous vertical ground movements, their causes, and important effects is given in Table 2.1.

2.1.3 Scope and Objectives

Scope

Ground movements considered in this chapter are caused by some internal change within the subsurface such as the extraction of fluids or solids, solution of rock or a cementing agent in soils, erosion, or physicochemical changes. Movements brought about by the application of surface loads from construction activity (that is, ground settlements resulting from embankments, buildings, etc.) will not be considered here.

Objectives

The objectives are to provide the basis for recognizing the potential for surface movements, and for preventing or controlling the effects.

TABLE 2.1
Summary of Hazardous Vertical Ground Movements

Movement	Description	Causes	Important Effects
Regional subsidence	Downward movement of ground surface over large area	Seismic activity,[a] groundwater extraction, oil and gas extraction	Flooding, growth faults, structure distortion
Ground collapse	Sudden downward movement of ground surface over limited area	Subsurface mining, limestone cavity growth, piping cavities in soils, leaching of cementing agents	Structure destruction, structure distortion
Soil subsidence	Downward movement of ground surface over limited area	Construction dewatering, compression under load applied externally, desiccation and shrinkage	Structure distortion
Ground heave	Upward movement of ground surface	Expansion of clays and rocks, release of residual stresses, tectonic activity,[b] ground freezing	Structure distortion, weakening of clay shale slopes

[a] See Section 3.3.3.
[b] See Section 3.3.3, Appendix A.

2.2 Groundwater and Oil Extraction

2.2.1 Subsurface Effects

Groundwater Withdrawal

Aquifer Compaction

Lowering the groundwater level reduces the buoyant effect of water, thereby increasing the effective weight of the soil within the depth through which the groundwater has been lowered. For example, for a fully saturated soil, the buoyant force of water is 62.4 pcf (1 t/m^3) and if the water table is lowered 100 ft (30 m), the increase in effective stress on the underlying soils will be 3.0 tsf (30 t/m^3), a significant amount. If the prestress in the soils is exceeded, compression occurs and the surface subsides. In an evaluation of the effect on layered strata of sands and clays, the change in piezometric level in each compressible stratum is assessed to permit a determination of the change in effective stress in the stratum. Compression in sands is essentially immediate; cohesive soils exhibit a time delay as they drain slowly during consolidation. Settlements are computed for the change in effective stress in each clay stratum from laboratory consolidation test data.

The amount of subsidence, therefore, is a function of the decrease in the piezometric level, which determines the increase in overburden pressures and the compressibility of the strata. For clay soils the subsidence is a function of time.

Construction Dewatering

Lowering the groundwater for construction projects has the same effect as "aquifer compaction," that is, it compresses soil strata because of an increase in effective overburden stress.

Oil Extraction

Oil extraction differs from groundwater extraction mainly because much greater depths are involved, and therefore much greater pressures. Oil (or gas) extraction results in a reduction of pore-fluid pressures, which permits a transfer of overburden pressures to the intergranular skeleton of the strata.

In the Wilmington oil field, Long Beach, California, Allen (1973) cites "compaction as taking place primarily by sand grain arrangement, plastic flow of soft materials such as micas and clays, and the breaking and sharding of grains at stressed points." Overall, about two thirds of the total compaction at the Wilmington field is attributed to the reservoir sands and about one third to the interbedded shales (Allen and Mayuga, 1969). During a period of maximum subsidence in 1951 to 1952, faulting apparently occurred at depths of 1500 to 1750 ft (450 to 520 m), shearing or damaging hundreds of oil wells.

2.2.2 Surface Effects

Regional Subsidence

General

Surface subsidence from fluid extraction is a common phenomenon and probably occurs to some degree in any location where large quantities of water, oil, or gas are removed. Short-term detection is difficult because surface movements are usually small, are distributed over large areas in the shape of a dish, and increase gradually over a span of many years.

Monitoring Surface Deflections

Traditionally, subsidence has been measured periodically using normal surveying methods. When large areas are involved the procedure is time-consuming, costly, and often incomplete. Since 1992, some major cities have been monitoring subsidence with InSAR (*Interferometric Synthetic Aperture Radar*) (Section 2.2.3). SAR images are presently obtained by the European Space Agency (ESA) satellites. Cities mapped include Houston, Phoenix, and Las Vegas in the United States, and Mexico City. The new InSAR maps provide new and significant aspects of the spatial pattern of subsidence not evident on conventional mapping (Bell et al., 2002),

Some Geographic Locations

Savannah, Georgia: The city experienced as much as 4 in. of subsidence over the 29 year period between 1933 and 1962 because of water being pumped from the Ocala limestone, apparently without detrimental effects (Davis et al., 1962).
Houston, Texas: A decline in the water table of almost 300 ft since 1890 has caused as much as 5 ft of subsidence with serious surface effects (see Section 2.2.4).
Las Vegas, Nevada, Tucson and Elroy, Arizona and the San Joaquin and Santa Clara Valleys, California: They have insignificant amounts of subsidence from groundwater withdrawal.
Mexico City: In the hundred years or so, between the mid-1800s and 1955, the city experienced as much as 6 m (20 ft) of subsidence from compression of the underlying soft soils because of groundwater extraction. By 1949, the rate was 35 cm/yr (14 in.). Surface effects have been serious (see Section 2.2.4).
London, England: A drop in the water table by as much as 200 ft has resulted in a little more than 1in. of subsidence, apparently without any detrimental effects because of the stiffness of the clays.
Long Beach, California: It has suffered as much as 30 ft of subsidence from oil extraction between 1928 and 1970 with serious effects (see Section 2.2.4).

Lake Maricaibo, Venezuela: The area underwent as much as 11 ft of subsidence between 1926 and 1954 due to oil extraction.
Po Valley, Italy: It has been affected by gas withdrawal. In the Adriatic Sea, close to the coast near Ravenna, gas withdrawal from 1971 to 1992 has resulted in 31 cm of land subsidence. Numerical predictions suggest that a residual subsidence of 10 cm may occur by 2042, 50 years after the field was abandoned (Bau et al., 1999).

Flooding, Faulting, and Other Effects

Flooding results from grade lowering and has been a serious problem in coastal cities such as Houston and Long Beach in the United States, and Venice in Italy. In Venice, although subsidence from groundwater withdrawal has been reported to be only 5.5 cm/year, the total amount combined with abnormally high tides has been enough to cause the city to be inundated periodically. The art works and architecture of the city have been damaged as a result. Flood incidence also increases in interior basins where stream gradients are affected by subsidence.

Faulting or *growth faults* occur around the periphery of subsided areas. Although displacements are relatively small, they can be sufficient to cause distress in structures and underground storm drains and sanitary sewers, and sudden drops in roadways. Oil extraction can cause movement along existing major faults.

Differential movement over large distances affects canal flows, such as in the San Joaquin Valley of California and over short distances causes distortion of structures, as in Mexico City.

Grade lowering can also result in the loss of head room under bridges in coastal cities and affect boat traffic, as in Houston (ENR, 1977).

Local Subsidence from Construction Dewatering

Drawdown of the water table during construction can cause surface subsidence for some distance from the dewatering system. Differential settlements reflect the cone of depression. The differential settlements can be quite large, especially when highly compressible peat or other organic soils are present, and the effect on adjacent structures can be damaging.

During the construction of the Rotterdam Tunnel, wellpoints were installed to relieve uplift pressures in a sand stratum that underlay soft clay and peat. The groundwater level in observation wells, penetrating into the sand, at times showed a drop in water level of 42 ft. Settlements were greatest next to the line of wellpoints: 20 in. at a distance of 3 ft and 3 in. at a distance of 32 ft. The water level was lowered for about 2.9 yr and caused an effective stress increase as high as 1.3 tsf (Tschebotarioff, 1973).

2.2.3 Physiographic Occurrence

General

Although subsidence can occur in any location where large quantities of fluids are extracted, its effects are felt most severely in coastal areas and inland basins. When groundwater depletion substantially exceeds recharge, the water table drops and subsidence occurs.

Coastal Areas

Many examples of coastal cities subsiding and suffering flooding can be found in the literature, and any withdrawal from beneath coastal cities with low elevations in reference to sea level must be performed with caution.

Interior Basins

In the semiarid to arid regions of the western United States, the basins are often filled with hundred of meters of sediments, which serve as natural underground reservoirs for the periodic rainfall and runoff from surrounding mountains. Subsidence can reach significant amounts: for example, as much as 30 ft in the San Joaquin Valley of California, 12 ft in the Santa Clara Valley in California, 15 ft in Elroy, Arizona, 6 ft in Las Vegas, Nevada, and 9 ft in Houston (Leake, 2004). Around the Tucson Basin, where the water level has dropped as much as 130 ft since 1947, it has been suggested that minor faulting is occurring and may be the reason for distress in some home foundations (Davidson, 1970, Pierce, 1972). Other effects will be increased flooding due to changes in stream gradients and the loss of canal capacity due to general basin lowering.

In 1989, the U.S. Department of Housing and Urban Development began requiring special subsidence hazard assessments for property located near subsidence features in Las Vegas (Bell et al., 1992). The requirement resulted from structural damage to a major subdivision in North Las Vegas that required the repair or displacement of more than 240 damaged or threatened homes at a cost of $12 to $13 million. The assessments included guidelines specifying detailed studies and specialized construction for all new developments within 150 m of a mapped fault.

The basin in which Mexico City is situated is filled with thick lacustrine sediments of volcanic origin, and groundwater withdrawal has resulted in serious consequences.

2.2.4 Significant Examples

Houston, Texas (Water Extraction: Flooding and Faulting)

Between 1906 and 1964, 5 ft of subsidence occurred, and reports place the subsidence at 9 ft at some locations (Civil Engineering, 1977). The cost of the subsidence, including flood damage, between 1954 and 1977 has been estimated to be $110 million and in 1977 was growing at the annual rate of $30 million (Spencer, 1977). The subsidence results in "growth" faults that cause distress in structures, large deflections of roadways, and rupture of utility lines; in flooding, resulting in homes being abandoned along Galveston Bay; and the lowering of bridges over the Houston Ship Canal.

The problem of growth faults in the Houston metropolitan area is severe. Activity has been recognized on more than 40 normal faults, which are prehistoric according to Van Siclen (1967). Major surface faults and the cumulative subsidence between 1906 and 1964 are given in Figure 2.1, and a profile of subsidence and groundwater decline for a distance of about 14 mi is given in Figure 2.2. The drop in water level of almost 300 ft causes an increase in overburden pressure of about 9 tsf, which is believed to be causing downward movement along the old faults as clay beds interbedded with sand aquifers consolidate (Castle and Youd, 1972).

Holdahl et al. (1991) report that the area has been divided into two zones, west and east of downtown Houston. In 1987, west zone subsidence was ranging up to 72 mm/year (2.8 in./year), but after 1978 the east zone has experienced 60 to 90% decreases in subsidence rates due to regulated groundwater withdrawal and the use of canals to carry water from Lake Houston. The east zone is the industrial zone. Subsidence patterns in Houston are being monitored with InSAR (Section 2.2.2).

Mexico City (Water Extraction: Subsidence and Foundation Problems)

Geologic Conditions

The basin of the valley of Mexico City, 2240 m above sea level, has been filled with 60 to 80 m of Pleistocene soils including interbedded sands, sands and gravels, and

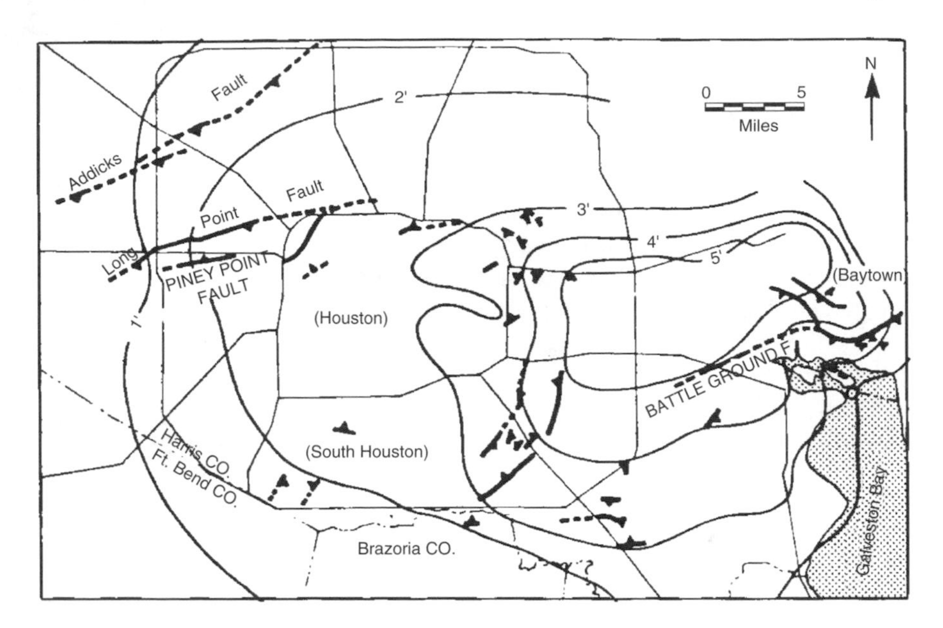

FIGURE 2.1
Surface faulting and cumulative subsidence (in ft) in the Houston area between 1906 and 1964. (From Castle, R. O. and Youd, T. L., *Bull. Assoc. Eng. Geol.*, 9, 1972. With permission.)

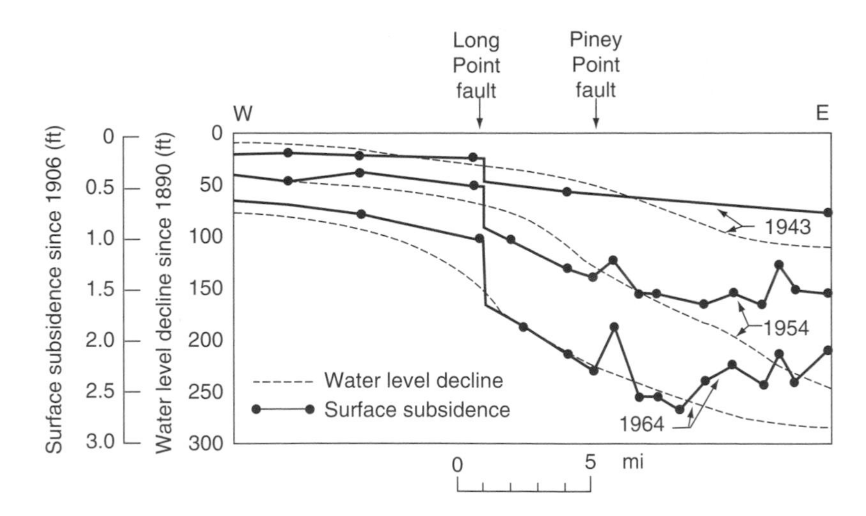

FIGURE 2.2
Profiles of subsidence and groundwater decline along a section trending due west from Houston. (From Castle, R. O. and Youd, T. L., *Bull. Assoc. Eng. Geol.*, 9, 1972. With permission.)

lacustrine volcanic clay, which overlie a thick deposit of compact sand and gravel. The soils to depths of about 33 m are very soft to medium-stiff clays. Void ratios as high as 13 and water contents as high as 400% indicate the very high compressibility of the soils. The clays are interbedded with thin sand layers, and thick sand strata are found at depths of 33, 45, and 73 m. Prior to the 1930s, groundwater beneath the city was recharged naturally.

Ground Subsidence

In the 100 years prior to 1955, a large number of water wells were installed between depths of 50 and 500 m in the sand and sand and gravel layers. Subsequent to the 1930s, groundwater withdrawal began to exceed natural recharge. The wells caused a large reduction in piezometric head, especially below 28 m, but the surface water table remained unaltered because of the impervious shallow clay formation. A downward hydraulic gradient was induced because of the difference in piezometric levels between the ground surface and the water-bearing layers at greater depths. The flow of the descending water across the highly compressible silty clay deposits increased the effective stresses, produced consolidation of the weak clays, and thus caused surface subsidence (Zeevaert, 1972).

In some places, as much as 7.6 m of subsidence had occurred in 90 years, with about 5 m occurring between 1940 and 1970. The maximum rate, with respect to a reference sand stratum at a depth of 48 m, was 35 cm/year, and was reached in 1949. About 80 to 85% of the subsidence is attributed to the soils above the 50 m depth. In 1955, the mayor of the city passed a decree prohibiting all pumping from beneath the city, and the rate of lowering of the piezometric levels and the corresponding rate of subsidence decreased considerably. As of 2001, however, subsidence was still continuing at rates of about 3 to 7 cm/year (Rudolph, 2001), and groundwater was being pumped. Even after pumping ceases, the compressible soils will continue consolidating under the increased effective stresses. Attempts are being made to bring water in from other watersheds. Subsidence patterns in Mexico City are being monitored with InSAR (Section 2.2.2).

Foundation Problems

Before the well shutdown program, many of the wells that were poorly sealed through the upper thin sand strata drew water from these strata, causing dish-shaped depressions to form around them. The result was the development of severe differential settlements, causing tilting of adjacent buildings and breakage of underground utilities. Flooding has become a problem as the city now lies 2 m below nearby Lake Texcoco (Rudolph, 2001).

A different problem occurs around the perimeter of the basin. As the water table drops, the weak bentonitic clays shrink by desiccation, resulting in surface cracks opening to as much as a meter in width and 15 m in depth. When the cracks open beneath structures, serious damage results.

The major problems, however, have been encountered within the city. Buildings supported on piles can be particularly troublesome. End-bearing piles are usually driven into the sand stratum at 33 m for support. As the ground surface tends to settle away from the building because of the subsidence, the load of the overburden soils is transferred to the piles through "negative skin friction" or "downdrag." If the piles do not have sufficient capacity to support both the building load and the downdrag load, settlements of the structure result. On the other hand, if the pile capacities are adequate, the structure will not settle but the subsiding adjacent ground will settle away from the building. When this is anticipated, utilities are installed with flexible connections and allowances are made in the first-floor design to permit the sidewalks and roadways to move downward with respect to the building.

Modern design attempts to provide foundations that enable a structure to settle at about the same rate as the ground subsides (Zeevaert, 1972). The "friction-pile compensated foundation" is designed such that downdrag and consolidation will cause the building to settle at the same rate as the ground subsidence. The Tower Latino Americano, 43 stories high, is supported on a combination of end-bearing piles and a compensated raft foundation (Zeevaert, 1957). The piles were driven into the sand stratum at 33 m, and a 13-m-deep excavation was made for the raft which had the effect of removing a substantial overburden load, subsequently replaced by the building load. The building was completed in 1951 and the settlements as of 1957 occurring from consolidation of the clay strata were as predicted, or about 10 cm/year.

Long Beach, California (Oil Extraction: Subsidence and Flooding)

Geologic Conditions

The area is underlain by 2000 ft of "unconsolidated" sediments of late Pliocene, Pleistocene, and Holocene age, beneath which are 4600 ft of oil-producing Pliocene and Miocene formations including sandstone, siltstone, and shale.

Ground subsidence began to attract attention in Long Beach during 1938 to 1939 when the extraction of oil began from the Wilmington field located primarily in the city (see site location map, Figure 2.3). A peak subsidence rate of over 20 in/year was reached in 1951 to 1952 (Section 2.2.1), By 1973, subsidence in the center of a large bowl-shaped area had reached 30 ft vertically, with horizontal movements as great as 13 ft. Flood protection for the city, which is now below sea level in many areas, is provided by extensive diking and concrete retaining walls. Deep-well recharging by salt-water injection has halted subsidence and some rebound of the land surface has occurred.

Baldwin Hills Reservoir Failure (Oil Extraction: Faulting)

Event

On December 14, 1963, the Baldwin Hills Reservoir, a pumped storage reservoir located in Los Angeles (Figure 2.3), failed and released a disastrous flood onto communities downstream (Jansen et al., 1967).

Background

The dam was located close to the Inglewood oil field and the Inglewood fault passed within 500 ft of the west rim of the reservoir. The Inglewood fault is part of the major Newport–Inglewood fault system (Figure 2.3). During construction excavation in 1948, two minor faults were found to pass through the reservoir area, but a board of consultants judged that further movement along the faults was unlikely. The dam was well instrumented with two strong-motion seismographs, tiltmeters, settlement measurement devices, and observation wells.

Surface Movements

Between 1925 and 1962, at a point about 0.6 mi west of the dam, about 10 ft of subsidence and about 6 in. of horizontal movement occurred. In 1957, cracks began to appear in the area around the dam. Six years later, failure occurred suddenly and the narrow breach in the dam was found to be directly over a small fault (Figure 2.4). It was judged that 6 in. of movement had occurred along the fault, which ruptured the lining of the reservoir and permitted the sudden release of water. It appears likely that subsidence from oil extraction caused the fault displacement, since there had been no significant seismic activity in the area for at least the prior month.

2.2.5 Subsidence Prevention and Control

Groundwater Extraction

General

Subsidence from groundwater extraction cannot be avoided if withdrawal exceeds recharge, resulting in significant lowering of the water table or a reduction in piezometric levels at depth, and if the subsurface strata are compressible.

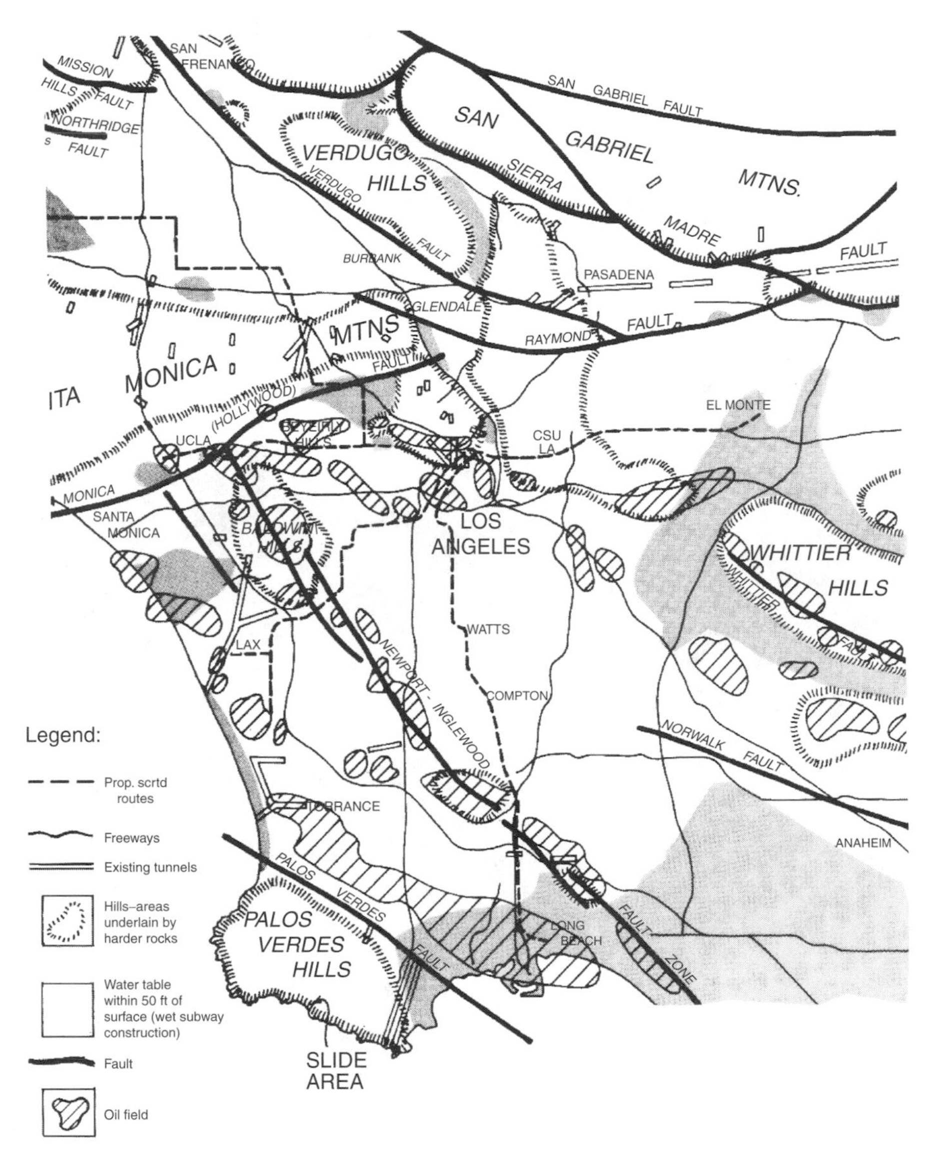

FIGURE 2.3
Water table, faults, and oil fields in the Los Angeles Basin. (From Proctor, R. J., *Geology, Seismicity and Environmental Impact,* Special Publication Association Engineering Geology, University Publishers, Los Angeles, 1973, pp. 187–193. With permission.)

Prediction

Prior to the development of a groundwater resource, studies should be made to determine the water-balance relationship and estimate magnitudes of subsidence. The water-balance relationship is the rate of natural recharge compared with the anticipated maximum rate of withdrawal. If recharge equals withdrawal, the water table will not drop and subsidence will not occur. If withdrawal significantly exceeds recharge, the water table will be lowered.

Estimates of the subsidence to be anticipated for various water-level drops are made to determine the maximum overdraft possible before surface settlement begins to be troublesome and cause flooding and faulting. By using concepts of soil mechanics it is possible to

FIGURE 2.4
Pumped storage reservoir failure 1968, Baldwin Hills, California. Failure occurred suddenly over a small fault.

compute estimates of the surface settlements for various well-field layouts, including differential deflections and both immediate and long-term time rates of settlements.

The increase in the flood hazard is a function of runoff and drainage into a basin or the proximity to large water bodies and their relative elevations, including tidal effects. The growth of faults is difficult to assess both in location and magnitude of displacement. In general, locations will be controlled by the locations of relict faults. New faults associated with subsidence are normally concentrated in a concentric pattern around the periphery of the subsiding area. The center of the area can be expected to occur over the location of wells where withdrawal is heaviest.

Prevention

Only control of overextraction prevents subsidence. Approximate predictions as to when the water table will drop to the danger level can be based on withdrawal, precipitation, and recharge data. By this time, the municipality must have provisions for an alternate water supply to avoid the consequences of overdraft.

Control

Where subsidence from withdrawal is already troublesome, the obvious solution is to stop withdrawal. In the case of Mexico City, however, underlying soft clays continued to consolidate for many years, even after withdrawal ceased, although at a much reduced rate. Artificial recharging will aid the water balance ratio.

Recharging by pumping into an aquifer requires temporary surface storage. The Santa Clara Valley Water District (San Jose, California) has been storing storm water for recharge pumping for many years (ENR, 1980). In west Texas and New Mexico, dams are to be built to impound flood waters that are to be used as pumped-groundwater recharge (ENR, 1980).

Where the locale lacks terrain suitable for water storage, deep-well recharging is not a viable scheme, and recharge is permitted to occur naturally. Venice considered a number of recharge schemes, in addition to a program to cap the city wells begun in 1965. Apparently, natural recharge is occurring and measurements indicate that the city appears to be rising at the rate of about 1 mm every 5 years (*Civil Engineering*, 1975).

Oil and Gas Extraction

Prediction of subsidence from oil and gas extraction is difficult with respect to both magnitude and time. Therefore, it is prudent to monitor surface movements and to have contingency plans for the time when subsidence approaches troublesome amounts.

Control by deep-well recharging appears to be the most practical solution for oil and gas fields. At Long Beach, water injection into the oil reservoirs was begun in 1956; subsidence has essentially halted and about 8 mi^2 of land area has rebounded, in some areas by as much as 1 ft (Allen, 1973; Testa, 1991).

Construction Dewatering

Prediction and Control

Before the installation of a construction dewatering system in an area where adjacent structures may be affected, a study should be made of the anticipated drop in water level as a function of distance, and settlements to be anticipated should be computed considering building foundations and soil conditions. Peat and other organic soils are particularly susceptible to compression. In many cases, condition surveys are made of structures and all signs of existing distress recorded as a precaution against future damage claims. Before dewatering, a monitoring system is installed to permit observations of water level and building movements during construction operations. The predicted settlements may indicate that preventive measures are required.

Prevention

Prevention of subsidence and the subsequent settlement of a structure are best achieved by placing an impervious barrier between the dewatering system and the structure, such as a slurry wall. Groundwater recharge to maintain water levels in the area of settlement-sensitive structures is considered to be less reliable.

Surcharging

Surcharging of weak compressible layers is a positive application of construction dewatering. If a clay stratum, for example, lies beneath a thickness of sands adequate to apply significant load when dewatered, substantial prestress can be achieved if the water table remains lowered for a long enough time. Placing a preload on the surface adds to the system's effectiveness.

2.3 Subsurface Mining

2.3.1 Subsidence Occurrence

General

Extraction of materials such as coal, salt, sulfur, and gypsum from "soft" rocks often results in ground subsidence during the mining operation or, at times, many years after operations have ceased. Subsidence can also occur during hard rock mining and tunneling operations.

In the United States, ground subsidence from mining operations has occurred in about 30 states, with the major areas located in Pennsylvania, Kansas, Missouri, Oklahoma, Montana, New Jersey, and Washington (Civil Engineering, 1978). Especially troublesome in terms of damage to surface structures are the Scranton-Wilkes-Barre and Pittsburgh areas of Pennsylvania. The approximate extent of coal fields in the eastern United States in the Pennsylvanian Formations is given in Figure 2.5.

Other troublesome areas: The midlands of England where coal has been, and is still being, mined. Paris, France, and surrounding towns have suffered surface collapse, which at times has swallowed houses, over the old underground limestone and gypsum quarries that were the source of building stone for the city in the 18th century (Arnould, 1970). In the Paris area, collapse has been intensified by groundwater pumping (see Section 2.4.2).

Metal Mining

During the 1700s and 1800s various ores, including iron (magnetite), copper, zinc, and lead, were extracted from relatively shallow underground mines in New Jersey and Pennsylvania. In recent years, the deterioration of these workings has resulted in the formation of surface sinkholes. Over 400 abandoned iron mines in Northern New Jersey have been identified (Cohen et al., 1996). A collapse resulting in a 30-m-diameter sinkhole jeopardizing a major road was reported. The Schuyler Copper Mine in North Arlington, New Jersey, is estimated to contain up to 55 shafts located in a 20-acre area within the town (Trevits and Cohen, 1996). In 1989, a collapse occurred and a large hole opened in the backyard of a home, and depressions formed at other shaft sites.

In Chester County, Pennsylvania, numerous shallow zinc and lead mines were developed in the 1800s. In general, the metals were deposited in thin, near-vertical formations. Mining proceeded along drifts that, in many cases, were extended upward toward the surface leaving shallow caps of overburden. The area periodically suffers the development of small sinkholes, particularly after heavy rains.

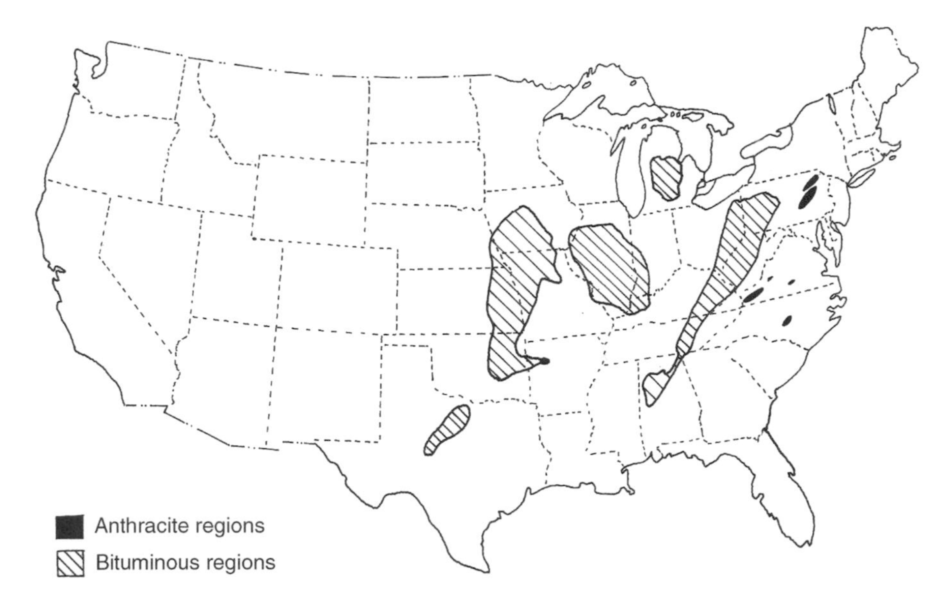

FIGURE 2.5
Approximate extent of coal fields of the Pennsylvanian formations in the eastern United States. (From Averitt, P., *USGS Bull, 1275,* U.S. Govt., Printing Office, Washington, DC, 1967.)

Salt Mines

In the United States, salt is mined from the subsurface in New York, Kansas, Michigan, Utah, Texas, Louisiana, and California.

Kansas

In Kansas, natural sinkholes, such as Lake Inman in McPherson County, are common on the solution front along the eastern edge of the Hutchinson Salt Member. Solution mining has resulted in subsidence in and around the city of Hutchinson. In 1974, a sinkhole 300 ft in diameter occurred at the Cargill Plant, southeast of the city.

In Russell County, I-70 crosses two active sinkholes that have been causing the highway to subside since its construction in the mid-1960s. In the area, sandstone and shale overlie a 270-ft-thick salt bed 1300 ft below the surface. It is believed that old, improperly plugged, abandoned oil wells allowed fresh water to pass through the salt bed resulting in dissolution of the salt and the creation of a large void. Subsidence has been of the order of 4 to 5 in./year, and in recent years a nearby bridge has subsided over another sinkhole. Over the years, the roadway grade has been raised and repaved. In 1978, a sinkhole 75 ft across and 100 ft deep, centered about an abandoned oil well, opened up about 20 mi from the I-70 sinks. Plugging old wells and performing cement injections have failed to stop the subsidence (Croxton, 2002).

New York

On March 12, 1994, a 650 × 650 ft panel 1180 ft under the Genesse Valley in Retsof, New York, failed catastrophically generating a magnitude 3.6 seismic event. A similar size panel failed a month later, and groundwater immediately entered the mine. Two sinkholes, 700 ft across and up to 75 ft deep, opened over the failed mine panels (William, 1995). Surface subsidence over a large area in the valley due to (1) dissolution of pillars and (2) dewatering of valley-fill sediments was occurring at rates measured in inches to feet per year in 1995. The salt bed, about 30 ft thick, was mined for many years by the room and pillar method where 40 to 50% of the salt was left in place. The collapse occurred in an area of the mine where smaller pillars, known as yield pillars, were used to maintain the mine roof (MSHA, 1995). Gowan and Trader (2000) suggest that an anomalous buildup of natural gas and brine pressure above the collapse area contributed to the failure.

Coal Mining

From the aspects of frequency of occurrence and the effects on surface structures, coal mining appears to be the most important subsurface mining operation. Mine collapse results in irregular vertical displacement, tilting, and horizontal strains at the surface, all resulting in the distortion of structures as illustrated in Figure 2.6. The incidence and severity of subsidence are a function of the coal bed depth, its thickness, percent of material extracted, tensile strength of the overburden, and the strength of pillars and other roof supports. InSAR is being used in Poland and other locations in Europe to monitor subsidence from coal mining.

General methods of extraction and typical subsidence characteristics include:

- The longwall panel method is used in Europe, and is finding increasing use in the United States. It involves complete removal of the coal. Where the mine is relatively shallow and the overlying materials weak, collapse of the mine, and surface subsidence progress with the mining operation. Subsidence is in the form of a large trough which mirrors, but is larger than, the plan dimensions of the mine.
- The room and pillar method has been used in the United States. Pillars are left in place to support the roof, but subsequent operations rob the pillars, weakening support. Collapse is often long term. Subsidence takes the forms of sinkholes (pits) or troughs (sags).

Most states have prepared detailed "Codes" relating to underground mining; for example, in Pennsylvania, Code 89.142a adopted in 1998 refers to "Subsidence Control: Performance Standards."

FIGURE 2.6
Building damaged by mine subsidence in Pittsburgh. (Photo courtesy of Richard E. Gray.)

2.3.2 Longwall Panel Extraction

Extraction Procedures

Most of the coal currently produced in Pennsylvania uses the longwall mining method (PaDEP, 2002). Mining begins with the excavation of hallways supported by pillars along the sides of the panel to be mined, generally 600 to 7000 ft or more in width (Figure 2.7). The hallways provide for access, ventilation, and the removal of the mined coal. Along the face of the panel a cutting head moves back and forth, excavating the coal, which falls onto a conveyor belt. A system of hydraulic roof supports prevents the mine from collapsing. As the excavation proceeds, the supports move with it and the mine roof behind the supports is permitted to collapse.

National Coal Board (NCB) Studies

Based on surveys of 157 collieries, the NCB of Great Britain (NCB, 1963, 1966) developed a number of empirical relationships for the prediction of the vertical component of surface displacements s and the horizontal component of surface strain e, associated with the trough-shaped excavation of the longwall panel method of coal extraction, illustrated in Figure 2.8.

Surveys were made over coal seams that were inclined up to 25° from the horizontal, were about 3 to 15 ft in thickness m, and ranged in depth h from 100 to 2600 ft. Face or panel width w varied from 100 to 1500 ft, and the panel width to depth ratio w/h varied from 0.05 to 4.0. Panel widths were averaged if the panel sides were nonparallel. The foregoing conditions assume that there is no zone of special support within the panel areas. The physical relationships are illustrated in Figure 2.8.

Subsidence Characteristics

Angle of the Draw

The area where the coal is mined is referred to as the "panel." Extraction results in surface subsidence termed "subsidence trough" or "basin." The width by which the subsidence trough exceeds the panel width is the "angle of the draw."

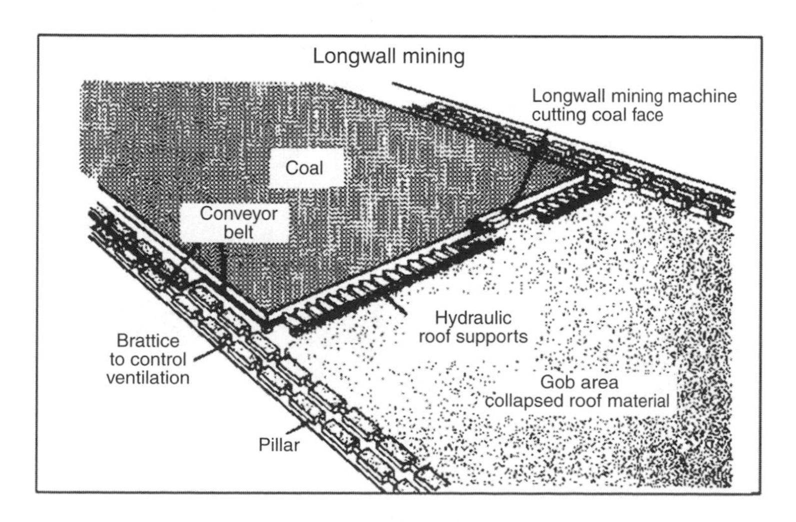

FIGURE 2.7
Longwall panel mining of coal beds. (From PaDEP, Commonwealth of Pennsylvania, Department of Environmental Protection, Bureau of Mining and Reclamation, Contract No. BMR-00-02, 2002. With permission.)

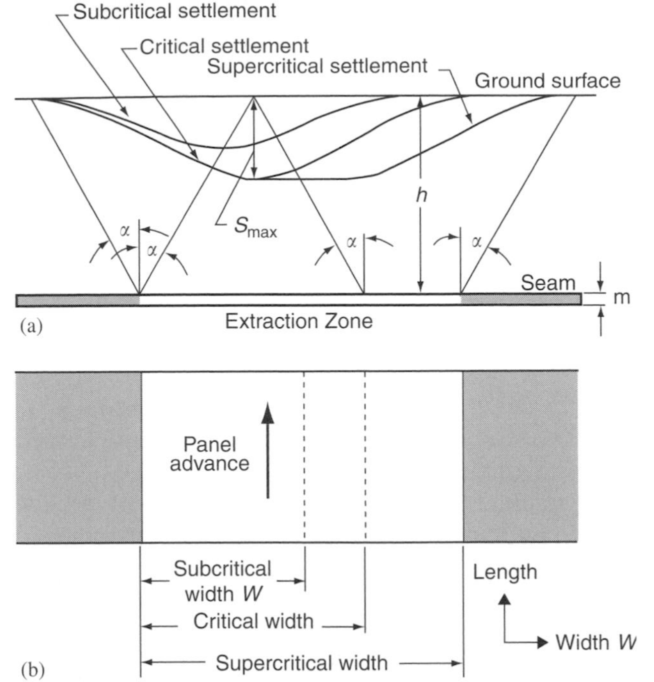

FIGURE 2.8
Mine subsidence vs. critical depth concept for longwall panel extraction developed by the National Coal Board of Great Britain: (a) section and (b) plan. (After Mabry, R. E., Proceedings of ASCE, 14th Symposium on Rock Mechanics, University Park, Pennsylvania, 1973.)

Most of the subsidence in the NCB studies was found to occur within a day or two of extraction in a dish-shaped pattern over an area bounded by lines projected upward from the limits of the collapsed area at the angle α as shown in Figure 2.8. The angle of the draw was found to be in the range of 25 to 35° for beds dipping up to 25°. The angle of the draw, however, could be controlled by defects in the overlying rock, such as a fault zone providing weak rock, and by surface topography. Residual subsidence may occur for several months to a year after mining (PaDEP, 2002).

Greatest maximum subsidence Smax possible was found to be approximately equal to 90% of the seam thickness, occurring at values of panel width w/depth h (w/h) greater than about 1.2. At values of $w/h < 0.2$, the maximum subsidence was less than 10% of the seam thickness.

Critical width is the panel width required to effect maximum subsidence (Figure 2.8); as the panel width is extended into the zone of supercritical width, additional vertical subsidence does not occur, but the width of the subsiding area increases accordingly. Detailed discussion and an extensive reference list are given in Voight and Pariseau (1970).

2.3.3 Room and Pillar Method (Also "Breast and Heading" Method)

Extraction

Early Operations

In the anthracite mines of the Scranton-Wilkes-Barre and Pittsburgh areas of Pennsylvania, "first mining" proceeded historically by driving openings, called breasts, in the up-dip direction within each vein, which were then connected at frequent intervals with heading openings. The usual width of the mine openings ranged from 15 to 25 ft and the distance between center lines of adjacent breasts varied generally from 50 to 80 ft with the shorter distances in the near-surface veins and the greater distances in the deeper veins. Pillars of coal were left in place to support the roof and the room-pillar configurations were extremely variable. One example is given in Figure 2.9.

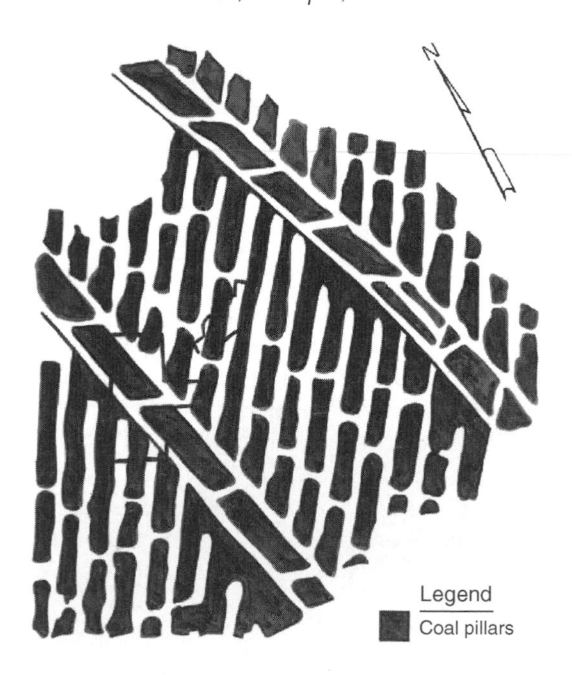

FIGURE 2.9
Plan of coal mine room and pillar layout, Westmoreland County, Pennsylvania. (From Gray, R. E. and Meyers, J. F., *Proc. ASCE J. Soil Mech. Found. Eng. Div.*, 96, 1970. With permission.)

"Robbing" occurred subsequent to first mining and consisted of removing the top and bottom benches of coal in thick veins and trimming coal from pillar sides.

Collapse occurs with time after mine closure. The coal is often associated with beds of clay shale, which soften and lose strength under sealed, humid mine conditions. Eventually failure of the pillars occurs, the roof collapses, and the load transfer to adjacent pillars causes them to collapse. If the mined area is large enough and the roof thin enough, subsidence of the surface results.

Old mines may contain pillars of adequate size and condition to provide roof support or robbed pillars, weakened and in danger of collapse. On the other hand, collapse may have already occurred.

Modern Operations

Two major coal seams of the Pennsylvanian underlie the Pittsburgh area, each having an average thickness of 6 ft: the Pittsburgh coal and the Upper Freeport coal. The Pittsburgh seam is shallow, generally within 200 ft of the surface, and has essentially been worked out. In 1970, the Upper Freeport seam was being worked to the north and east of the city where it lay at depths of 300 to 600 ft (Gray and Meyers, 1970).

In the new mines complete extraction is normally achieved. A system of entries and cross-entries is driven initially to the farthest reaches of the mine before extensive mining. Rooms are driven off the entries to the end of the mine, and when this is reached, a second, or retreat phase, is undertaken. Starting at the end of the mine, pillars are removed and the roof is permitted to fall.

In active mines where surface development is desired either no extraction or partial extraction proceeds within a zone beneath a structure determined by the "angle of the draw" (Gray and Meyers, 1970). The area of either no extraction or partial extraction is determined by taking an area 5 m or more in width around the proposed structure and projecting it downward at an angle of 15 to 25° from the vertical to the level of the mine (angle of the draw). With partial extraction, where 50% of coal pillars remain in place there is a small risk of surface subsidence, whereas with no extraction there should be no risk.

Mine Collapse Mechanisms

General

Three possible mechanisms which cause mine collapse are roof failure, pillar failure, or pillar foundation failure.

Roof Failure

Roof stability depends upon the development of an arch in the roof stratum, which in turn depends on the competency of the rock in relation to span width. In weak, fractured sedimentary rocks, this is often a very difficult problem to assess, since a detailed knowledge of the engineering properties and structural defects of the rock is required, and complete information on these conditions is difficult and costly to obtain. If the roof does have defects affecting its capability, it is likely that it will fail during mining operations and not at some later date, as is the case with pillars.

Roof support does become important when the pillars weaken or collapse, causing the span length to increase, which in turn increases the loads on the pillars. Either roof or pillars may then collapse.

Pillar Failure

The capability of a pillar to support the roof is a function of the compressive strength of the coal, the cross-sectional area of the pillar, the roof load, and the strength of the floor and roof. The cross-sectional area of the pillar may be reduced in time by weathering and spalling of its walls, as shown in Figure 2.10, to a point where it cannot support the roof and failure occurs.

FIGURE 2.10
Spalling of a coal pillar in a mine room (Pittsburgh). (Photo courtesy of Richard E. Gray.)

Pillar Punching

A common cause of mine collapse appears to be the punching of the pillar into either the roof or the floor stratum. Associated with coal beds, clay shale strata are often left exposed in the mine roof or floor. Under conditions of high humidity or a flooded floor in a closed mine, the clay shales soften and lose their supporting capacity. The pillar fails by punching into the weakened shales, and the roof load is transferred to adjacent pillars, which in turn fail, resulting in a lateral progression of failures. If the progression involves a sufficiently large area, surface subsidence can result, depending upon the type and thickness of the overlying materials.

Earthquake Forces

In January 1966, during the construction of a large single-story building in Belleville, Illinois, settlements began to occur under a section of the building, causing cracking. It was determined that the settlements may have started in late October or early November 1965. An earthquake was reported in Belleville on October 20, 1965. The site was located over an old mine in a coal seam 6 to 8 ft thick at a depth of 130 ft, which was closed initially in about 1935, then reworked from 1940 to 1943. Mansur and Skouby (1970) considered that building settlements were the result of pillar collapse and mine closure initiated by the earthquake. Some investigators, however, consider that the collapsing mine was the shock recorded in Belleville.

Subsidence over Abandoned Mines

Two types of subsidence occurring above abandoned mines have been classified; sinkholes (pits) and troughs (sags). The subsidence form usually relates to mine depth and geologic conditions.

Sinkholes (Pits)

A sinkhole is a depression on the ground surface resulting from the collapse of overburden into a limited mine opening, such as a room or entry. They usually develop where the cover over the mine is less than 50 to 100 ft. Competent strata above the mine will limit sinkhole development, but sinkholes developed over a mine in Illinois where the overburden was 164 ft deep.

Troughs

Where a pillar or pillars fail by crushing or punching into the mine roof, troughs develop. Subsidence troughs usually resemble those that form above active mines but often do not conform to the mine boundaries. Trough diameters above abandoned mines in the Northern Appalachian Coal Field commonly measure 1.5 to 2.5 times the overburden thickness.

2.3.4 Strength Properties of Coal

General

Pillar capacity analysis requires data on the strength properties of coal. A wide range of values has been obtained by investigators either by testing specimens in the laboratory, or by back-analysis in which the strength required to support an existing roof is calculated for conditions where failure has not occurred. For the determination of the stability of a working mine, the strength of fresh rock specimens governs, whereas for problems involving stability after a lapse of many years, the strength of weathered specimens of the pillar and its roof and floor support pertain.

Typical coals contain a system of orthogonal discontinuities consisting of horizontal bedding planes and two sets of vertical cracks called "cleats," which are roughly perpendicular to each other as shown in Figure 2.11. This pattern makes the recovery of undisturbed specimens difficult. Some unconfined compression strength values for coals from various locations are given in Table 2.2.

Triaxial Compression Tests

A series of triaxial compression tests was performed in the laboratory on specimens from Eire, Colorado; Sesser, Illinois; and Bruceton, Pennsylvania, by Ko and Gerstle (1973). Confining pressures of 50, 100, 250, and 600 psi were used, with the maximum pressures considered as the overburden pressure on a coal seam at a depth of 600 ft. While confining pressure was held constant, each specimen was loaded axially to failure while strains and stresses were recorded. The test specimens were oriented on three perpendicular axes (α, β, and γ), and the failure load was plotted as a function of the confining pressure P_c to obtain the family of curves presented in Figure 2.12.

It was believed by the investigators that the *proportional limit* represents the load level at which microcracking commences in the coal. The safe limit is an arbitrary limit on the applied load set by the investigators to avoid internal microcracking.

2.3.5 Investigation of Existing Mines

Data Collection

Collection of existing data is a very important phase of investigation for projects that are to be constructed over mines. The data should include information on local geology, local

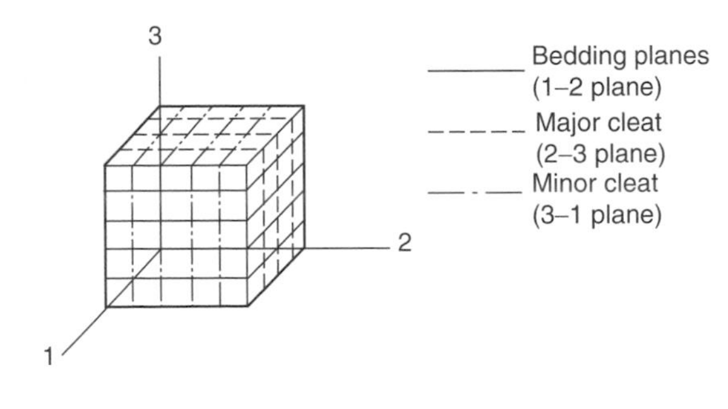

FIGURE 2.11
Bedding and cleat orientation of coal and the principal material axis. (From Ko, A. Y. and Gerstle, K. H., *Proceedings of ASCE, 14th Symposium on Rock Mechanics*, University Park, Pennsylvania, 1973, pp. 157–188. With permission.)

TABLE 2.2
Unconfined Compression Strength Values for Coal

Specimen Source	Description	Strength kg/cm^2	Comments	Reference
South Africa	2 ft cube	19	Failure sudden	Voight and Pariseau (1970)
Not given	Not given	50–500	None	Fanner (1968)
Pittsburgh coal	Sound pillar	~57	Pillar with firm bearing. $H=3$ m, sides, 4.8×4.8 m	Greenwald et al. (1941)
Anthracite from Pennsylvania	1 in. cubes	200,[a] 404[a]	First crack appears Crushing strength	Mabry (1973) from Griffith and Conner (1912)

[a] Values selected by Mabry (1973), from those obtained during a comprehensive study of the strength of 116 cubic specimens reported by Griffith and Conner (1912).

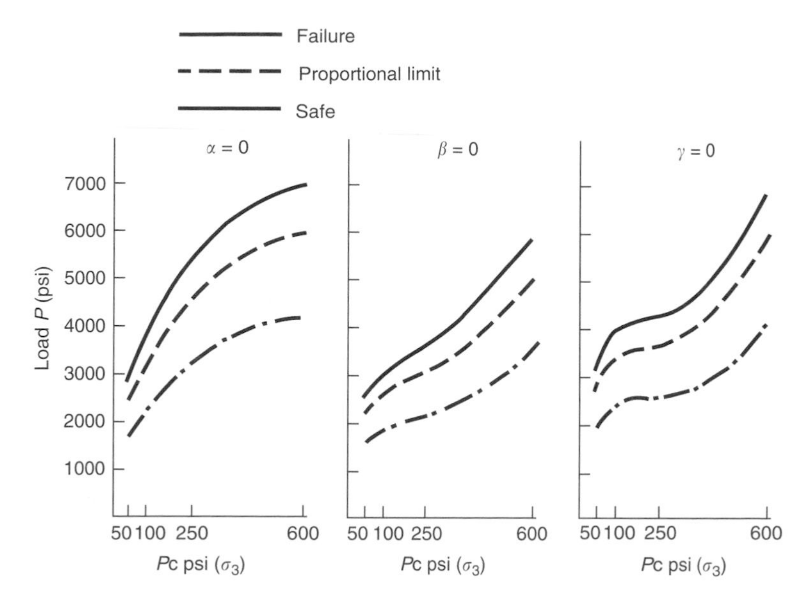

FIGURE 2.12
Triaxial compression tests results from coal specimens. (From Ko, A. Y. and Gerstle, K. H., *Proceedings of ASCE, 14th Symposium on Rock Mechanics,* University Park, Pennsylvania, 1973, pp. 157–188. With permission.)

subsidence history, and mining operations beneath the site. Many states and municipalities have extensive catalogs of abandoned mine maps. For example, the Ohio Division of Geological Survey has maps for 4138 mines, most of which are coal mines.

If possible, data on mining operations beneath the site should be obtained from the mining company that performed the extraction. The data should include information on the mine limits, percent extraction, depth or depths of seams, pillar dimensions, and the closure date. Other important data that may be available include pillar conditions, roof and floor conditions, flooding incidence, the amount of collapse that has already occurred, and accessibility to the mine.

Explorations

Exploration scopes will vary depending upon the comprehensiveness of the existing data and the accessibility of the mine for examination. Actual inspection of mine conditions is extremely important, but often not possible in old mines.

Preliminary explorations where mine locations and collapse conditions are unknown or uncertain may include the use of:

- Gravimeters to detect anomalies indicating openings.
- Rotary probes, if closely spaced, to detect cavities and indicate collapse conditions.
- Borehole cameras to photograph conditions, and borehole TV cameras, some of which are equipped with a zoom lens with an attached high-intensity light, to inspect mines remotely.
- Acoustical emissions devices, where mines are in an active collapse state, to locate the collapse area and monitor its growth.
- Electrical resistivity using the pole–dipole method was used in a study in Scranton, Pennsylvania, where the abandoned coal seam was 45 to 75 ft below the surface. Three signatures were interpreted: intact rock, caved rock, and voids. Test borings confirmed the interpretations.

Detailed study of conditions requires core borings to obtain cores of the roof, floor, and pillars, permitting an evaluation of their condition by examination and laboratory testing.

Finite-Element Method Application

General

The prediction of surface distortions beneath a proposed building site caused by the possible collapse of old mines by use of the finite element method is described by Mabry (1973). The site is in the northern anthracite region of Pennsylvania near Wilkes-Barre, and is underlain by four coal beds at depths ranging from 260 to 600 ft, with various percentages of extraction *R*, as illustrated in Figure 2.13.

Finite Element Model

Pillar analysis revealed low safety factors against crushing and the distinct possibility of subsidence. To evaluate the potential subsidence magnitude, a finite-element model was prepared incorporating the geometry of the rock strata and mines to a depth of 700 ft, and engineering properties including density, strength, and deformation moduli of the intervening rock and coal strata. The finite element mesh is given in Figure 2.14.

Analysis and Conclusions

Gravity stresses were imposed and the ground surface subsidence due to the initial mining in the veins was determined. Subsequent analysis was made of the future surface distortions, such as those would be generated by pillar weathering and eventual crushing. Pillar weathering was simulated by reducing the joint stiffness, and pillar collapse or yield was simulated by setting the joint stiffness in the appropriate intervals of the coal seams to zero. After pillar weathering in the coal seams was evaluated by changing joint stiffness values, the intervals of the veins were "collapsed" in ascending order of computed safety factors for several extraction ratios. The results are summarized in Table 2.3.

After an evaluation of all of the available information, it was the judgment of the investigator that the more realistic case for plant design was Case 2 (Table 2.3), and that the probability of Cases 3 and 4 developing during the life of the structure was very low.

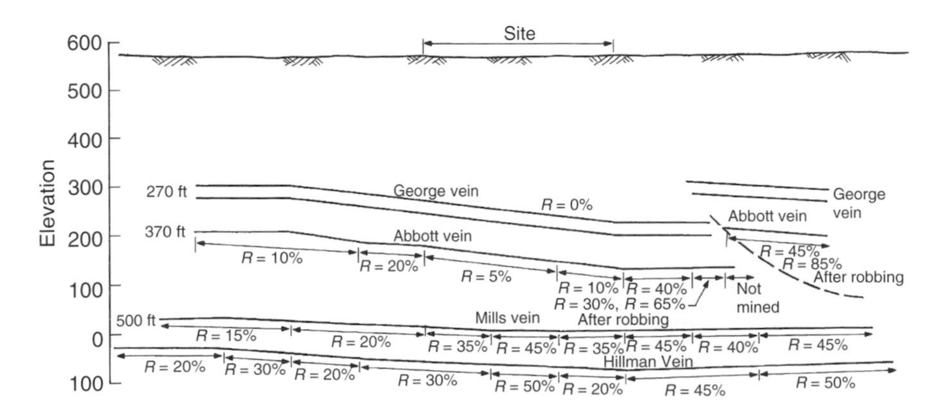

FIGURE 2.13
Section illustrating coal mines and percent extraction *R* beneath a proposed construction site near Wilkes-Barre, Pennsylvania. (From Mabry, R. E., *Proceedings of ASCE, 14th Symposium on Rock Mechanics*, University Park, Pennsylvania, 1973. With permission.)

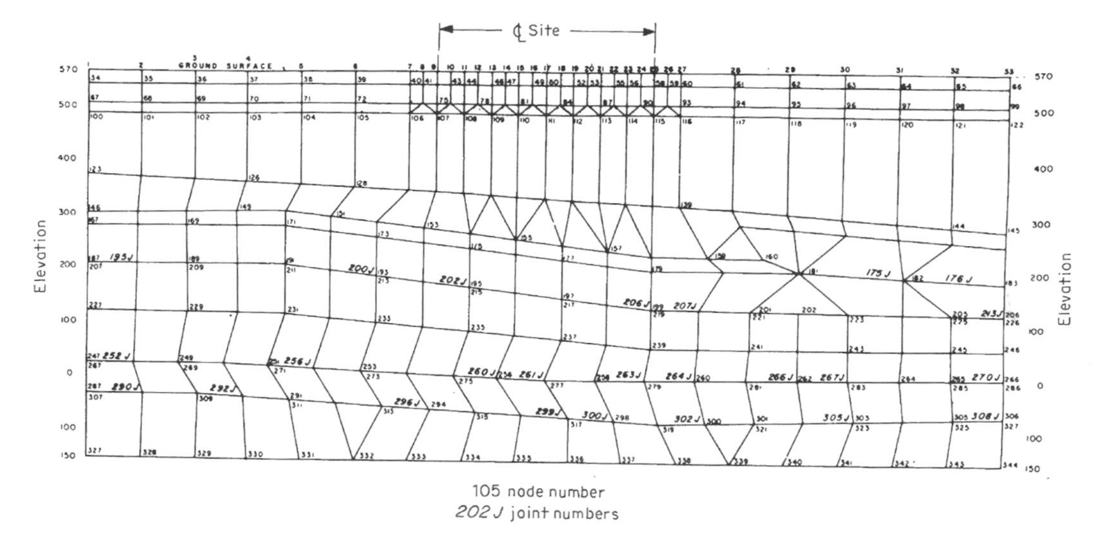

FIGURE 2.14
Finite-element mesh for coal mine conditions beneath site near Wilkes-Barre, Pennsylvania. (Mabry, R. E., *Proceedings of ASCE, 14th Symposium on Rock Mechanics,* University Park, Pennsylvania, 1973. With permission.)

TABLE 2.3
Summary of Finite-Element Analysis of Coal Mine Study[a]

Condition	Safety Factor against Pillar Failure	Cumulative Settlement (cm)	Maximum Distortion within Plant Site	
			Angular Rotation (rad)	Horizontal Strain (+ = tension)
Weathering in all veins		1.0	20.0×10^{-6}	$\pm12.0=10^{-6}$
Collapse in Hillman, R=50%	0.80	4.0	3.6×10^{-4}	$+1.7\times10^{-4}$
Collapse in Mills and Hillman, R=40%	0.89 to 0.99	29.0	3.0×10^{-3}	$+10.1\times10^{-4}$
Collapse in Mills, R=30%	1.1	63.2	4.8×10^{-3}	$+19.2\pm10^{-4}$

[a] From Mabry, R. E., *Proceedings of ASCE, 14th Symposium on Rock Mechanics*, University Park, Pennsylvania, June 1972. With permission.

2.3.6 Subsidence Prevention and Control and Foundation Support

New Mines

In general, new mines should be excavated on the basis of either total extraction, permitting collapse to occur during mining operations (if not detrimental to existing overlying structures), or partial extraction, leaving sufficient pillar sections to prevent collapse and resulting subsidence at some future date. Legget (1972) cites the case where the harbor area of the city of Duisburg, West Germany, was purposely lowered 1.75 m by careful, progressive longwall mining of coal seams beneath the city, without damage to overlying structures.

Old Mines

Solutions are based on predicted distortions and their probability of occurrence.

Case 1

No or small surface distortions are anticipated when conditions include adequate pillar support, or complete collapse has occurred, or the mined coal seam is at substantial depth overlain by competent rock. Foundations may include mats, doubly reinforced continuous footings, or articulated or flexible design to allow compensation for some differential movements of structures.

Case 2

Large distortions are anticipated or small distortions cannot be tolerated, when pillar support is questionable, collapse has not occurred, and the mine is at relatively shallow depths. Solutions may include:

- Relocate project to a trouble-free area.
- Provide mine roof support with construction of piers in the mine or installation of grout columns, or completely grout all mine openings from the surface within the confines of a grout curtain installed around the site periphery.
- Install drilled piers from the surface to beneath the mine floor.

2.4 Solution of Rock

2.4.1 General

Significance

Ground subsidence and *collapse* in soluble rock masses can result from nature's activities, at times aided by humans, or from human-induced fluid or solid extraction. Calcareous rocks, such as limestone, dolomite, gypsum, halite, and anhydrite, are subject to solution by water, which causes the formation of cavities of many shapes and sizes. Under certain conditions, the ground surface over these cavities subsides or even collapses, in the latter case forming sinkholes.

The Hazard

Geographic distribution is widespread, and there are many examples in the literature of damage to structures and even deaths caused by ground collapse over soluble rocks. Examples are the destruction of homes in central Florida (Sowers, 1975) (Figure 2.15); the sudden settlement of a seven-story garage in Knoxville, Tennessee (ENR, 1978); and a foundation and structural failure in an Akron, Ohio, department store that resulted in 1 dead and 10 injured (ENR, 1969). Collapses resulting in substantial damage and in some cases deaths have also been reported for locations near Johannesburg and Paris (see Section 2.4.3). Subsidence and sinkholes associated with the removal of halite have been reported for areas around Detroit, Michigan; Windsor, Ontario; and Hutchinson, Kansas.

Collapse incidence is much less than that for slope failures, but nevertheless the recognition of its potential is important, especially since the potential may be increasing in a given area. Collapse does occur as a natural phenomenon, but the incidence increases substantially in any given area with an increase in groundwater withdrawal.

2.4.2 Solution Phenomenon and Development

Characteristics of Limestone Formations

General

Limestone, the most common rock experiencing cavity development, is widely distributed throughout the world, and is exposed in large areas of the United States, as shown in Figure 2.16. The occurrence, structure, and geomorphology of carbonate rocks are briefly summarized in this section.

FIGURE 2.15
Collapse of two houses into a funnel-shaped sink in Bartow, Florida in 1967. Cause was ravelling of medium to fine sand into chimney-like cavities in limestone at depths of 50 to 80 ft below the surface. (Photo courtesy of George F. Sowers.)

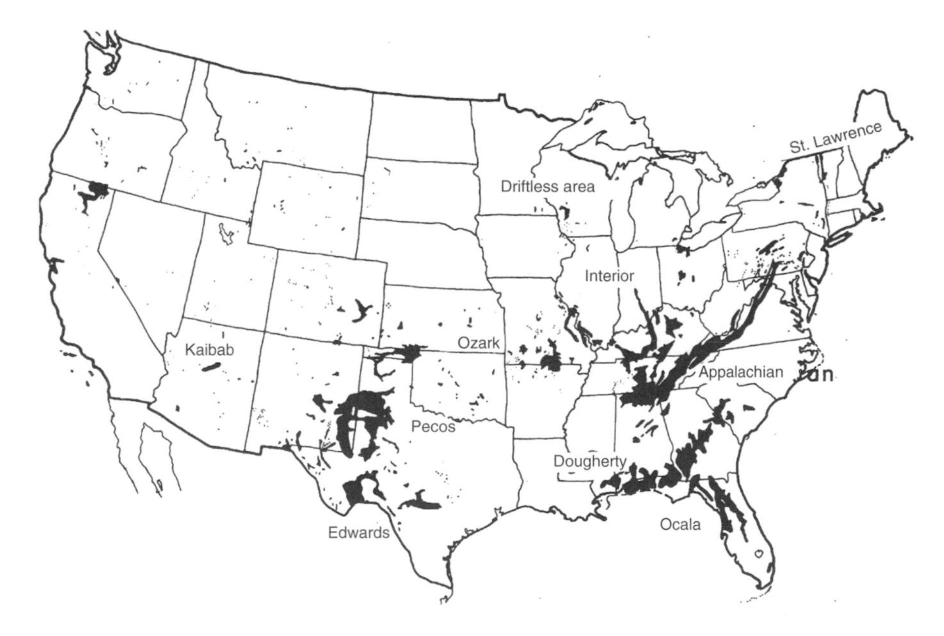

FIGURE 2.16
Distribution of karst regions in the United States. (Compiled by William E. Davies; from White, W. B., *Encyclopedia of Geomorphology*, Dowden Hutchinson & Ross Publ., Stroudsburg, Pennsylvania, 1968, pp. 1036–1039. With permission.)

Rock Purity and Cavity Growth

Purer limestone, normally found as thick beds of dense, well-indurated rock, is the most susceptible to cavity growth. At least 60% of the rock must be made up of carbonate materials for karst development, and a purity of 90% or more is required for full development (Corbel, 1959).

Impure limestone is characteristically thinly bedded and interbedded with shale and is resistant to solution.

Jointing: Groundwater moves in the rock along the joints, which are usually the result of strain energy release (residual from early compression) that occurs during uplift and rebound subsequent to unloading by erosion. This dominant origin causes most joints to be normal to the bedding planes. Major joints, cutting several beds, usually occur in parallel sets and frequently two sets intersect, commonly at about 60°, forming a conjugate joint system.

Cavity Growth, Subsidence, and Collapse

Solution

Groundwater moving through the joint system at depth and rainfall entering the joint system from above result in the solution of the rock. As rainwater passes through the surface organic layer, it becomes a weak acid that readily attacks the limestone. Solution activity is much greater, therefore, in humid climates with heavy vegetation than in dry climates with thin vegetation.

Geologic Conditions and Cavity Growth Form

Horizontal beds develop cavities vertically and horizontally along the joints, which grow into caverns as the solution progresses. Cavern growth is usually upward; surface subsidence occurs when the roof begins to deflect, or when broken rock in the cavern provides partial support, preventing a total collapse. When a cavern roof lacks adequate arch to support overburden pressures, collapse occurs and a sinkhole is formed.

Horizontal beds overlain by thick granular overburden are also subject to sudden raveling into cavities developing along joints. An example of a large sink developing under these conditions is given in Figure 2.15. Sowers (1975) states, "Raveling failures are the most widespread and probably the most dangerous of all the subsidence phenomena that are associated with limestone." The author considers this statement to apply to conditions like those in central Florida, that is, relatively thick deposits of granular alluvium overlying limestone undergoing cavity development from its surface.

Dipping beds develop cavities along joint dips, as shown in Figure 2.17, creating a very irregular rock surface, characterized by pinnacles. As the cavity grows, the overburden moves into the void, forming a soil arch. With further growth the arch collapses and a sinkhole results. In granular soils, the soil may suddenly enter a cavity by raveling, wherein the arch migrates rapidly to the surface, finally collapsing. A pinnacled rock surface is typical of eastern Pennsylvania, as illustrated in Figure 2.18. In addition to sinkholes resulting from soil raveling into cavities, the very irregular rock surface poses difficult foundation conditions.

Natural Rate of Cavity Growth

It has been estimated that the rainfall in the sinkhole region of Kentucky will dissolve a layer of limestone 1 cm thick in 66 years (Flint et al., 1969). Terzaghi (1913), reporting on a geologic study that he made in the Gacka region of Yugoslavia, observed that solution proceeded much more rapidly in heavily forested areas than in areas covered lightly by grass or barren of vegetation. In an analysis he assumed that 60% of the annual rainfall, or 700 mm/year, entered the topsoil and that the entire amount of carbon dioxide developed

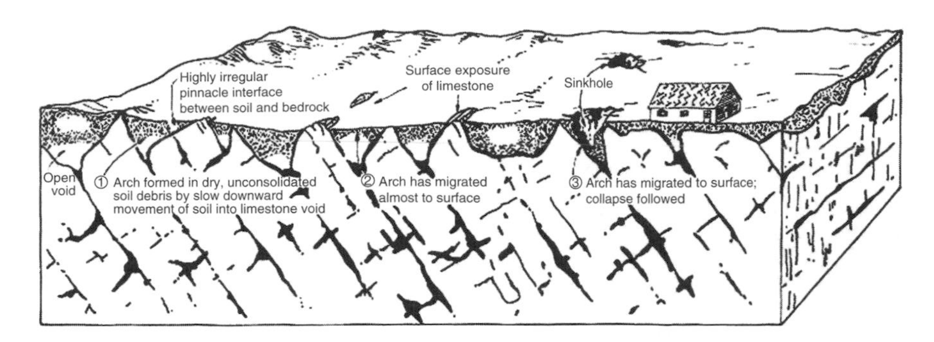

FIGURE 2.17
Hypothetical section through a carbonate valley showing stages of sinkhole development. (From Pennsylvania Geological Survey, *Engineering Characteristics of the Rocks of Pennsylvania*, Pennsylvania Geological Survey, 1972. With permission.)

FIGURE 2.18
Very irregular rock surface (pinnacles) and cavities in quarry wall. House in upper left gives scale (Ledger dolomite, Upper Merion, Pennsylvania).

in the topsoil was used up in the process of solution, which removed the limestone at the rate of 0.5 mm/year, or 1 cm in 20 years.

Collapse Causes Summarized

Collapse of limestone cavities can result from:

- Increase in arch span from cavity growth until the strength is insufficient to support the overburden weight.
- Increase in overburden weight over the arch by increased saturation from rainfall or other sources, or from groundwater lowering, which removes the buoyant force of water.
- Entry of granular soils by raveling into a cavity near the rock surface.
- Applications of load to the surface from structures, fills, etc.

Geomorphic Features of Karst

Karst refers in general to the characteristic, readily recognizable terrain features that develop in the purer limestone. The important characteristics of karst topography are its predominantly vertical and underground drainage, lack of surface drainage systems, and the development of circular depressions and sinks. At times, streams flow a short distance and suddenly disappear into the ground.

Youthful karst is characterized by numerous sinkholes and depressions as well as deranged and intermittent drainage.

Mature karst is characteristic of humid tropical climates. The landform consists of numerous rounded, steep-sided hills ("haystacks" or "pepinos").

Buried karst is illustrated by the ERTS image of Florida, showing numerous lakes that have filled subsidence depressions. In the Orlando region, the limestone is often buried under 60 to 100 ft of alluvium.

Groundwater Pumping Effects

Significance

Groundwater withdrawal greatly accelerates cavity growth in soluble rocks, and lowering of the water table increases overburden pressures. The latter activity, which substantially increases the load on a naturally formed arch, is probably the major cause of ground subsidence and collapse in limestone regions (Prokopovich, 1976). Even if groundwater withdrawal is controlled with the objective of maintaining a water balance and preserving the natural water table, the water table drops during severe and extensive droughts and collapse activity increases significantly, as occurred in central Florida during the spring of 1981.

Examples

Pierson, Florida: The sink illustrated in Figure 2.19, 20 m in diameter and 13 m deep, formed suddenly in December 1973 after 3 days of continuous pumping from nearby irrigation wells. The limestone is about 30 m in depth.

Johannesburg, South Africa: A large pumping program was begun in 1960 to dewater an area, underlain by up to 1000 m of Transvaal dolomite and dolomitic limestone, for gold mining operations near Johannesburg. In December 1962, a large sinkhole developed suddenly under the crushing plant adjacent to one of the mining shafts, swallowed the entire plant, and took 29 lives. In 1964, the lives of five persons were lost when their home suddenly fell into a rapidly developing sinkhole. Between 1962 and 1966, eight sinkholes larger than 50 m in diameter and 30 m in depth had formed in the mine area (Jennings, 1966).

Paris, France: Groundwater withdrawal has increased the solution rate and cavern growth in old gypsum quarries beneath the city and some suburban towns (Arnould, 1970). Ground collapse has occurred, causing homes to be lost and industrial buildings to be damaged. In one case, it was estimated that pumping water from gypsum at the rate of about 85 ft^3/min removed 136 lb of solids per hour.

Hershey, Pennsylvania: Increased dewatering for a quarry operation caused groundwater levels to drop over an area of 5000 acres and soon resulted in the appearance of over 100 sinkholes (Foose, 1953). The original groundwater levels were essentially restored after the quarry company sealed their quarry area by grouting.

Round Rock, Texas: The Edwards limestone outcrops in the area and is known to be cavernous in some locations. During a study for new development, interpretation of stereo-pairs of air photos disclosed several sink holes and depressions in one area of the

FIGURE 2.19
Sink 65 ft in diameter and 42 ft deep formed suddenly in December 1973 in Pierson, Florida, after 3 days of continuous pumping from nearby irrigation wells. The limestone is about 100 ft in depth. (Photo: Courtesy of George F. Sowers.)

site as illustrated in Figure 2.20 (Hunt, 1973). During reconnaissance of the site in 1972, the discovery of a sink 20 ft wide and 6 ft deep (Figure 2.21) that did not appear on the aerial photos, dated 1969, indicated recent collapse activity. Since the Edwards overlies the major aquifer in the area and groundwater withdrawals have increased along with area development, it may be that the incidence of collapse activity is increasing.

2.4.3 Investigation

Preliminary Phases

Data Collection

The existing geologic data are gathered to provide information on regional rock types and their solubility, bedding orientation and jointing, overburden types and thickness, and local aquifers and groundwater withdrawal.

Landform Analysis

Topographic maps and remote-sensing imagery are interpreted by landform analysis techniques. Some indicators of cavernous rock are:

- *Surface drainage*: Lack of second- and third-order streams; intermittent streams and deranged drainage; and streams ending suddenly.
- *Landform*: Sinks and depressions or numerous dome-shaped, steep-sided hills with sinks in between.
- *Photo tone*: Soils in slight depressions formed over cavity development will have slightly higher moisture contents than those in adjacent areas and will show as slightly darker tones on black-and-white aerial photos such as Figure 2.20. Infrared is also useful in detecting karstic features because of differences in soil moisture.

FIGURE 2.20
Stereo-pair of aerial photos of area near Round Rock, Texas, showing sinks and depressions from partial collapse of shallow limestone. (From Hunt, R. E., *Bull. Assoc. Eng. Geol.*, 10, 1973. With permission.)

FIGURE 2.21
Recent sink, not shown in aerial photos dated 1969 (Figure 2.19). Photo taken in 1972. (From Hunt, R. E., *Bull. Assoc. Eng. Geol.*, 10, 1973. With permission.)

Preliminary Evaluation

From a preliminary evaluation of the data, judgments are formulated regarding the potential for or the existence of cavity development from natural or human causes, and the possible locations, type, and size of ground subsidence and collapse. From this data base explorations are programmed.

Explorations

General

The location of all important cavities and the determination of their size and extent is a very difficult and usually impossible task. Explorations should never proceed without completion of data collection and landform analysis.

Geophysics

Explorations with geophysical methods can provide useful information, but the degrees of reliability vary. Seismic refraction surveys may result in little more than "averaging" the depth to the limestone surface if it is highly irregular. Seismic-direct cross-hole surveys may indicate the presence of cavities if they are large. Electrical resistivity, at times, has indicated shallow cavity development.

Gravimeter surveys were found to be more useful than other geophysical methods (seismic cross-hole surveys or electrical resistivity) in a study for a nuclear power plant in northwestern Ohio (Millet and Moorhouse, 1973).

Ground-probing radar (GPR) may disclose cavities, but interpretation of the images is difficult.

Test Borings and Pits

Test and core borings are programmed to explore anomalies detected by geophysical explorations and terrain analysis as well as to accomplish their normal purposes. Voids are disclosed by the sudden drop of the drilling tools and loss of drilling fluid. The material in a sinkhole is usually very loose compared with the surrounding materials, and may overlie highly fractured rock where a roof has collapsed. One should be suspicious when low SPT N values, such as 2 or 3, are encountered immediately above the rock surface, following significantly higher N values recorded in the overlying strata.

Test pits are useful to allow examination of the bedrock surface. Although the normal backhoe reach is limited to 10 to 15 ft, on important projects, such as for dams or construction with heavy foundation loads, deeper excavations, perhaps requiring dewatering, may be warranted.

Rotary Probes and Percussion Drilling

If cavity presence has been confirmed, it is usually prudent to make either core borings, rotary probes, or pneumatic percussion drill holes at the location of each footing before final design, or before construction. The objective is to confirm that an adequate thickness of competent rock is present beneath each foundation.

Percussion drilling with air track rigs is a low-cost and efficient method of exploring for cavities and the rock surface, especially in dipping limestone formations. Core borings usually drill 20 to 40 ft/day, whereas air track probes readily drill 300 to 500 ft/day. During investigation several core borings should be drilled immediately adjacent to air track holes for rock-quality correlations.

Proof testing with air drills is an alternate to rotary probes at each footing location, particularly for drilled piers where installations are relatively deep. Proof testing the bottom of each pier founded on rock with air drills is much less costly than rotary probes or core borings.

Evaluation and Analysis

Basic Elements

The following elements should be considered:

- Rock bedding, i.e., horizontal vs. dipping.
- Overburden thickness and properties and thickness variations, which are likely to be substantial.
- Bedrock surface characteristics, i.e., weathered, sound, relatively sound and smooth with cavities following joint patterns from the rock surface, or highly irregular in configuration and soundness (see Figure 2.17).
- Cavities within the rock mass — location, size, and shape.
- Arch characteristics, i.e., thickness, span, soundness, and joint characteristics and properties.
- Groundwater depth and withdrawal conditions, i.e., present vs. future potential.

Analysis

Analysis proceeds in accordance with a normal foundation study (except where foundations may overlie a rock arch, then rock-mechanics principles are applied to evaluate the minimum roof thickness required to provide adequate support to foundations). A generous safety factor is applied to allow for unknown rock properties.

2.4.4 Support of Surface Structures

Avoid the High Hazard Condition

Project relocation should be considered where cavities are large and at relatively shallow depths, or where soluble rock is deep but overlain by soils subject to raveling. The decision is based on the degree of hazard presented, which is directly related to the occurrence of groundwater withdrawal or its likelihood in the future. Groundwater withdrawal represents very high hazard conditions. In fact, the probability and effects of groundwater withdrawal are the most important considerations in evaluating sites underlain by soluble rock.

Foundation Treatments

Dental Concrete

Cavities that can be exposed by excavation can be cleaned of soil and filled with lean concrete, which provides suitable support for shallow foundations.

Grouting

Deep cavities that cannot be reached by excavation are often filled by grout injection, but the uncertainty will exist that not all cavities and fractures have been filled, even if check explorations are made subsequent to the grouting operations. Grouting has the important advantage of impeding groundwater movement and therefore cavity growth, even where pumping is anticipated.

Deep Foundations

Deep, heavily loaded foundations, or those supporting settlement-sensitive structures, when founded on, or rock-socketed in, soluble rock, should be proof-tested, whether grouted or not (see discussion of explorations in Section 2.4.3).

2.5 Soil Subsidence and Collapse

2.5.1 General

Causes

Subsidence in soils results from two general categories of causes:

1. Compression refers to the volume reduction occurring under applied stress from grain rearrangement in cohesionless soil or consolidation in a cohesive soil. The phenomenon is very common and always occurs to some degree under foundation loadings.
2. Collapse is the consequence of a sudden closure of voids, or a void, and is the subject of this chapter. Collapsible or metastable soils undergo a sudden decrease in volume when internal structural support is lost; piping soils are susceptible to the formation of large cavities, which are subject to collapse.

The Hazard

Subsidence from compression or soil collapse is a relatively minor hazard, resulting in structural distortions from differential settlements.

Piping erosion forms seepage channels in earth dams and slopes and in severe cases results in the collapse of the piping tunnel, which can affect the stability of an earth dam or a natural or cut slope.

2.5.2 Collapsible or Metastable Soils

Collapse Mechanisms

Temporary Internal Soil Support

Internal soil support, which is considered to provide temporary strength, is derived from a number of sources. Included are capillary tension, which provides temporary strength in partially saturated fine-grained cohesionless soils; cementing agents, which may include iron oxide, calcium carbonate, or clay in the clay welding, of grains; and other agents, which include silt bonds, clay bonds, and clay bridges, as illustrated in Figure 2.22.

Collapse Causes

Wetting destroys capillary bonds, leaches out cementing agents, or softens clay bonds and bridges in an open structure. Local shallow wetting occurs from surface flooding or broken pipelines, and subsidence can be substantial and nonuniform. Intense, deep local wetting from the discharge of industrial effluents or irrigation can also result in substantial and nonuniform subsidence. A slow and relatively uniform rise in the groundwater level usually results in uniform and gradual subsidence.

Increased saturation under an applied load can result in gradual settlement, or in a sudden collapse as the soil bonds are weakened.

Applied load of critical magnitude can cause a sudden collapse of the soil structure when the bonds break in a brittle type of failure, even at natural moisture content.

Susceptible Soils

Loess

Loess is a fine-grained aeolian deposit.

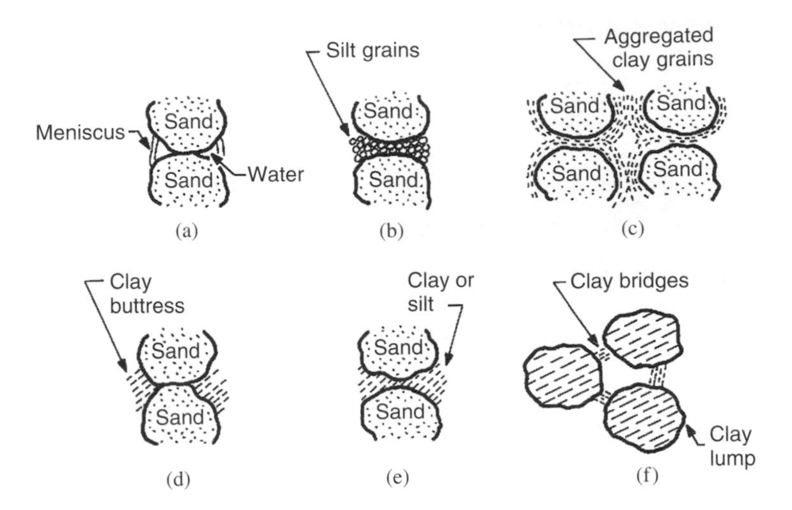

FIGURE 2.22
Typical collapsible soil structures: (a) capillary tension; (b) silt bond; (c) aggregated clay bond; (d) flocculated clay bond; (e) mudflow type of separation; (f) clay bridge structure. (From Clemence, S. P. and Finbarr, A. O., *Proceedings of ASCE*, Preprint 80–116, 1980, 22 pp. Adapted from Barden, L. et al., *Engineering Geology*, 1973, pp. 49–60.)

Valley Alluvium: Semiarid to Arid Climate

In arid climates, occasional heavy rains carry fine soils and soluble salts to the valley floor to form temporary lakes. The water evaporates rapidly leaving a loosely structured, lightly cemented deposit susceptible to subsidence and erosion. Often characteristic of these soils are numerous steep-sided gulleys as shown in the oblique aerial photo (Figure 2.23) taken near Tucson, Arizona. Tucson suffers from the collapsing soil problem, and damage to structures has been reported (Sultan, 1969).

Studies of subsidence from ground collapse were undertaken by the U.S. Bureau of Reclamation (USBR) to evaluate problems and solutions for the construction of the California aqueduct in the San Joaquin Valley of California (Curtin, 1973). Test sites were selected after extensive ground and aerial surveys of the 2000 mi^2 study area. It is interesting to note that the elevation of the entire valley had undergone *regional subsidence* and had been lowered by as much as 30 ft by groundwater withdrawal since the early 1920s.

USBR test procedures to investigate subsidence potential involved either inundating the ground surface by ponding or filling bottomless tanks with water as shown in Figure 2.24. One of the large ponds overlays 250 ft of collapsible soils. Water was applied to the pond for 484 days, during which an average settlement of 11.5 ft occurred. Benchmarks had been set at the surface and at 25 ft intervals to a depth of 150 ft. A plot of the subsidence and compaction between benchmarks as a function of time is given in Figure 2.25. It is seen that the effects of the test influenced the soils to a depth of at least 150 ft. The subsidence appears to be the summation of soil collapse plus compression from increasing overburden pressure due to saturation. The test curves show an almost immediate subsidence of 1 ft upon saturation within the upper 25 ft; thereafter, the shape of the curves follows the curve expected from normal consolidation.

San Joaquin Valley soils are described by Curtin (1973). They have a texture similar to that of loess, characterized by voids between grains held in place by clay bonds, with bubble cavities formed by entrapped air, interlaminar openings in thinly laminated sediments, and unfilled polygonal cracks and voids left by the disintegration of entrapped vegetation. The classification ranges from a poorly graded silty sand to a clay, in general with more

FIGURE 2.23
Fine-grained valley fill near Tucson, Arizona. The steep-sided gullies form in easily eroded collapsible soils.

FIGURE 2.24
Subsidence after 3 months caused by ground saturation around test plot in San Joaquin Valley, California. (From Curtin, G., *Geology, Seismicity and Environmental Impact, Special Publication* Association Engineering, Geology, University Publishers, Los Angles, 1973. With permission.)

than 50% passing the 220 sieve. Dry density ranges from 57 to 110 pcf with a porosity range of 43 to 85%. The predominant clay mineral is montmorillonite, and the clay content of collapsing soils was reported to be from 3 to 30%. The observation was made that soils

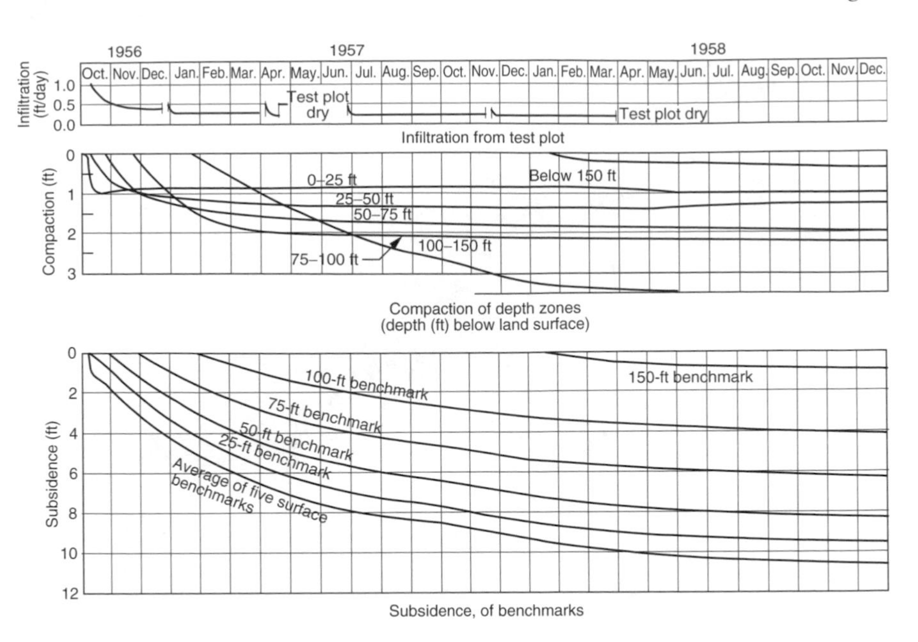

FIGURE 2.25
Subsidence measured by benchmarks, test pond B, west side of San Joaquin, California. (From Curtin, G., *Geology, Seismicity and Environmental Impact, Special Publication* Association Engineering, Geology, University Publishers, Los Angles, 1973. With permission.)

with a high clay content tended initially to swell rather than collapse. During field testing, the benchmarks on the surface first rose upon inundation as the montmorillonite swelled; soon subsidence overcame the swelling and the benchmarks moved downward. A laboratory consolidation test curve showing soil collapse upon the addition of water is given in Figure 2.26.

In describing the soils in western Fresno County, Bull (1964) notes that maximum subsidence occurs where the clay amounts to about 12% of solids; below 5% there is little subsidence and above 30% the clay swells.

Residual Soils

Collapse has been reported to occur in residual soils derived from granite in South Africa and northern Rhodesia (Brink and Kantey, 1961), and from sandstone and basalt in Brazil (Vargas, 1972).

Porous clays of Brazil (argila porosa) occur intermittently in an area of hundreds of square kilometers ranging through the central portions of the states of Sao Paulo, Parana, and Santa Catarina. The terrain consists of rolling savannah and the annual rainfall is about 1200mm (47 in.), distributed primarily during the months of December, January, and February. The remaining 9 months are relatively dry. Derived from Permian sandstone, Triassic basalt, and Tertiary sediments, the soils are generally clayey. Typical gradation curves are given in Figure 2.27, and plasticity index ranges in Figure 2.28.

The upper zone of these formations, to depths of 4 to 8 m, typically yields low SPT values, ranging from 0 to 4 blows/ft; void ratios of 1.3 to 2.0 are common. Below the "soft" (dry) upper zone a hard crust is often found. A typical boring log is given in Figure 2.29; it is to be noted that groundwater was not encountered, which is the normal condition. Some laboratory test data are also included in the figure.

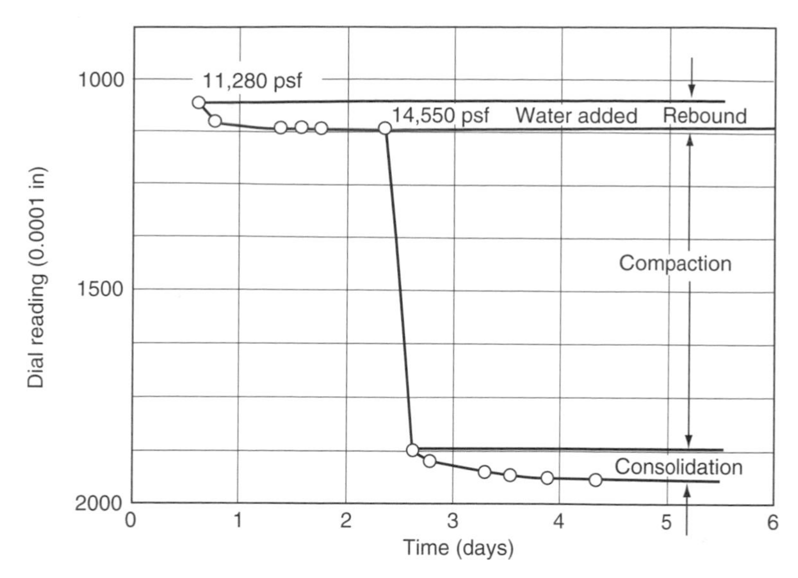

FIGURE 2.26
Laboratory consolidation test curve of compression vs. time for a collapsible soil from the San Joaquin Valley. (From Curtin, G., *Geology, Seismicity and Environmental Impact, Special Publication* Association Engineering, Geology, University Publishers, Los Angles, 1973. With permission.)

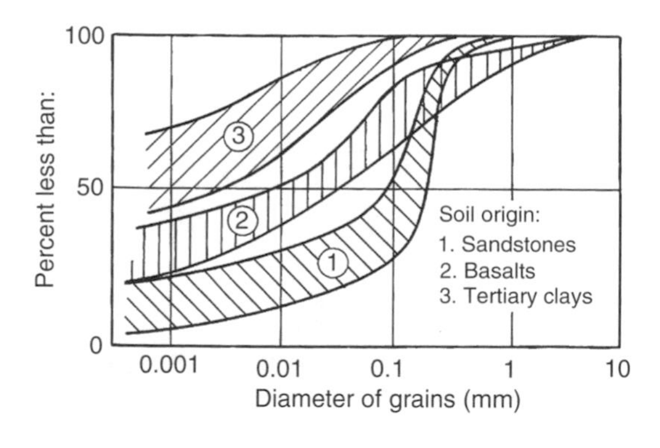

FIGURE 2.27
Gradation curves for typical porous clays of Brazil. (From Vargas, 1972.)

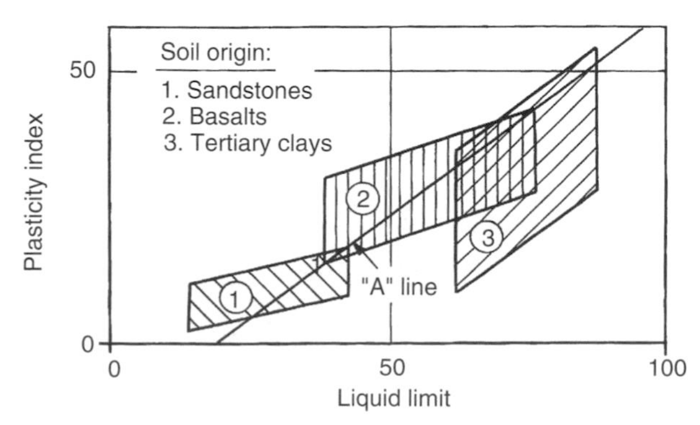

FIGURE 2.28
Relationship between plasticity index and liquid limit, porous clays of Brazil. (From Vargas, 1972.)

Although the soil is essentially a clay, its open, porous structure provides for high permeability and the rapid compression characteristics of a sand, hence the term "porous clay." The porous characteristics are the result of the leaching out of iron and other minerals that are carried by migrating water to some depth where they precipitate to form

Depth (m)	"N"	Soil type	Laboratory test data				
0			LL	PI	w (%)	γ	e
1 ■	3	Very soft to soft	45	21	32	1.25	1.94
2 ■	2	Red silty clay	46	19	31	1.19	2.01
3 ■	5	(CL)	48	24	34	1.51	1.53
4 ■	5		43	16	32	1.39	1.39
5	10	Stiff tan to brown					
6	12	Very silty clay					
7	16						
8	14						
9	11						
10	14						
11	14						
12	36	Becoming hard					
13	50						
14	31						
15	5/1 cm (refusal)						

Note:

■ – Block sample

γ_t – gm/cm^3

N – Blows per foot (SPT)

GWL – Groundwater table not encountered

FIGURE 2.29
Test boring log and laboratory test data for a porous clay derived from basalt (Araras, Sao Paulo, Brazil).

the aforementioned hard zone, which often contains limonite nodules. The effect of saturation on a consolidation test specimen is given in Figure 2.30, a plot of void ratio vs. pressure. The curve is typical of collapsing soils. In the dry condition strengths are high, and excavation walls will stand vertical for heights greater than 4 or 5 m without support in the same manner as loess.

The recognition of porous clays can often be accomplished by terrain analysis. Three factors appear to govern the development of the weak, open structure: a long relatively dry period followed by heavy summer rains, a relatively high ground elevation in rolling, hilly terrain with a moderately deep water table, and readily leachable materials. An examination of aerial photos, such as Figure 2.31, reveals unusual features for a clay soil: characteristically thin vegetation and lack of any surface drainage system, both indicative of the open porous structure. Here and there, where terrain is relatively level, bowl-shaped areas, often 3 to 4 m deep and 20 m across, with no apparent existing drainage, seem to indicate areas of possible natural collapse (Figure 2.31). These may have occurred during periods of very heavy rains, which either created ponds or fell on zones that had been very much weakened by leaching.

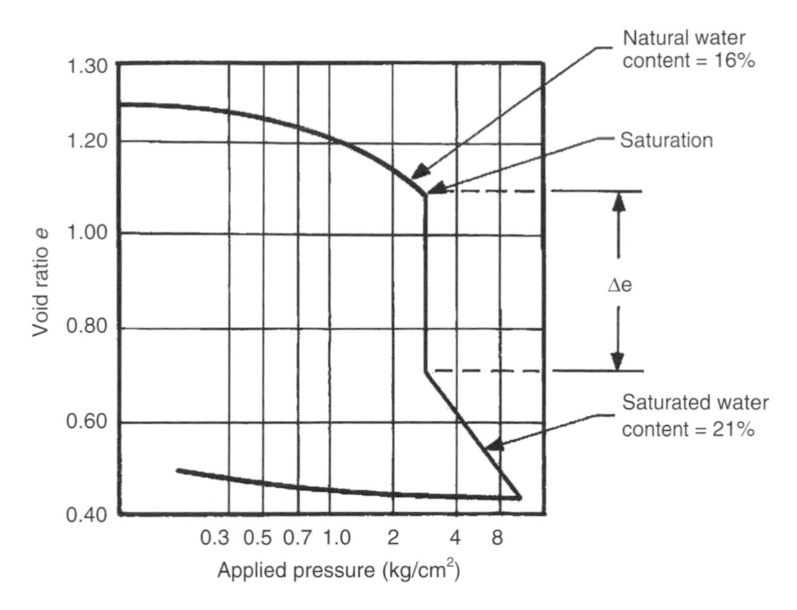

FIGURE 2.30
Effect of saturation on the pressure vs. void ratio curve of a porous clay from Brazil. (From Vargas, M., *VIII Seminario Nacional de Grandes Barragems*, São Paulo, 1972. With permission.)

FIGURE 2.31
Stereo-pair of aerial photos of area of porous clays showing probable collapse zones (state of São Paulo, Brazil). Although residual soils are clays, the lack of drainage patterns indicates high infiltration. (From Hunt, R. E. and Santiago, W. B., *Proceedings 1, Congresso Brasileiro de Geologia de Engenharia,* Rio de Janeiro, 1976, pp. 79–98. With permission.)

2.5.3 Predicting Collapse Potential

Preliminary Phases

Data Collection

A preliminary knowledge of the local geology from a literature review aids in anticipating soils with a collapse potential, since they are commonly associated with loess and other fine-grained aeolian soils, and fine-grained valley alluvium in dry climates. The susceptibility of residual soil is difficult to determine from a normal literature review. Rolling terrain with a moderately deep water table in a climate with a short wet season and a long dry season should be suspect.

Landform Analysis

Loess and valley alluvium are identified by their characteristic features. Residual soils with collapse potential may show a lack of surface drainage channels indicating rainfall infiltration rather than runoff, especially where the soils are known to be clayey. Unexplained collapse depressions may be present.

Explorations

Test Borings and Sampling

Drilling using continuous-flight augers (no drilling fluid) will yield substantially higher SPT values in collapsible soils than will drilling with water, which tends to soften the soils. A comparison of the results from both methods on a given project provides an indication of collapse potential. Undisturbed sampling is often difficult in these materials. Representative samples, suitable for laboratory testing, were obtained with the Denison core barrel in the Negev Desert of Israel.

In residual soils, a "soft" upper zone with low SPT values will be encountered when drilling fluids are used, often underlain by a hard zone or crust which may contain limonite nodules or concretions.

Test pits are useful for close examination and description of the soils in the undisturbed state, *in situ* natural density tests, and the recovery of block samples for laboratory testing.

Simple Hand Test

A hand-size block of the soil is broken into two pieces and each is trimmed until the volumes are equal. One is wetted and molded in the hand and the two volumes then compared. If the wetted volume is obviously smaller, then collapsibility may be suspected (Clemence and Finbarr, 1980).

Field Load Tests

Ground saturation by ponding or using bottomless tanks is useful for evaluating collapse and subsidence where very large leakage may occur, as through canal linings.

Full-scale or plate load tests, for the evaluation of foundation settlements, should be performed under three conditions to provide comparative data at founding level:

1. Soils at natural water content when loads are applied.
2. Loads applied while soil is at the natural water content until the anticipated foundation pressure is reached. The ground around and beneath the footing is then wetted by pouring water into auger holes. (In very dry climates and clayey soils, several days of treatment may be required to achieve an adequate level of

saturation. In all cases, natural soil densities and water contents should be measured before and after wetting).

3. Soils at the test plot are wetted prior to any loading, and then loads are applied.

Laboratory Testing

Natural Density

When the natural density of fine-grained soils at the natural moisture content is lower than normal, about 87 pcf or less, collapse susceptibility should be suspected.

Density vs. liquid limit as a collapse criterion is given by Zur and Wiseman (1973) as follows:

$D_o / D_{LL} < 1.1$, soil prone to collapse

$D_o / D_{LL} > 1.3$, soil prone to swell

where D_0 is the *in situ* dry density and D_{LL} the dry density of soil at full saturation and at moisture content equal to the liquid limit.

Normal Consolidation Test

Loads are applied to the specimen maintained at its natural water content: the specimen is wetted at the proposed foundation stress, and compression is measured as shown in Figure 2.30.

Double Consolidation Test

The double consolidation test is used to provide data for the calculation of estimates of collapse magnitude (Jennings and Knight, 1975; Clemence and Finbarr, 1980).

Two specimens of similar materials (preferably from block samples) are trimmed into consolidometer rings and placed in the apparatus under a light 0.01 tsf seating load for 24 h. Subsequently, one specimen is submerged and the other kept at its natural water content for an additional 24 h. Load applications on each specimen are then carried out in the normal manner.

The e–log p curve for each test is plotted on the same graph, along with the overburden pressure p_o and the preconsolidation pressure p_c for the saturated specimen. The curves for a normally consolidated soil ($p_c/p_o = 0.8 - 1.5$) are given in Figure 2.32, and for an overconsolidated soil ($p_c/p_o > 1.5$) in Figure 2.33. The curve for the natural water content condition is relocated to the point (e_o, p_o) given in the figures.

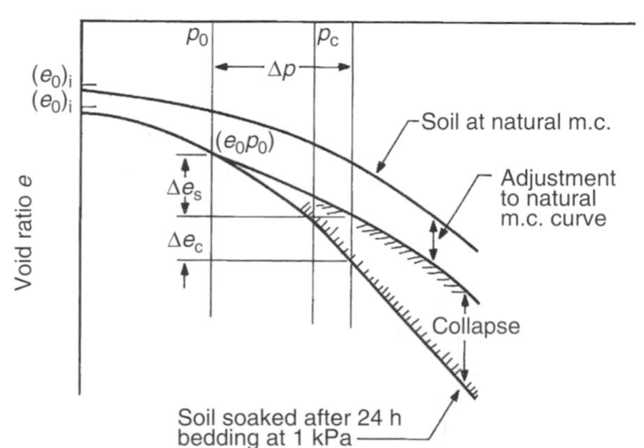

FIGURE 2.32
Double consolidation test *e–p* curves and adjustments for a normally consolidated soil. (From Clemence, S. P. and Finbarr, A. O., *Proceedings of ASCE*, Preprint 80–116, 1980, 22 pp. With permission.)

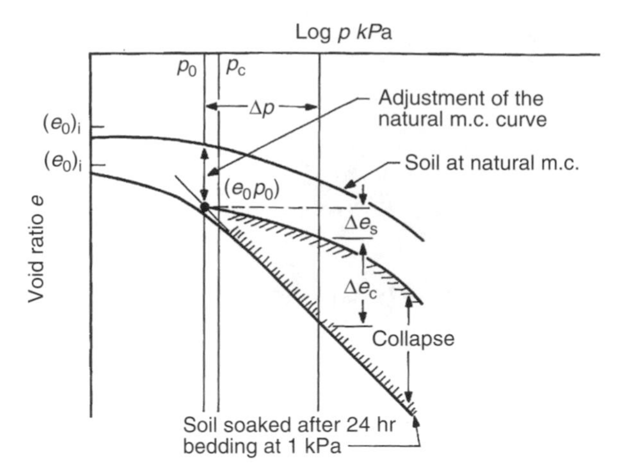

FIGURE 2.33
Double consolidation test *e–p* curves and adjustments for an overconsolidated soil. (From Clemence, S. P. and Finbarr, A. O., *Proceedings of ASCE*, Preprint 80–116, 1980, 22 pp. With permission.)

For an increase in foundation stress ΔP, and wetted soil, the settlement ρ is estimated from the expression

$$\rho = (\Delta e_s / 1 + e_o) + \Delta e_c / (1 + e_o) \tag{2.1}$$

2.5.4 Treatment and Support of Structures

Evaluate the Degree of Hazard and Risk

Degree of hazard is basically a function of the probability of significant ground wetting and of the magnitude of the potential collapse, if the critical pressure that will cause collapse at the natural water content is not approached. Sources of ground wetting have been given in Section 2.5.2 in the discussion of collapse causes.

- *Low-hazard* conditions exist, where potential collapse magnitudes are small and tolerable, or the probability for significant ground wetting is low.
- *Moderate-hazard* conditions exist, where the potential collapse magnitudes are undesirable but the probability of substantial ground wetting is low.
- *High-hazard* conditions exist, where the potential collapse magnitudes are undesirable and the probability of occurrence is high.

Degree of risk relates to the sensitivity of the structure to settlement and to the importance of the structure.

Reduce the Hazard

Prevent ground wetting and support structures on shallow foundations designed for an allowable bearing value sufficiently below the critical pressure to avoid collapse at natural water content. The critical pressure is best determined by *in situ* plate-load tests, and the allowable soil pressure is based on FS = 2 to 3, depending upon the settlement tolerances of the structure.

In some cases, such as large grain elevators, where the load and required size of a mat foundation impose bearing pressures of the order of the critical pressure, the structures have been permitted to settle as much as 1 ft (30 cm), provided that tilting is avoided. This solution has been applied to foundations on relatively uniform deposits of loess with natural water contents of the order of 13% in eastern Colorado and western Kansas. In collapsible soils derived from residual soils, however, such solutions may not be applicable because of the likelihood that a variation in properties will result in large differential settlements. Adequate site drainage should be provided to prevent ponding and all runoff should be collected and directed away from the structure. Avoid locating septic tanks and leaching fields near the structure, and construct all utilities and storm drains carefully to ensure tightness. Underground water lines near the structures have been double-piped or encased in concrete to assure protection against exfiltration. In a desert environment, watering lawns, which is commonly accomplished by flooding, should be avoided, and areas should be landscaped with natural desert vegetation.

Lime stabilization has been used in Tucson, Arizona, to treat collapsible soils that have caused detrimental settlements in a housing development (Sultan, 1969). A water-lime mixture was pumped under high pressure into 2-in.-diameter holes to depths of about 5 ft, and significant movements were arrested.

Hydrocompaction to preconsolidate the collapsible soils was the solution used by the California Department of Water Resources for the construction of the California aqueduct (Curtin, 1973). Dikes and unlined ditches were constructed and flooded along the canal route to precompact the soils at locations where collapse potential was considered high. A section of the canal being constructed over areas both precompacted and not precompacted is shown in Figure 2.34; the subsidence effects on the canal sides are evident on the photo. When hydrocompaction is used, however, the possibility of long-term settlements from the consolidation of clay soils under increased overburden pressures should be considered.

Vibroflotation was experimented with before the construction of the California aqueduct, but adequate compaction was not obtained in the fine-grained soils along the alignment.

FIGURE 2.34
Mendoza test plot showing prototype canal section along the California aqueduct. Crest width is 168ft and length is 1400ft. Note that both the lined and unlined sections of the canal are subsiding where the land was not precompacted. Concentric subsidence cracks indicate former locations of large test ponds. (From Curtin, G., *Geology, Seismicity and Environmental Impact, Special Publication* Association Engineering, Geology, University Publishers, Los Angles, 1973. With permission. Photo courtesy of California Department of Water Resources.)

Dynamic compaction involves dropping 8- to 10-ton tamping blocks from heights of 30 to 120 ft (9 to 36 m). The drops, made by a crane in carefully regulated patterns, produce high-energy shock waves that have compacted soils to depths as great as 60 ft (18 m) (ENR, 1980).

Avoid the Hazard

Settlement-sensitive structures may be supported on deep foundations that extend beyond the zone of potential collapse or, if the collapsible soils extend to limited depths, shallow foundations may be established on controlled compacted earth fill after the collapsible soils are excavated.

In Brazil, buildings constructed over collapsible residual soils are normally supported on piers or piles penetrating to the hard stratum at depths of 3 to 5 m. Floors for large industrial buildings are often supported on porous clays or small amounts of fill, and protection against ground wetting is provided. The risk of subsidence is accepted with the understanding that some relevelling by "mudjacking" may be required in the future.

2.5.5 Piping Soils and Dispersive Clays

General

Soils susceptible to piping erosion and dispersion are not a cause of large-scale subsidence. Ground collapse can occur, however, when the channels resulting from piping and dispersion grow to significant dimensions.

Piping Phenomena

Piping refers to the erosion of soils caused by groundwater flow when the flow emerges on a free face and carries particles of soil with it.

Occurrence in natural deposits results from water entering from the surface, flowing through the soil mass along pervious zones or other openings, finally to exit through the face of a stream bank or other steep slope as illustrated in Figure 2.35. As the water flows, the opening increases in size, at times reaching very large dimensions as shown in Figure 2.36,

FIGURE 2.35
Piping erosion in road cut, Tucson, Arizona. (1972 photo courtesy of Robert S. Woolworth.)

FIGURE 2.36
Tunnel about 7 m high formed from piping in colluvial–lacustrine clayey silts, near La Paz, Bolivia. The vertical slopes are about 15 m in height.

forming in the remains of the enormous ancient mudflow shown in Figure 1.55. The massive movement of the flow destroyed the original structure of the formation and it came to rest in a loose, remolded condition in which fissures subsequently developed. Rainwater entering surface cracks passes along the relict fissures and erodes their sides.

Dispersive Clays

Occurrence

Erosion tunnels from piping in earth dams constructed with certain clay soils are a relatively common occurrence that can seriously affect the stability of the embankment (Sherard et al., 1972). It was originally believed that the clay soils susceptible to dispersion erosion were limited to dry climates, but in recent years these soils and their related problems have been found to exist in humid climates. Sherard et al. (1972) cite examples in the United States from Oklahoma and Mississippi and from western Venezuela.

The Phenomenon

Dispersive clays erode in the presence of water by dispersion or deflocculation. In certain clay soils in which the electrochemical bond is weak, contact with water causes individual particles to detach or disperse. Flowing fresh water readily transports the dispersed particles in the form of piping erosion, in time creating voids or tunnels in the clay mass as illustrated in Figure 2.37 and Figure 2.38. Any fissure or crack from desiccation or settlement can provide the initial flow channel. In expansive clays, the cracks will close by swelling and dispersion will not occur. The failure of many small dams and dikes has been attributed to dispersive clays.

Soil Susceptibility

The main property governing susceptibility to dispersion appears to be the quantity of dissolved sodium cations in the pore water relative to the quantities of other main cations

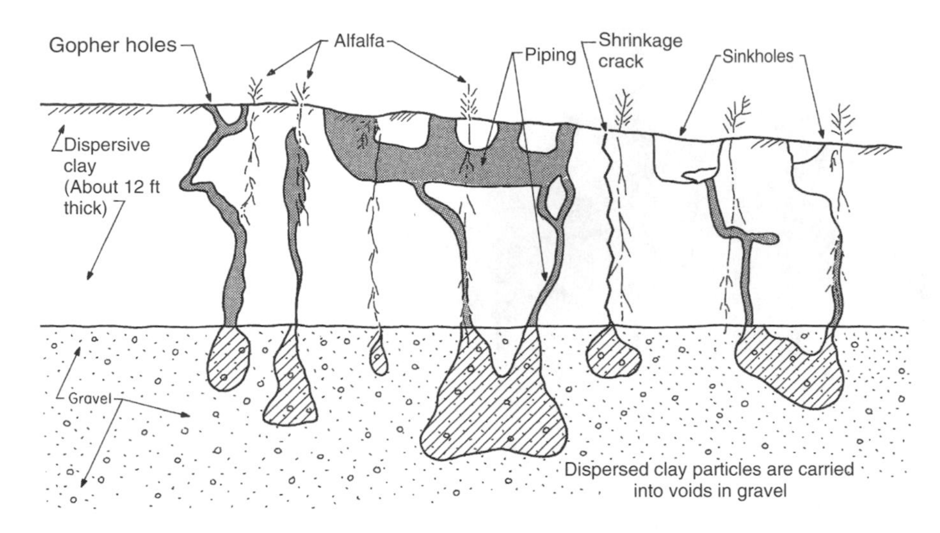

FIGURE 2.37
Damage to agricultural fields in Arizona from piping and sinkhole formation in dispersive clays. (From Sherard, J. L. et al., *Proceedings ASCE*, Purdue University, Vol. I, 1972. With permission.)

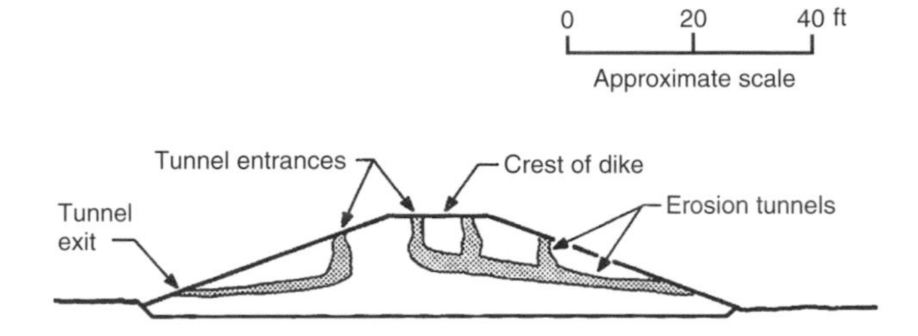

FIGURE 2.38
Schematic of typical rainfall erosion tunnels in clay flood-control dike in badly damaged section. (From Sherard, J. L. et al., *Proceedings ASCE*, Purdue University, Vol. I, 1972. With permission.)

(calcium and magnesium), i.e., the higher the percentage of sodium cation, the higher the susceptibility to dispersion. Soil scientists refer to this relationship as "exchangeable sodium percentages" (ESP).

As of 1976, there appeared to be no good relationship between the ESP and the index tests used by the geotechnical engineer to classify soils, except for the identification of expansive clays that are nonsusceptible. Highly dispersive clays frequently have the same Atterberg limits, gradation, and compaction characteristics as nondispersive clays. These clays plot above the *A* line on the plasticity chart and are generally of low to medium plasticity (LL = 30–50%; CL-CH classifications), although cases of dispersion have been reported in clay with a high liquid limit.

The pinhole test is a simple laboratory test developed to identify dispersive clays (Sherard et al., 1976).

Prevention of Piping and Dispersion

Piping in natural deposits is difficult to prevent. The formation of large tunnels is rare and limited to loose, slightly cohesive silty soils where they are exposed in banks and steep slopes and seepage can exit from the slope. A practical preventive measure is to place a filter at the erosion tunnel outlet to reduce flow exit velocities, but in some cases such as shown in Figure 2.36, the area should be avoided since tunnels are likely to open at other locations.

Piping in dispersive clays used in embankment construction is prevented by the proper design of filters to control internal seepage and by the use of materials that are not susceptible to the phenomenon. Where piping is already occurring it may be necessary to reconstruct the embankment using proper design and materials, if placing a filter at the outlet is not effective.

2.6 Heave in Soil and Rock

2.6.1 General

Origins of Ground Heave

Heave on a regional basis occurs from tectonic activity (see Section 3.3.1 and the Appendix). On a local basis heave occurs from stress release in rock excavations, expansion from freezing, and expansion from swelling in soil or rock.

The Swelling Hazard

Swelling in Geologic Materials

Clay soils and certain minerals readily undergo volume change, shrinking when dried, or swelling when wet. When in the dry state, or when less than fully saturated, some clays have a tremendous affinity for moisture, and in some cases may swell to increase their volume by 30% or more. Pressures in excess of 8 tsf can be generated by a swelling material when it is confined. Once the material is permitted to swell, however, the pressures reduce. There is a time delay in the swelling phenomenon. Noticeable swell may not occur for over a year after the completion of construction, depending on the soil's access to moisture, but may continue for 5 years or longer.

Damage to Structures

Ground heave is a serious problem for structures supported on shallow foundations, or deep foundations if they are not isolated from the swelling soils. Heave results in the uplift and cracking of floors and walls, and in severe cases, in the rupture of columns. It also has a detrimental effect on pavements. Shrinkage of soils also causes damage to structures. Damages to structures in the United States each year from swelling soils alone have been reported to amount to $700 million (ENR, 1976).

Geographic Distribution

Swelling Soils

Swelling soils are generally associated with dry climates such as those that exist in Australia, India, Israel, the United States, and many countries of Africa. In the United States, foundation problems are particularly prevalent in Texas, Colorado, and California, and in many areas of residual soils, as shown on the map given in Figure 2.39.

Swelling in Rocks

Swelling in rocks is associated primarily with clay shales and marine shales. In the United States, it is particularly prevalent in California, Colorado, Montana, North and South Dakota, and Texas.

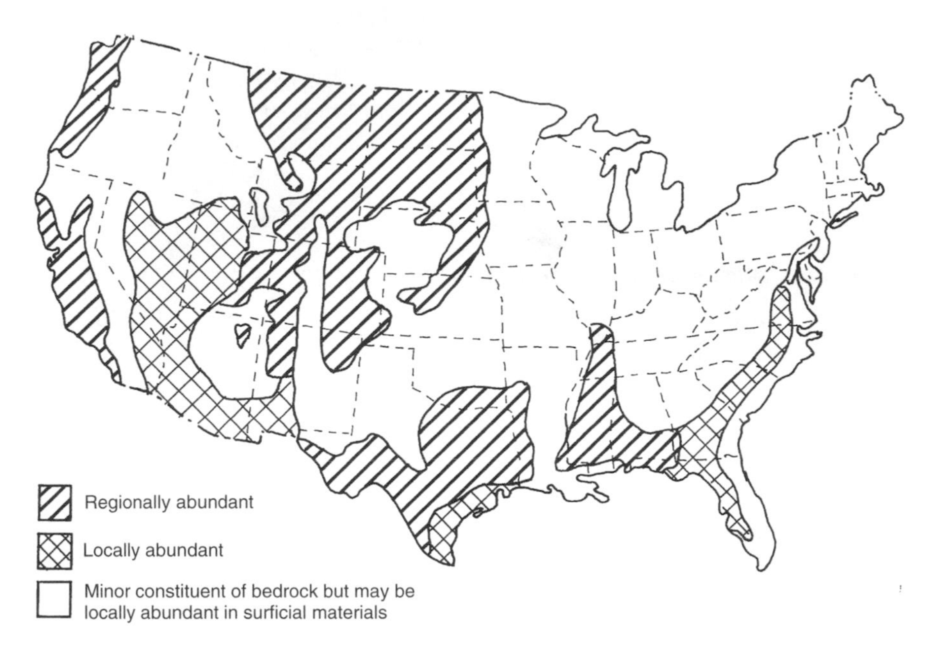

FIGURE 2.39
Distribution of expansive soils in the United States. They are most widespread in areas labeled "regionally abundant," but many locations in these areas will have no expansive soils, while in some unshaded portions of the map, some expansive soils may be found. (From Godfrey, K. A., *Civil Engineering, ASCE,* 1978, pp. 87–91. With permission.)

2.6.2 Swelling in Soils

Determining Swell Potential

Basic Relationships

The phenomenon of adsorption and swell is complex and not well understood, but it appears to be basically physicochemical in origin. Swell potential is related to the percentage of the material in the clay fraction (defined as less than 2 μm, 0.002 mm), the fineness of the clay fraction, the clay structure, and the type of clay mineral. Montmorillonite has the highest potential for swell, followed by illite, with kaolinite being the least active. Thus, mineral identification is one means of investigating swell potential.

Clay Activity

Swell potential has been given by Skempton (1953) in terms of activity defined as the ratio of the plasticity index to the percent finer by weight than 2 μm. On the basis of activity, soils have been classified as inactive, normal, and active. The activity of various types of clay minerals as a function of the plasticity index and the clay fraction is given in Figure 2.40. It is seen that the activity of sodium montmorillonite is many times higher than that of illite or kaolinite.

Prediction from Index Tests

Any surface clay with plasticity index PI > 25 (CH clays) and a relatively low natural moisture content approaching the plastic limit must be considered as having swell potential. The colloid content (percent minus 0.001 mm), the plasticity index, and shrinkage limit are used

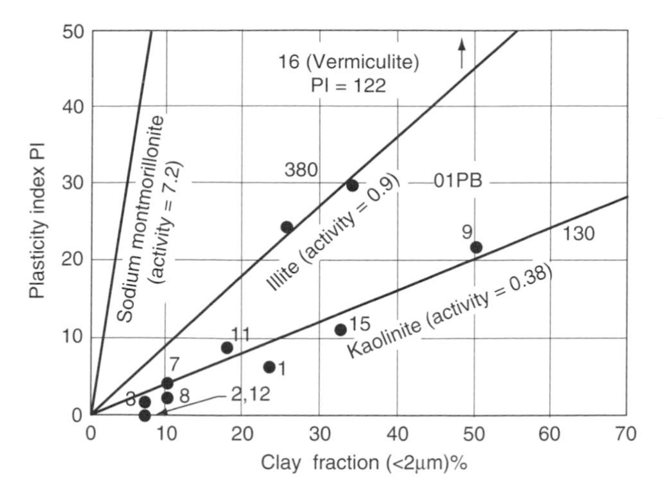

FIGURE 2.40
Clay fraction and plasticity index of natural soils in relation to activity of natural soils in relation to activity chart of Skempton, A. W., *Proceedings of the 3rd International Conference on Soil Mechanics and Foundation Engineering,* Zurich, Vol. I, 1953, pp. 57–61. (After Basu, R. and Arulanandan, K., *Proceedings of the 3rd International Conference on Expensive Soils, Haifa, Israel,* 1973.)

TABLE 2.4
Relation of Soil Index Properties to Probable Volume Changes for Highly Plastic Soils[a]

Data from Index Tests[b]				
Colloid Content (%) <0.001 mm	**Plasticity Index**	**Shrinkage Limit (%)**	**Probable Expansion, Percent Probable Total Volume Change (dry to saturated condition)**	**Degree of Expansion**
>28	>35	<11	>30	Very high
20–31	25–41	7–12	20–30	High
13–23	15–28	10–16	10–20	Medium
<15	<18	>15	<10	Low

[a] From USBR, Earth Manual, U.S. Bureau of Reclamation, Federal Center, Denver, Colorado, 1974. With permission.
[b] All three index tests should be considered in estimating expensive properties.
[c] Based on a vertical loading of 1.0 psi as for concrete canal lining. For higher loadings the amount of expansion is reduced, depending on the load and day characteristics.

by the USBR (1974) as criteria for estimating the probable total volume change from the dry to the saturated condition, as given in Table 2.4.

In the method developed by Seed et al. (1962), expansion was measured as percent swell by placing samples at 100% maximum density and optimum water content in a Standard AASHO compaction mold under a surcharge of 1psi and then soaking them. A family of curves given in Figure 2.41 was developed to describe the percent swell potential for various clay types in terms of activity and clay fraction present. The chart given in Figure 2.42 provides another basis for estimating potential expansiveness.

Index tests have limitations. They do not always identify the swell potential for all natural deposits, nor provide information on the true amount of heave or pressures that may develop. Because the tests are made on remolded specimens, the natural structure of the material is destroyed and other environmental factors are ignored.

Laboratory Tests

Tests in the laboratory on undisturbed samples trimmed into the consolidometer will provide an indication of potential heave and the swelling pressures that may develop.

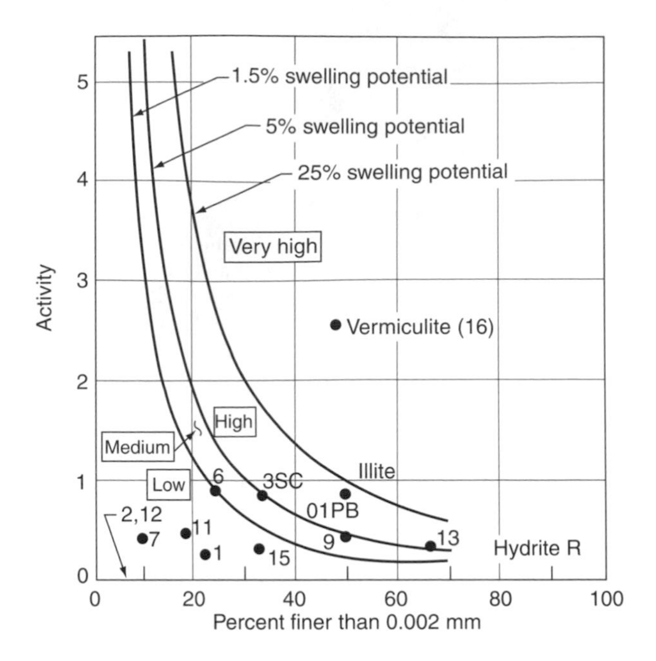

FIGURE 2.41
Classification chart for swelling potential. (After Seed, H. B. et al., *Proc. ASCE J. Soil Mech. Found. Eng. Div.*, 88, 1962. With permission. From Basu, R. and Arulanandan, K., *Proceedings of the 3rd International Conference on Expensive Soils, Haifa, Israel*, 1973.)

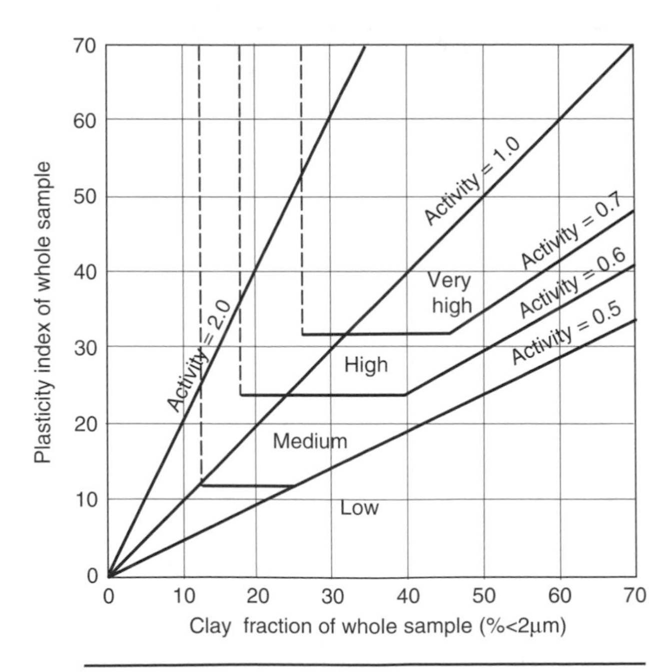

Potential expansiveness	Inch per foot of soil
Very high	1.0
High	0.5
Medium	0.25
Low	0

FIGURE 2.42
Proposed modified chart for determining expansiveness of soils. (From Williams, A. A. and Donaldson, G. W., *Proceedings of the 4th International Conference on Expansive Soils*, Denver, Colorado, Vol. II, 1980, pp. 834–844. With permission.) The Table is after Van der Merwe, U. H., *Proceedings of the 6th Regional Conference for Africa on Soil Mechanics and Foundation Engineering*, Durban, Vol. 2, 1975, pp. 166–167.

Environmental Factors

Basic Factors

Water-table depth: For a soil to develop substantial heave it must lie above the static water table in a less than saturated state and have moisture available to it. Moisture can originate from capillary action or condensation as well as in the form of free water.

Climate: Highly expansive soils are found in climates from hot to cold. Long periods with little or no rainfall permit the water table to drop and the soils to decrease in moisture. The soils dry and shrink, large cracks open on the surface, and fissures develop throughout the mass, substantially increasing its permeability. In Texas, during the dry season, these cracks can extend to depths of 20 ft. Rainfall or other moisture then has easy access for infiltration to cause swelling.

Topography affects runoff and infiltration. Poorly drained sites have a higher potential for ground heave than slopes.

Environmental Changes Cause Surface Movements

Decrease in moisture, causing shrinkage and fissuring, results from:

- Prolonged dry spells or groundwater pumping producing a drop in the water table.
- Growth of trees and other vegetation producing moisture loss by transpiration.
- Heat from structures such as furnaces, boiler plants, etc., producing drying.

Increase in moisture, causing swelling and heave, results from:

- Rainfall and a rise in the water table.
- Drilling holes, such as for pile foundations, through a perched water table that permits permeation into a lower clay stratum (ENR, 1969).
- Retarding evaporation by covering the ground with a structure or a pavement.
- Thermo-osmosis, or the phenomenon by which moisture migrates from a warm zone outside a building area to the cool zone beneath the building.
- Condensation from water lines, sewers, storm drains, and canals.
- Removal of vegetation that increases susceptibility to fissuring and provides access for water.

Time Factor

Usually, a year or more passes after construction is completed before the effects of heave are apparent, although heave can occur within a few hours when the soil has sudden access to free water as from a broken water main or a clogged drain.

A plot of yearly rainfall and heave as a function of time for a house in the Orange Free State in South Africa is given in Figure 2.43. It shows almost no heave for the first year, then a heave of 11.6 cm occurring over a period of 4 years, after which movement essentially stops. The long-term effect is the result of the slow increase, due to natural events, of moisture content beneath a covered area.

2.6.3 Swelling in Rock Masses

Marine and Clay Shale

Characteristics

Montmorillonite is a common constituent of marine and clay shale; therefore, these shales have a high swell potential. They are commonly found disintegrated and badly

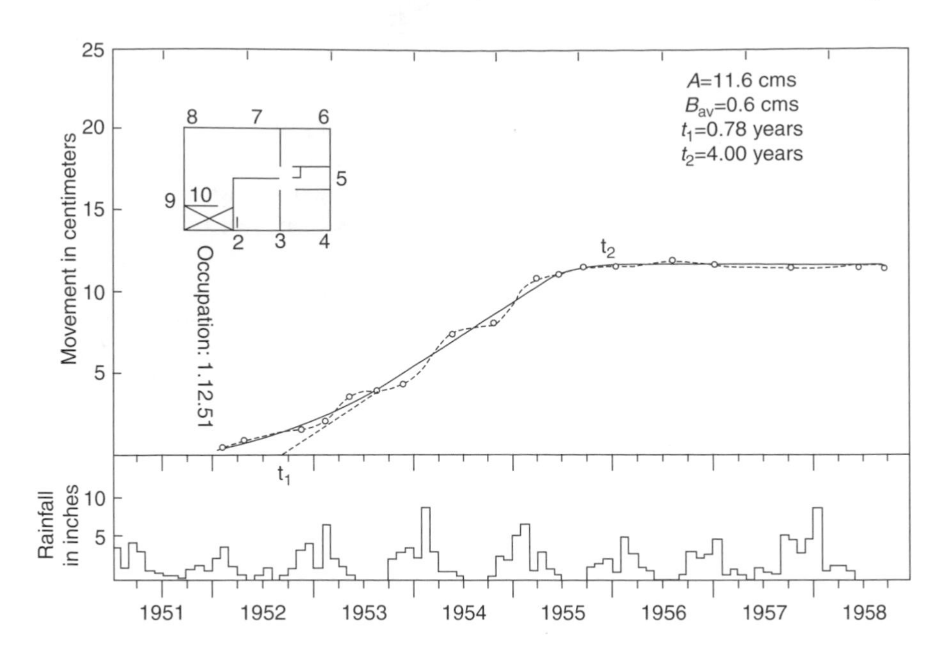

FIGURE 2.43
Typical heave record of single-story brick house in Orange Free State gold fields. (From Jennings, J. E., *Trans. South African Inst. Civil Eng.*, 1969. With permission.)

broken and microfissured from the weathering and expansion of clay minerals. As a result, their shallower portions consist of a mass of hard fragments in a soil matrix. The hard fragments can vary from a medium-hard rock to a very hard indurated clay. Where the weathering is primarily mechanical from the swelling of the clay minerals the disintegrated shale can be found extending downward nearly from the surface, or where chemical weathering has dominated it can be found under a residue of highly expansive clay.

Classification of these materials as either a residual soil or a weathered rock is difficult because of their properties and characteristics. Clay and marine shale are often described as soils in engineering articles and as rock in most geologic publications. They are truly transitional materials, even referred to at times as claystone.

Example: Menlo Park, California

Project: A large urban development suffered substantial and costly damage from the expansion of clay shale (Meehan et al., 1975).

Geologic conditions: Prior to development, the area was virgin hilly terrain underlain by interbedded sandstone and claystone dipping typically at 40 to 60°. In the lower elevations, the claystones are overlain by 3 to 10 ft of black clayey residual or colluvial soils. The weathered portions of the claystone are yellow to olive brown in color, becoming olive gray in fresh rock. The fresh claystone, a soft rock, has two major joint systems, the master system spaced at about 3 ft and the secondary system from 1 to 3 in. Groundwater is generally about 30 ft below the surface. Typical claystone properties are:

- Index properties: LL = 70%, PI = 20, Gs = 2.86, γ_t = 135 pcf
- Clay mineral: Montmorillonite
- Swell pressures: Measured in the consolidometer, 4 to 9 tsf

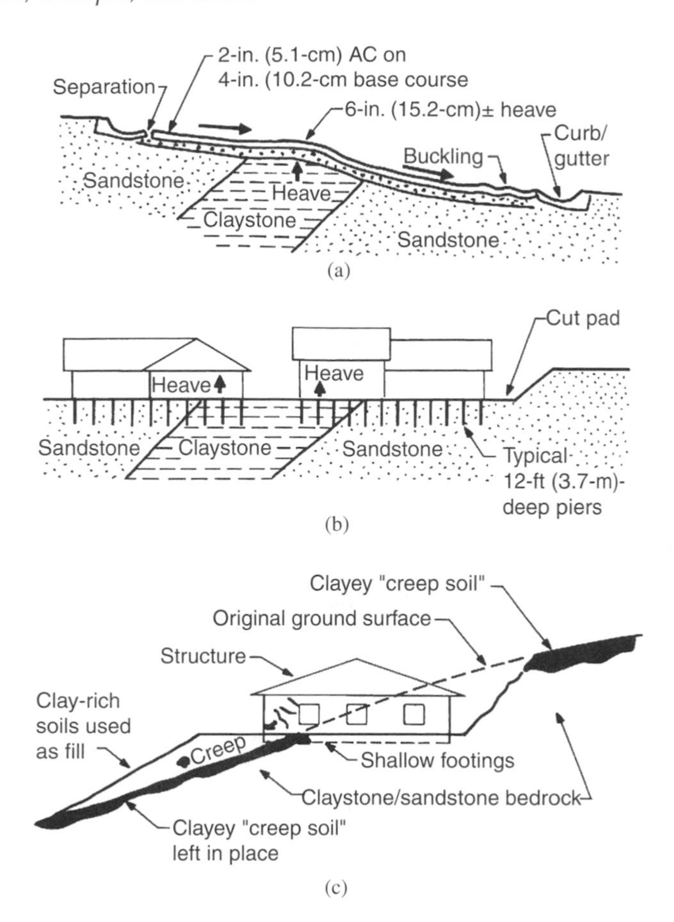

FIGURE 2.44
Problems of heave and creep in the interbedded claystones and sandstones at Menlo Park, California: (a) typical pavement damage; (b) damage to houses on shallow piers; (c) conditions resulting in creep damage. (From Meehan, R. L. et al., *Proc. ASCE J. Geotech. Eng. Div.*, 101, 1975. With permission.)

The problems: Damage from heave to houses and street pavements had been severe, with movements of the order of 4 to 6 in. The major problem was that of differential movements because of the alternating surface exposure of the dipping beds of sandstone and claystone as illustrated in Figure 2.44a and b. Claystone swelling where structures have been placed in cuts typically occurs over several years following construction. It may not be observed for 1 or 2 years, but generally continues to be active for 5 to 7 years after construction. Slopes are unstable, even where gentle. In dry weather the heavy black clays exhibit shrinkage cracks a few feet or so deep. During the winter rains, the cracks close and the clay becomes subject to downhill creep at about 1/2 in. (1.2 cm) per year. Downhill movement of fills has occurred, causing severe distress in structures located partly in cut and partly on fill as shown in Figure 2.44c. Attempted and successful solutions are described in Section 2.6.4.

Other Rocks

Pyritic Shales

In Kansas City, the Missouri limestone has been quarried underground and the space left from the operations has been used since 1955 for warehousing, manufacturing, office, and laboratory facilities. Underlying the limestone and exposed in the floors of many facilities is a black, pyritic shale of Upper Pennsylvanian age. Sulfide alteration of the pyrite results in swelling, and in subsequent years as much as 3 to 4 in. (8 to 11 cm) of floor heave occurred, causing severe floor cracking as well as cracking of the mine pillars left to provide roof support (Coveney and Parizek, 1977). Possible solutions are discussed in Section 2.6.4.

Heaving from the swelling of a black, pyritic carbonaceous shale is reported to have caused damage to structures in Ottawa, Canada (Grattan-Bellew and Eden, 1975). In some instances, the concrete of floors placed in direct contact with the shale has turned to "mush" over a period of years. Apparently, the pyrite oxidizes to produce sulfuric acid, which reacts with calcite in the shale to produce gypsum, the growth of which results in the heave. The acid builds up in the shale to lower the pH to an observed value of 3, leaching the cement from the concrete. The phenomenon does not appear to affect the more deeply embedded footing foundations.

Gneiss and other metamorphic rocks may contain seams of montmorillonite that can be troublesome to deep foundations, tunnels, and slopes.

2.6.4 Treatments to Prevent or Minimize Swelling and Heave

Foundations

Excavations

Sections as small as practical should be opened in shales, and water infiltration prevented. The opening should be covered immediately with foundation concrete, cyclopic concrete, or compacted earth. The objective is to minimize the exposure of the shales to weathering, which occurs very rapidly, and is especially important for dams and other large excavations.

Deep foundations, which generally are drilled piers extending below the permanent zone of saturation, eliminate the heave hazard. The piers, grade beams, and floors must be protected against uplift from swelling forces.

Shallow rigid mat or "rigid" interconnected continuous footings may undergo heave as a unit, but they provide protection against differential movements when adequately stiff.

Other methods, often unsatisfactory, include:

- Preflooding to permit expansion, then designing to contend with settlements of the softened material and attempting to maintain a balance between swell and consolidation.
- Injection with lime has met with some success, but in highly active materials it may aggravate swelling, since the lime is added with water.
- Excavation of the upper portion of the swelling clays, mixing that portion with lime, which substantially reduces their activity, and then replacing the soil–lime mixture as a compacted fill is a costly, and not always satisfactory, procedure.

Floors

Floors should consist of a structural slab not in contact with the expansive materials or supported on a free-draining gravel bed that permits breathing. Infiltration of water must be prevented.

Pavements

Foundations for pavements are prepared by excavating soil to some depth, which depends on clay activity and environmental conditions, and replacing the materials with clean granular soils, with the same soil compacted on the wet side of optimum, or with the same

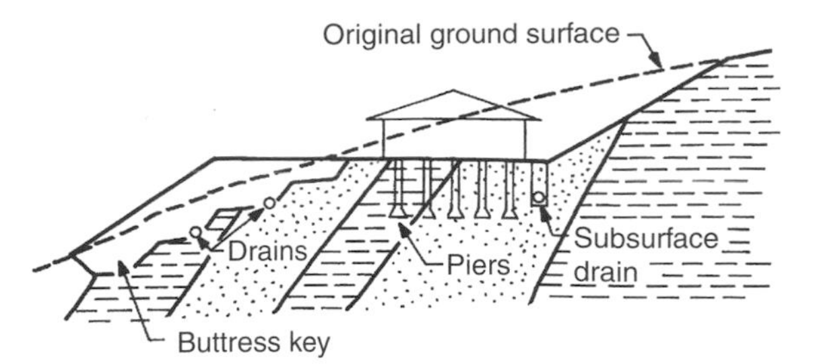

FIGURE 2.45
Site grading and foundation solutions at Menlo Park, California. (From Meehan, R. L. et al., *Proc. ASCE J. Geotech. Eng. Div.*, 101, 1975. With permission.)

soil mixed with lime. Protection against surface water infiltration is provided at the pavement edges, adequate underdrainage is provided, and pavement cracks are sealed.

Solutions at Menlo Park

Site Grading

The procedures cited in Meehan et al. (1975) to correct the problems described in Section 2.6.3 called for the stripping of the shallow, surface soils, and keying fills into slopes in a series of steps. Subsurface drains were provided to prevent water migration into the expansive weathered rock as illustrated in Figure 2.45.

Foundation support was provided by drilled piers taken to depths of 20 to 30 ft, or to the sandstone if shallower — a satisfactory but costly solution for homes.

Lime injected into closely spaced holes was not successful in reducing heave, and at times the situation was aggravated because the lime was added with water.

Pavements were treated by ripping the subgrade to a depth of about 5 ft, injecting a lime slurry into the loosened claystone, then recompacting the surface before paving or repaving. New streets were constructed with "full-depth" asphaltic concrete with typical thicknesses of 5 in. laid directly on the subgrade. When Meehan et al. (1975) published their paper, the success history of the treatments for pavements was about 2 years.

Possible Solutions for Kansas City

For the problems in the pyritic shales in Kansas City (see Section 2.6.3), several possible solutions appear feasible:

- Immediate coating of the shale after excavation with bitumen or a comparable airtight substance.
- Removal of the shales beneath the floor areas and replacement with concrete to some moderately substantial depth, but with consideration given to the phenomenon reported for the shales in Ottawa (see Section 2.6.3).
- Bypass the old mine pillars as roof support with concrete supports founded beyond the active zone of the shale.

References

Allen, D. R., Subsidence, Rebound and Surface Strain Associated with Oil Producing Operations, Long Beach, California, *Geology, Seismicity and Environmental Impact*, Special Publishing, Association of Engineering Geology, University Publishers, Los Angeles, 1973.

Allen, D. R. and Mayuga, M. N., The Mechanics of Compaction and Rebound, Wilmington Oil Field, Long Beach, CA, *Land Subsidence*, Vol. II, Pub. No. 89, IASH-UNESCO-WHO, 1969, pp. 410–423.

Amould, M., Problems Associated with Underground Caverns in the Paris Region, *Proc. Annual Mtg., Assoc. Engrg. Geol.*, 1970.

Averitt, P., Coal Resources of the United States, January, 1967, *USGS Bull. 1275*, U.S. Govt. Printing Office, Washington, DC, 1967.

Barden, L., McCown, A., and Collins, K., The collapse mechanism in partially saturated soil, *Eng. Geol.*, 49–60, 1973.

Basu, R. and Arulanandan, K., A New Approach to the Identification of Swell Potential for Soils, *Procedings of the 3rd International Conference on Expansive Soils*, Haifa, Israel, 1973.

Bau, D., Gambolati, G., and Teatini, P., Residual Land Subsidence Over Depleted Gas Fields in the Northern Adaiatic Basin, *Environmental and Engineering Science*, Vol. V, 1999, pp. 389–405, Joint Pub., AEG and GSA.

Bell, J. W., Amelung, F., Ramelli, A. R., and Blewitt, G., Land Subsidence in Las Vegas, Nevada, 1935–2000. New Geodetic Data Show Evolution, Revised Spatial Patterns, and Reduced Rates, *Environmental and Engineering Science*, Vol. VIII, 2002, pp. 389–405, Joint Pub, AEG and GSA.

Bell, J. W., Price, J. G., and Mifflin, M. D., Subsidence-induced fissuring along preexisting faults in Las Vegas Valley, NV, *Proceedings. AEG*, 35th Annual Mtg., Los Angeles, 1992, pp. 66–75.

Brink, A. B. A. and Kantey, B. A., Collapsible Grain Structure in Residual Granite Soils in Southern Africa, *Proceedings of the 5th International Conference on Soil Mechanics and Foundation Engineering*, Paris, Vol. 1, 1961, pp. 611–614.

Bull, W. B., Alluvial Fans and Near-surface Subsidence in Western Fresno County, CA, Geological Survey Prof. Paper 437-A, Washington, DC, 1964.

Castle, R. O. and Youd, T. L., Discussion: the Houston fault problem, *Bull. Assoc. Engrg. Geol.* 9, 1972.

Civil Engineering, Venice Is Actually Rising? ASCE, November 1975, p. 81.

Civil Engineering, Houston Area Tackles Subsidence, ASCE, March 1977, p. 65.

Civil Engineering, Ground Surface Subsidence Is a World-Wide Problem, ASCE, June 1978, p. 18.

Cohen, K. K., Trevits, M. A., and Shea, T. K., Abandoned Metal Mine Reclamation in New Jersey Applies Geophysics to Target Subsurface Hazards for Safety, Remediation and Land Use AEG National Mtg., Abstracts, 1996.

Corbel, J., Erosion en terrain calcaire, *Ann. Geog., 68*, 97–120, 1959.

Coveney, R. M. and Parizek, E. J., Deformation of mine floors by sulfide alteration, *Bull. Assoc. Engrg. Geol.*, 14, 131–156, 1977.

Clemence, S. P. and Finbarr, A. O., Design Considerations and Evaluation Methods for Collapsible Soils, *Proc. ASCE*, Preprint 80-116, April 1980, 22 pp.

Croxton, N. M., Subsidence on I-70 in Russell County, Kansas related to salt dissolution — a history, *The Professional Geologist*, 2002, pp. 2–7.

Curtin, G., Collapsing Soil and Subsidence, *Geology, Seismicity and Environmental Impact*, Spec. Pub., Assoc. Eng. Geol., University Publishers, Los Angeles, 1973.

Davidson, E. S., Geohydrology and Water Resources of the Tucson Basin, Arizona, Open File Report, Geological Survey, U.S. Dept. Of Interior, Tucson, Arizona, May 1970.

Davis, G. H., Small, J. B., and Counts, H. B., Land Subsidence Related to Decline of Artesian Pressure in the Ocala Limestone at Savannah, Georgia, Engineering Geology Case Histories No. 4, The Geological Society of America, Engineering Geology Div., New York, 1964, pp. 185–192.

ENR, Sinkhole Causes Roof Failure, *Engineering News-Record*, Dec. 11, 1969, p. 23.

ENR, Expansive Soils Culprit in Cracking, *Engineering News-Record*, Nov. 4, 1976, p. 21.

ENR, Houston Land Subsidence Sinks 1970 Bridge Specs, *Engineering News-Record*, June 9, 1977a, p. 11.

ENR, USGS Bets on Las Vegas: Sinking Is No Emergency, *Engineering News-Record*, Dec. 1, 1977b, p. 17.

ENR, Limestone Cavity Sinks Garage, *Engineering News-Record*, Feb. 9, 1978, p. 11.

ENR, Groundwater Bottled Up to Check Moving Piles, *Engineering News-Record*, Oct. 18, 1979, p. 26.

ENR, Flood Control Plan Recycles Water, *Engineering News-Record*, April 24, 1980, pp. 28-29.

ENR, Reclaimed Site Pays Off, *Engineering News-Record*, Jan. 3, 1980, p.14.

Farmer, I. W., *Engineering Properties of Rocks*, E. & F. N. Spon, London, 1968.

Flint, R. F., Longwell, C. R., and Sanders, J. E., *Physical Geology*, John Wiley & Sons, New York, 1969.

Foose, R. M., Groundwater behavior in the Hershey Valley, *Bull. Geol. Soc. Amer.*, 64, 1953.
Godfrey, K. A., Expansive and Shrinking Soils-Building Design Problems Being Attacked, *Civil Engineering*, ASCE, October 1978, pp. 87–91.
Gowan, S.W. and Trader, S.M., Mine Failure Associated with a Pressurized Brine Horizon: Retsof Salt Mine, Western New York, *Environmental and Engineering Science*, 6, 2000, pp. 57–70.
Grattan-Bellew, P. E. and Eden, W. J., Concrete deterioration and floor heave due to biogeochemical weathering of underlying shale, *Canadian Geotech. J.*, 12, 372–378, 1975, National Research Council of Canada, Ottawa.
Gray, R. E. and Meyers, J. F., Mine subsidence and support methods in the Pittsburgh area, *Proc. ASCE J. Soil Mech. Found. Eng. Div.*, 96, 1970.
Greewald, H. P., Howarth, H. C., and Hartmann, I., Experiments on Strength of Small Pillars of Coal in the Pittsburgh Bed, U.S. Bureau of Mines, Report of Investigations 3575, June 1941.
Griffith, W. and Conner, E. T., Mining Conditions Under the City of Scranton, Pennsylvania, *Bull. 25,* Bureau of Mines, U.S. Dept. of Interior, Washington, DC, 1912.
Holdahl, S. R., Zilkoski, D. B., and Holzschuh, J. C., Subsidence in Houston, Texas, 1973–1987, *Procedings of the 4th International Symposium on Land Subsidence*, Johnson, A.I., Ed., IAHS Publ. No. 200, E. 1991.
Hunt, R. E., Round rock, Texas New Town: geologic problems and engineering solutions, *Bull. Assoc. Engrg. Geol.*, 10, 1973.
Hunt, R. E. and Santiago, W. B., A fundo critica do engenheiro geologo em estudos de implantaclo de ferrovias, *Proc. 1, Cong. Brasileiro de Geologia de Engenharia*, Rio de Janeiro, August, Vol. 1, 1976, pp. 79–98.
Jansen, R. B., Dukleth, G. W., Gordon, B. B., James, L. B., and Shields, C. E., Earth Movement at Baldwin Hills Reservoir, *Proc. ASCE J. Soil Mechs. and Found. Eng. Div.*, 93, 1976; Jansen, R.B., Dukleth, G. W., Gordon, B. B., James, L. B., and Shields, C. E., Subsidence-A Geological Problem with a Political Solution, *Civil Engineering* ASCE, 1980, pp. 60–63.
Jennings, J. E., Building on Dolomites in the Transvaal, *Trans. South African Inst. of Civ. Eng.*, January 1966.
Jennings, J. E., The Engineering Problems of Expansive Soils, *Proceedings of the 2nd International Research and Engineering Conference on Expansive Soils*, Texas A&M Univ., College Station, Texas, 1969.
Jennings, J. E. and Knight, K., A Guide to Construction on or with Materials Exhibiting Additional Settlement Due to 'Collapse' of Grain Structure, *6th Regional Conference for Africa on Soil Mechs. Found. Engrg.*, September 1975, pp. 99–105.
Ko, H. Y. and Gerstle, K. H., Constitutive Relations of Coal, *New Horizons in Rock Mechanics*, Proceedings ASCE, 14th Symposium Rock Mechanics, Univ. Park, PA. June 1972, 1973, pp. 157–188.
Leake, S. A., Land Subsidence from Groundwater Pumping USGS, 2004, from the Internet
Legget, R. F., Duisburg Harbour Lowered by Coal Mining, *Canadian Geotech. J.*, 9, 1972.
Mabry, R. E., An Evaluation of Mine Subsidence Potential, *New Horizons in Rock Mechanics, Proceedings ASCE, 14th Symposium on Rock Mechanics.*, Univ. Park, PA, June 1972, 1973.
Mansur, C. I. and Skouby, M. C., Mine grouting to control building settlement, *Proc. ASCE, J. Soil Mechs. Found. Engrg. Div.*, 96, 1970.
Meehan, R. L., Dukes, M. T. and Shires, P. O. A case history of expansive clay stone damage, *Proc. ASCE J. Geotech. Engrg. Div.* 101, 1975.
Millet, R. A. and Moorhouse, D. C., Bedrock Verification Program for Davis-Besse Nuclear Power Station, *Proc. ASCE, Special Conference Structural Design Nuclear Power Plant Facilities*, Chicago, December 1973, pp. 89–113.
MSHA, Report Released on Collapse, Flooding of New York Salt Mine, News Release No. 95-032, Mine Safety and Health Admin., Aug. 8, 1995.
NCB Principles of Subsidence Engineering, Information Bull. 63/240, National Coal Board of the United Kingdom, Production Department, 1963.
NCB *Subsidence Engineers Handbook*, National Coal Board of the United Kingdom, Production Department, 1966.
PaDEP, Final Report, Effects of Longwall Mining on Real Property Value and the Tax Base of Green and Washington Counties, Pennsylvania, Commonwealth of PA, Department of Environmental Protection, Bur. Of Mining and Reclamation, Contract No. BMR-00-02, Dec. 13, 2002.

Penn. Geol. Survey, *Engineering Characteristics of the Rocks of Pennsylvania*, Pennsylvania Geological Survey, 1972.

Pierce, H. W., Geologic Hazards and Land-Use Planning, *Earth Science and Mineral Resources in Arizona*, Vol. 2, No. 3, September, Ariz. Bur. Mines, 1972.

Proctor, R. J., Geology and Urban Tunnels-Including a Case History of Los Angeles, *Geology, Seismicity and Environmental Impact*, Spec. Pub. Assoc. Engrg. Geol., University Publishers, Los Angeles, 1973, pp.187–193.

Prokopovich, N. P., Some geologic factors determining land subsidence, *Bull. Assoc. Engrg. Geol.*, 75–81, 1976.

Rudolph, M., Urban Hazards: Sinking of a Titantic City, *Geotimes*, July 2001, p. 10.

Seed, H. B., Woodward, R. J., and Lundgren, R. Prediction of swelling potential for compacted clays, *Proc. ASCE J. Soil Mechs. Found. Eng. Div.*, 88, 1962.

Sherard, J. L., Decker, R. S., and Ryker, N. L., Piping in Earth Dams of Dispersive Clay, *Performance of Earth and Earth Supported Structures, Proc. ASCE*, Purdue Univ., Vol. I, Part 1, 1972.

Sherard, J. L., Dunnigan, L. P., Decker, R. S., and Steele, E. F. Pinhole test for identifying dispersive soils, *Proc. ASCE J. Geotech.Eng. Div.*, 102, 69–85, 1976.

Skempton, A. W., The Colloidal Activity of Clays, *Proceedings of the 3rd International Conference on Soil Mechanics Foundational Engineering* Zurich, Vol. I, 1953, pp. 57–6l.

Sowers, G. F., Failures in limestones in humid subtropics, *Proc. ASCE J. Geotech. Eng. Div.*, 101, August 1975.

Spencer, G. W., The Fight to Keep Houston from Sinking, *Civil Engineering*, ASCE, September, 1977 pp. 69–71.

Sultan, H. A., Foundation Failures on Collapsing Soils in the Tucson, Arizona Area, *Proceedings of the 2nd International Research and Engineering Conference on Expansive Clay Soils*, Texas A & M Univ., College Station, Texas, 1969.

Terzaghi. K., Landforms and Subsurface Drainage in the Gacka Region in Yugoslavia, reprinted in *Theory to Practice in Soil Mechanics*, Wiley, NY, 1960, 1913.

Testa, S. M., Elevation Changes Associated with Groundwater Withdrawal and Reinjection in the Wilmington Area, Los Angeles Coastal Plain, California, *Proceedings of the 4th International Conference on Land Subsidence*, Houston, TX, May 1991, *IAHS* Pub. No. 200

Trevits, M. A. and Cohen, K. K., Remediation of Problematic Shafts Associated with the Abandoned Schuyler Copper, Mine, North Arlington, NJ, Abstracts, AEG National Mtg., 1996.

Tschebotarioff, G. P., *Foundations, Retaining and Earth Structures*, 2d ed., McGraw-Hill, New York, 1973.

USBR, *Earth Manual*, U.S. Bureau of Reclamation, Federal Center, Denver, CO, 1974.

Van der Merwe, U. H., Contribution to Specialty Session B, Current Theory and Practice for Building on Expansive Clays, *Proceedings of the 6th Regional Conference for Africa on Soil Mechanics and Foundation Engineering*, Durban, Vol. 2, 1975, pp. 166–167.

Van Siclen, D. C., The Houston Fault Problem, Amer. Inst. Prof. Geol., Proc. 3rd Annual Meeting Texas Section, Dallas, 1967.

Vargas, M., Fundañao de Barragems de Terra Sobre Solos Porosos, *VIII Seminario Nacional de Grandes Barragems*, São Paulo, Nov. 27, 1972.

Voight, B. and Pariseau, W., State of Predictive Art in Subsidence Engineering, *Proc. ASCE J. Soil Mechs. Found. Eng.* Div. 96, 1970.

White, W. B., Speleology, *Encyclopedia of Geomorphology*, Dowden, Hutchinson & Ross Publ., Stroudsburg, PA, 1968, pp. 1036–1039.

William, K., Geology, Mining History, and Recent Developments at the Akzo-Nobel Salt Mine, Retsof, NY, AIPG Abstracts, Oct. 24, 1995.

Williams, A. A. B. and Donaldson, G. W., Building on Expansive Soils in South Africa: 1973–1980, *Proceedings of 4th International Conference on Expansive Soils*, Denver, CO, Vol. II, 1980, pp. 834–844.

Zeevaert, L., Foundation Design and Behavior of Tower LatinoAmericano in Mexico City, *Geotechnique*, 7, 1957.

Zeevaert, L., *Foundation Engineering for Difficult Subsoil Conditions*, Van Nostrand Reinhold Co., New York, 1972, pp. 245–281.

Zur, A. and Wiseman, G., A Study of Collapse Potential of an Undisturbed Loess, *Proceedings of the International Conference on Soil Mechanics and Foundation Engineering* Moscow, Vol. 2.2, 1973, p. 265.

Further Reading

Braun, W., Aquifer Recharge to Lift Venice, Underground Services, 1973.

Christie, T. L., Is Venice sinking? in *Focus on Environmentol Geology*, R. Tank, Ed., Oxford University Press, New York, 1973, pp. 121–123.

Dahl, H. D. and Choi, D. S., Some Case Studies of Mine Subsidence and its Mathematical Modeling, Applications of Rock Mechanics, *Proceedings ASCE, 15th Symposium on Rock Mechanics.*, Custer State Park, ND, 1975, pp. 1–22.

Davies, W. E., Caverns of West Virginia, West Vir. Geol. Econ. Surv. Pub. 19A, 1958.

Dudley, I. H., Review of Collapsing Soils, *Proc. ASCE, J. Soil Mech. Found. Eng. Div.*, 96, 1970.

ENR, Texas Puts Land Subsidence Cost at $110 million, *Engineering News-Record*, Oct. 31, 1974, p. 18.

Fuqua, W. D. and Richter, R. C., Photographic Interpretation as an Aid in Delineating Areas of Shallow Land Subsidence in California, *Manual of Photographic Interpretation*, Amer. Soc. of Photogrammetry, Washington, DC, 1960.

Goodman, R., Korbay, S., and Buchignani, A., Evaluation of Collapse Potential over Abandoned Room and Pillar Mines, *Bull. Assoc. Engrg. Geol.*, 17, 27–38, 1980.

Green, J. P., An Approach to analyzing multiple causes of subsidence, in *Geology, Seismicity and Environmental Impact*, Spec Pub. Assoc. Engrg. Geol., University Publishers, Los Angeles, 1973.

Greenfield, R. J., Review of the Geophysical Approaches to the Detection of Karst, *Bull. Assoc. Eng. Geol.*, 16, 393–408, 1979.

Herak, M. and Stringfield, V. T., *Karst*, Elsevier, New York, 1972.

Holtz, W. G. and Gibbs, H., Engineering properties of expansive clays, *Trans. ASCE*, 120, 1956.

Lamoreaux, P. E., Remote Sensing Techniques and the Detection of Karst, *Bull. Assoc. Eng. Geol.*, 16, 383–392, 1979.

Legget, R. F., *Cities and Geology*, McGraw-Hill Book Co., New York, 1973.

Liszkowski, J., The influence of karst on geological environment in regional and urban planning, *Bull. Assoc. Eng. Geol.*, December 1975.

Livneh, M. and Greenstein, J., The Use of Index Properties in the Design of Pavements on Loess and Silty Clays, Transportation Research Inst. Pub. No. 77–6, Technion-Israel University, Haifa, 1977.

Mathewson, C. C., Dobson, B. M., Dyke, L. D., and Lytton, R. L., System interaction of expansive soils with light foundations, *Bull. Assoc. Eng. Geol.*, 17, 55–94, 1980.

Mayuga, M. N. and Allen, D. R., Long Beach Subsidence, in *Focus on Environmental Geology*, Tank, R., Ed., Oxford University Press, New York, 1973, p. 347.

Meyer, K. T. and Lytton, A. M., Foundation Design in Swelling Clays, paper presented to Texas Section, ASCE, October 1966.

Mitchell, J. K., Influences of Mineralogy and Pore Solution Chemistry on the Swelling and Stability of Clays, *Proceedings of the International Conference on Expansive Soils*, Haifa, Israel, 1973.

Poland, J. F., Land Subsidence in the Western United States, in *Focus on Environmental Geology*, Tank, R., Ed., Oxford University Press, New York, 1973, p. 335.

Price, D. G., Malkin, A. B., and Knill, J. L., Foundations of multi-story blocks on the coal measures with special reference to old mine workings, *Q. J. Eng. Geol. London*, 1969.

Prokopovich, N. P., Past and future subsidence along San Luis Drain, San Joaquin Valley, California, *Bull. Assoc. Eng. Geol.*, 12, 1975.

Soderberg, A. D., Expect the unexpected: foundations for dams in karst, *Bull. Assoc. Eng. Geol.*, 16, 409–426, 1979.

Sowers, G. F., Shallow foundations, in *Foundation Engineering*, Leonards, G. A., Ed., McGraw-Hill Book Co., New York, 1962, Chap. 6.

Spanovich, M., Construction over Shallow Mines: Two Case Histories, ASCE Annual and National Meeting on Structural Engineering, Pittsburgh, September 1968, Preprint 703.

Turnbull, W. J., Expansive Soils — Are We Meeting the Challenge? *Proceedings of the 2nd International Conference on Expansive Clay Soils*, Texas A&M Univ., College Station, TX, 1969.

Van Siclen, D. C. and DeWitt, C., Reply: The Houston Fault Problem, *Bull. Assoc. Eng. Geol.*, 9, 1972.

Zeevaert, L., Pore Pressure Measurements to Investigate the Main Source of Subsidence in Mexico City, *Proceedings of the 3rd International Congress: Soil Mechanics Foundation Engineering*, Zurich, Vol. II, 1953, p. 299.

Zeitlen, J. G., Some Approaches to Foundation Design for Structures in Expansive Soil Areas, *Proceedings of the 2nd International Research and Engineering Conference on Expansive Clay Soils*, Texas A&M Univ., College Station, TX, 1969.

3

Earthquakes

3.1 Introduction

3.1.1 General

The Hazard

Earthquakes are the detectable shaking of the Earth's surface resulting from seismic waves generated by a sudden release of energy from within the Earth. Surface effects can include damage to or destruction of structures; faults and crustal warping, subsidence and liquefaction, and slope failures offshore or onshore; and tsunamis and seiches in water bodies.

Seismology is the science of earthquakes and related phenomena (Richter, 1958). The Chinese began keeping records of earthquakes about 3000 years ago, and the Japanese have kept records from about 1600 A.D. Scientific data, however, were lacking until the first seismographs were built in the late 1800s. Strong-motion data, the modern basis for aseismic design, did not become available until the advent of the accelerograph, the first of which was installed in Long Beach, California, in 1933.

Engineering Aspects

Objectives of engineering studies are to design structures to resist earthquake forces, which may have a wide variety of characteristics requiring prediction and evaluation.

Important elements of earthquake studies include:

- Geographic distribution and recurrence of events.
- Positions as determined by focus and epicenter.
- Force as measured by intensity or magnitude.
- Attenuation of the force with distance from the focus.
- Duration of the force.
- Characteristics of the force as measured by: (1) amplitude of displacement in terms of the horizontal and vertical acceleration due to gravity and (2) its frequency component, i.e., ground motion.
- Response characteristics of engineered structures and the ground.

Earthquake damage factors to be considered include:

- Magnitude, frequency content, and duration of the event.
- Proximity to populated areas.
- Local geologic conditions.
- Local construction practices.

3.1.2 Geographic Distribution

Worldwide

General Distribution

The relationship between earthquake zones and tectonic plates (see Appendix A.3) is given in Figure 3.1. It is seen that concentrations are along the boundaries of subducting plates and zones of seafloor spreading, the concept of which is illustrated in Figure 3.2. The great Precambrian shields of Brazil, Canada, Africa, India, Siberia, and Australia are generally aseismic although their margins are subjected to earthquake activity.

Earthquake occurrence predominates in three major belts: island chains and land masses forming the Pacific Ocean; the mid-Atlantic ridge; and an east–west zone extending from China through northern India, Turkey, Greece, Italy, and western North Africa to Portugal. Countries with a high incidence of damaging earthquakes include Chile, China, Greece, India, Indonesia, Italy, Iran, Japan, Mexico, Morocco, Peru, the Philippines, Rumania, Spain, Turkey, Yugoslavia, the East Indies, and California in the United States.

Important worldwide events are summarized in Table 3.1. Selection is based on consideration of the number of deaths; the effect on the land surface in terms of faulting, subsidence, or landslides; and the contribution to the knowledge of seismology and earthquake engineering. The table also serves to illustrate the partial distribution of events by city and country.

During the 20 years since the first edition of this book there have been many significant, damaging earthquakes. Of particular note, outside of the United States, were the Izmit, Turkey event of 1999 (Section 3.3.1) and the Kobe, Japan (Hanshin event), of January 17, 1995. With $M=6.9$, Hanshin resulted in the deaths of 5500 persons. The deep focus depth was 10 mi, located along the Nojima strike-slip fault, and the duration less than 1 min. Major damage occurred where structures were located over soft ground; elevated highway and rail lines and buildings generally five to ten stories collapsed. Liquefaction destroyed port facilities and caused other buildings to collapse.

Two General Classes

Plate-edge earthquakes: The boundaries of the lithospheric plates are defined by the principal global seismic zones in which about 90% of the world's earthquakes occur (Figure 3.1 and Figure A.2). Note that in Figure A.2 additional plates have been identified, such as the Juan de Fuca plate in Northern California and Oregon.

Intraplate earthquakes: Areas far from the plate edges are characterized by fewer and smaller events, but large destructive earthquakes occur from time to time such as those of New Madrid, Charleston, and northern China (Table 3.1). These events and others indicate that the lithospheric plates are not rigid and free of rupture.

Continental United States

The distribution of the more damaging earthquakes in the continental United States through 1966 is given in Figure 3.3. California has by far the largest incidence of damaging events. The largest events recorded, however, include New Madrid, Missouri (1811 and 1812), Charleston, South Carolina (1886), and Anchorage, Alaska (1964), all of which had magnitudes estimated to be greater than 8. The New Madrid quakes had tremendous effects on the central lowlands, causing as much as 15 to 20 ft of subsidence in the Mississippi valley, forming many large lakes, and were felt from New Orleans to Boston (Guttenberg and Richter, 1954).

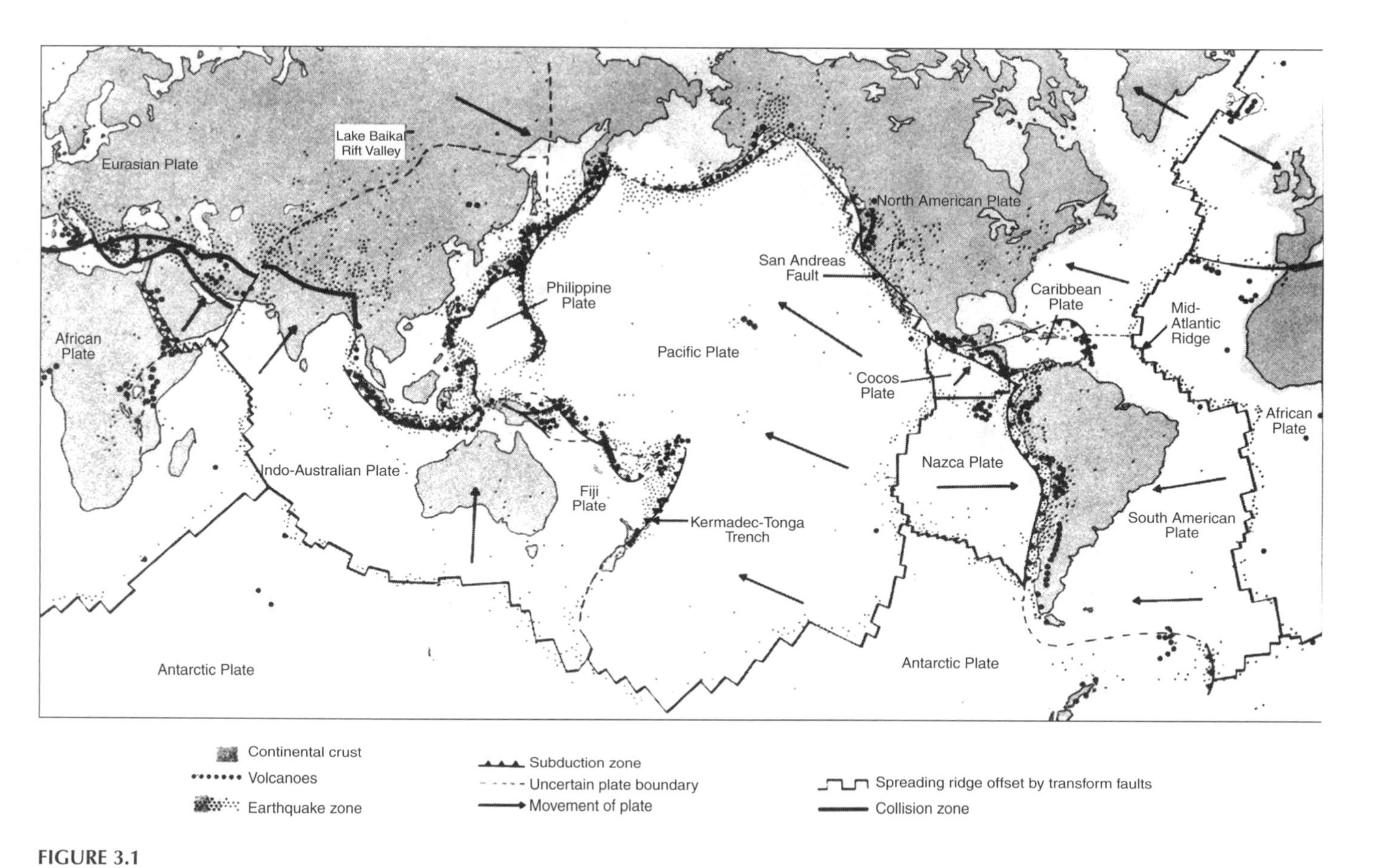

FIGURE 3.1
Worldwide distribution of earthquakes and volcanoes in relation to the major tectonic plates. (From Bolt, B. A. et al., *Geological Hazards*, Springer, New York, 1975. Reprinted with permission of Springer.)

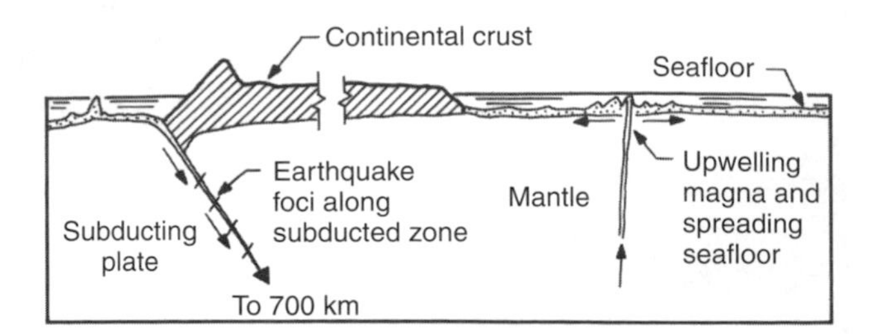

FIGURE 3.2
The concept of a subducting plate and the spreading seafloor. (From Deitz and Holden, 1970.)

TABLE 3.1
Important Earthquakes of the World

Location	Date	Magnitude[a]	Importance
Corinth, Greece	856		45.000 dead
Chihli, China	1290		100,000 dead
Shensi, China	1556		830,000 dead
Three Rivers, Quebec	Feb. 5, 1663		Strongest quake, NE North America (est. I=X)
Calcutta, India	1737		300,000 dead
Lisbon, Portugal	Nov. 1, 1755	≈8.7	60,000 dead; caused large tsunamis, and seiches in lakes to distances of 3500km
Calabria, Italy	Feb. 5, 1788		50,000 dead
New Madrid, Missouri	Dec. 16, 1811 and Feb. 7, 1812	≈8	Three great quakes occurred in this period that affected an area of over 8000 km^2 and caused large areas to subside as much as 6m
Charleston, South Carolina	Aug. 31, 1886	≈8	Large-magnitude shock in low-seismicity area
San Francisco, California	Apr. 18, 1906	≈8.3	About 450 dead, great destruction, especially from fire. San Andreas fault offset for 430 km or more; one of the longest surface ruptures on record. First "microzonation map" prepared by Wood
Colombia-Ecuador border	Jan. 31, 1906	8.9	Largest magnitude known, in addition to one in Japan, 1933
Messina, Italy	Dec. 28, 1908	7.5	120,000 dead
Kansu, China	Dec. 16, 1920	8.5	200,000 dead or more. Perhaps the most destructive earthquake. Entire cities destroyed by flows occurring in loess
Tokyo, Japan	Sept. 1, 1923	8.2	The Kwanto earthquake; 143,000 dead, primarily from fires
Attica, New York	Aug. 12, 1929	7.0	Largest recent event in NE United States and eastern Canada
Grand Banks, Newfoundland	Nov. 18, 1929	≈7.5	Undersea quake caused turbidity currents which broke 12 undersea cables in 28 locations
Long Beach, California	Mar. 10, 1933	6.3	Small shock caused much destruction to poorly designed and constructed buildings. Resulted in improved building code legislation regarding schools. First shock recorded on an accelerograph
Quetta, India	May 30, 1935	7.5	30,000 dead
Western Turkey	Dec. 26, 1939	8.0	20,000 to 30,000 dead
El Centro, California	May 18, 1940	6.5	The Imperial valley event. The first quake to provide good strong motion data from an accelerograph. Provided the basis for seismic design for many years
Kern Co., California	July 21, 1952	7.7	First major shock in California after earthquake-resistant construction began; showed the value of resistant design

(Continued)

TABLE 3.1
(*Continued*)

Location	Date	Magnitude[a]	Importance
Churchill Co., Nevada	Dec. 16, 1954	7.1	Caused much surface faulting in an area 32 × 96 km, with as much as 6 m vertical and 36 m horizontal displacement
Mexico City, Mexico	July 28, 1957	7.5	Maximum acceleration only 0.05 to 0.1g in the city, but caused the collapse of multistory buildings because of weak soils
Hebgen Lake, Montana	Aug. 17, 1959	7.1	Triggered large landslide in mountainous region which took 14 lives
Agadir, Morocco	Feb. 29, 1960	5.8	Small shallow-focus event destroyed the poorly constructed city and caused 12,000 deaths of 33,000 population. Previous heavy shock was in 1751
Central Chile	May 21, 1960	8.4	Strong, deep-focus quake was felt over large area, and generated one of the largest tsunamis on record. Much damage in Hilo, Hawaii, from a 10-m-high wave and in Japan from a 4 m high wave
Skopje, Yugoslavia	July 26, 1963	6.0	2,000 dead, city 85% destroyed by relatively small shock in poorly constructed area
Anchorage, Alaska	Mar. 27, 1964	8.6	3-min-long acceleration caused much damage in Anchorage, Valdez, and Seward, particularly from landsliding
Niigata, Japan	June 16, 1964	7.5	City founded on saturated sands suffered much damage from subsidence and liquefaction. Apartment houses overturned
Parkfield, California	June 27, 1966	5.6	Low magnitude, short duration, but high acceleration shock (0.5 g) caused little damage. The San Andreas broke along a 37 km length and displacement continued for months after the shock. An accelerograph located virtually on the fault obtained good strong-motion data
Caracas, Venezuela	July 29, 1967	6.3	277 dead, much damage but occurred selectively
Near Chimbote, Peru	May 31, 1970	7.8	50,000 dead, including 18,000 from avalanche triggered by the quake
San Fernando, California	Feb. 9, 1971	6.5	Strongest damaging shock in Los Angeles area in 50 years resulted in 65 deaths from collapsing buildings, caused major damage to modern freeway structures. Gave highest accelerations yet recorded: 0.5 to 0.75*g* with peaks over 1.0*g*. (see Section 3.2.4)
Managua, Nicaragua	Dec. 23, 1972	6.2	6000 dead; shallow focal depth of 8 km beneath the city
Guatemala City, Guatemala	Feb. 3, 1976	7.9	22,000 dead, great damage over 125 km radius
Tangshan, China	July 28, 1976	7.8	655,000 dead (Civil Engineering, November 1977)
Mindanao, Philippines	Aug. 18, 1976	7.8	Over 4000 dead in northern provinces
Vrancea, Rumania	Mar. 4, 1977	7.2	2000 dead; much damage to Bucharest
El Asnam, Algeria	Oct. 10, 1980	7.2	Over 3000 dead
Eboli, Italy	Nov. 30, 1980	6.8	Over 10,000 dead

[a] Magnitude of 4.0 is usually given as the threshold of damage.

Some recent earthquakes of significance that have occurred in the United States include:

Loma Prieta, California: October 17, 1990, $M=7.0$, duration=8 sec, occurred along the San Andreas Fault, 47 mi south of San Francisco. Deep focus depth was about 11 mi. Major damage occurred where structures were located over soft ground, and 63 deaths were recorded.

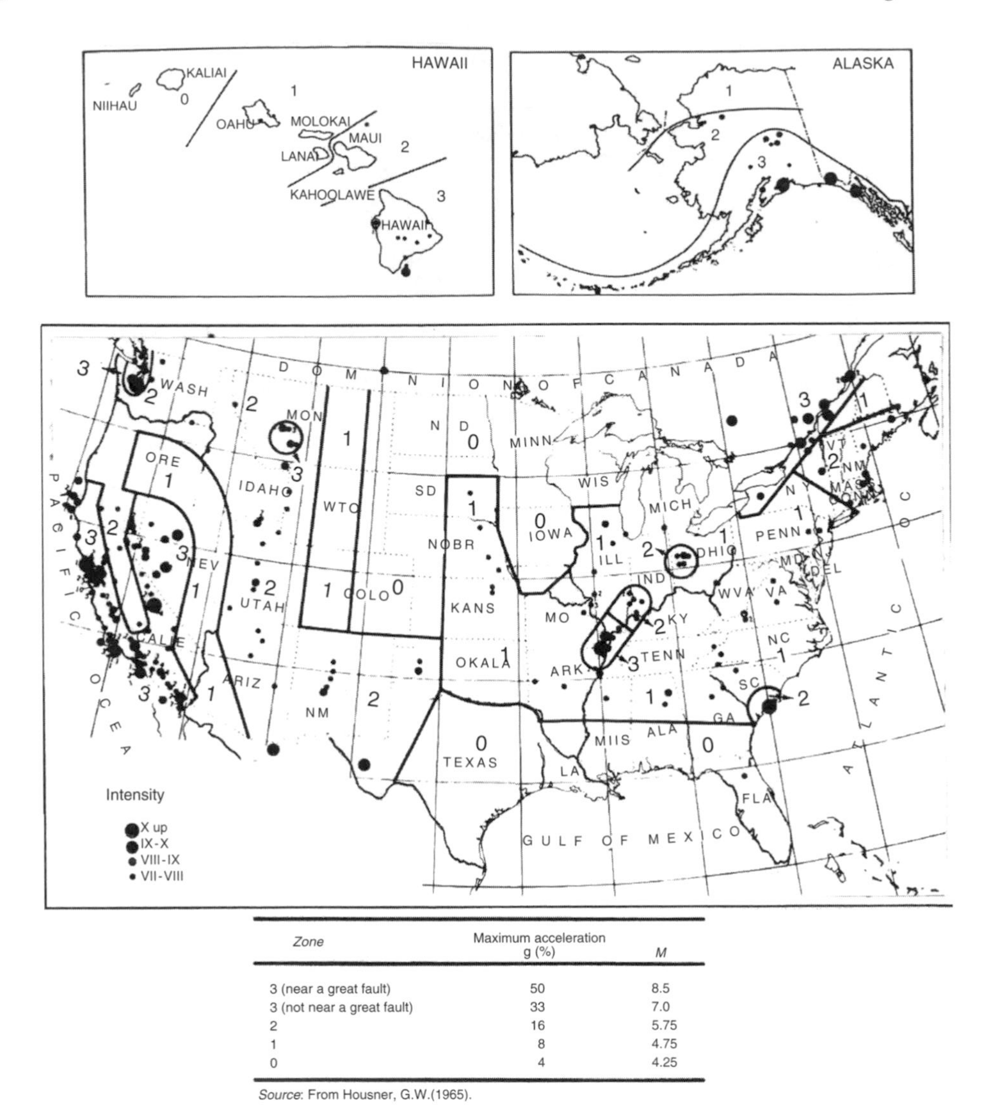

Zone	Maximum acceleration g (%)	M
3 (near a great fault)	50	8.5
3 (not near a great fault)	33	7.0
2	16	5.75
1	8	4.75
0	4	4.25

Source: From Housner, G.W.(1965).

FIGURE 3.3
Seismic risk zones and location of damaging earthquakes in the United States through 1966. See table above for significance of zones. Compare with Figure 3.13. (Map prepared by the U.S. Geological Society.)

Apartment buildings in the San Francisco, Marina District, collapsed and burned, a portion of the Oakland Bay Bridge separated and the upper deck fell onto the lower roadway, and a two-level elevated structure of highway I-880 collapsed. Of particular interest was the lack of surface rupture. Instrumentation showed that the Pacific Plate slipped 2 m northwest past the North American Plate (Figure 3.1) and rode upward about 1 m.

Northridge, Los Angeles, California: January 17, 1994, M=6.7, occurred along an unknown, buried (blind) thrust fault. Apartment buildings, concrete parking garages, and ten highway bridges collapsed, and 60 deaths were recorded.

Pymatuning Reservoir, Pennsylvania: September 25, 1998, M=5.2, was the last significant earthquake to have occurred in the northeastern U.S. (Geotimes [2001], excerpted from USGS Fact Sheet FS-006-01." Earthquakes in and near the northeastern U.S., 1638–1998.") Damage was minor and no deaths or injuries were reported.

3.1.3 Objectives and Scope

Objectives

The objectives of this chapter are to summarize and interrelate all of the aspects of earthquakes including their causes, characteristics, and surface effects, to provide a basis for recognizing the in hazard potential, for investigating quakes comprehensively, and for minimizing their consequences.

There is no field in geotechnical engineering in which the state of the art is changing more rapidly, and it is expected that some of the concepts and methodology presented may quickly become obsolete.

Scope

The earthquake phenomenon is described in terms of its geographic distribution, its location as determined by focus and epicenter, its force as measured by intensity and magnitude, attenuation of the force with distance, and its causes and predictability. In addition, ground and structural response to its forces, including effects on the geological environment such as faults and crustal warping, liquefaction and subsidence, slope failures, tsunamis, and seiches are also considered. Structural response is treated only briefly in this text as a background to the understanding of those elements of earthquake forces that require determination for the analysis and design of structures during investigation.

3.2 Earthquake Elements

3.2.1 The Source

Tectonic Earthquakes

General

Tectonic earthquakes are those associated with the natural overstress existing in the crust as described in Appendix A. This overstress is evidenced by crustal warping, faults, and residual stresses in rock masses as well as earthquakes. It is generally accepted that large earthquakes are caused by a rupture in or near the Earth's crust that is usually associated with a fault or series of faults, but primarily along one dominant fault termed the *causative fault*.

Most earthquakes result from motion occurring along adjacent plates comprising the Earth's crust or lithosphere, such as the Pacific Plate "subducting" (Figure 3.2) beneath the North American Plate. The plates are driven by the convection motion of the material within the Earth's mantle, which in turn is driven by heat generated within the Earth's core. Motion along adjacent plates is constricted by friction, which causes strain energy to accumulate (Scawthorn, 2003).

Elastic rebound theory is described by Richter (1958): "The energy source for tectonic earthquakes is potential energy stored in the crustal rocks during a long growth of strain. When the accompanying elastic stress accumulates beyond the competency of the rocks, there is fracture; the distorted blocks then snap back toward equilibrium, and this produces an earthquake." Earthquakes at very shallow focus may be explained by the elastic rebound theory, but the theory does not explain deep-focus events.

Plastic Yielding

At a depth of about 3 mi or so, the lithostatic pressure is approximately equal to the strength of massive rock at the temperature (500°C) and pressure present. Rock deformation under stress, therefore, would be expected to be plastic yielding rather than the brittle rupture needed for a large release of energy. The cause of earthquakes that originate with deeper foci is not clearly understood. The dilantancy theory has been used to explain rupture at substantial depths (Section 3.2.8).

Deep-focus earthquakes appear to be generally associated with tectonic plates and spreading seafloor movements.

Volcanic Activity

The worldwide distribution of volcanic activity is shown in Figure 3.1, where it is seen that volcanoes are generally located near plate edges. Large earthquakes were at one time attributed to volcanic activity but there is usually a separation of about 120 mi (200 km) or more between belts of active volcanoes and major tectonic activity. The seismic shocks occurring before, during, and after eruptions are referred to by Richter (1958) as volcanic tremors. The volcano hazard is described in Section 3.3.6.

Other Natural Causes

Minor Earth shaking over a relatively small surface area can occasionally be attributed to the collapse of mines or caverns, to large slope failures such as avalanches, or to meteorites striking the Earth.

Human-Induced Causes

Reservoirs

Filling reservoirs behind dams, forming lakes of the order of 100 m or more in depth, creates stress changes in the crust which may be of sufficient magnitude over a large area to induce earthquakes, especially where faults are near, or within, the reservoir area. The cause of reservoir-induced earthquakes is not clearly understood but seems to be more closely associated with an increase in pore- and cleft-water pressures in the underlying rocks than with the reservoir weight. Artificial reservoirs associated with seismic activity and some of their characteristics are given in Table 3.2. Of the 52 reservoirs over 100 m in height in the United States (1973), only about 20% caused seismic activity from water impounding (Bolt et al., 1975). Over 10,000 shocks have been recorded in the area of Lake Mead behind Hoover Dam (H=221 m) since its impoundment in 1935, with the largest having an intensity of MM=IV occurring in 1939. The 236-m-high Oroville Dam in California had not caused detectable seismicity within 10 km from the date of its impoundment in 1968 through early 1975. Following a series of small shocks, an event of magnitude M=5.7 occurred on August 1, 1975.

At Nurek Dam, Tadzhikistan, U.S.S.R., during construction in a seismically active area, the number of yearly events increased significantly since 1972 when the water level reached 116 m (ENR, 1975). Completed in 1980, Nurek Dam is the world's highest at 315 m. Seismic events are being monitored and studied. At the Hsinfengkaing Dam (H=105 m), 160 km from Canton, China, constructed in an area that had no record of damaging earthquakes, a shock of M=6.1 occurred 7 months after the reservoir was filled, causing a crack 82 m long in the upper dam structure.

Accelerographs (Section 3.2.3) are used to instrument large dams to monitor reservoir-induced seismicity. The practice is also being applied to dams lower than 100 m where there is substantial risk to the public if failure should occur.

TABLE 3.2

Artificial Reservoirs with Induced Seismicity[a]

Location (Dam, Country)	Dam Height (m)	Capacity ($m^3 \times 10^9$)	Basement Geology	Date Impounded	Date of First Earthquake	Seismic Effect
L'Oued Fodda, Algeria	101	0.0002	Dolomitic marl	1932	1/33	Felt
Hoover, United States	221	38.3	Granites and Precambrian shales	1935	9/36	Noticeable (M=5)
Talbingo, Australia	176	0.92				Seismic (M<3.5)
Hsinfengkiang, China	105	11.5	Granites	1959		High activity (M=6.1)
Grandval, France	78	0.29		1959–60	1961	MM intensity V in 1963
Monteynard, France	130	0.27	Limestone	1962	4/63	M=4.9
Kariba, Zimbabwe	128	160	Archean gneiss and Karoo sediments	1958	7/61	Seismic (M<6)
Vogorno, Switzerland	230	0.08		8/64	5/65	
Koyna, India	103	2.78	Basalt flows of Deccan trap	1962	1963	Strong (M=6.5); 177 people killed
Benmore, New Zealand	110	2.04	Greywackes and argillites	12/64	2/65	Significant (M=5.0)
Kremasta, Greece	160	4.75	Flysch	1965	12/65	Strong (M=6.2); 1 death, 60 injuries
Nuzek Tadzik, USSR	300	10.5		1972 (to 100 m)		Increased activity (M=4.5)
Kurobe, Japan	186			1960–69		Seismic (M=4.9)

[a] From Bolt, B. A. et al., *Geological Hazards*, Springer, New York, 1975. Reprinted with permission of Springer.

Deep-Well Withdrawal and Injection

Faulting and minor tremors occurred in the Wilmington oil field, Long Beach, California, associated with the extraction of oil (see Section 2.2.2).

Pumping waste fluids down a borehole to depths of 2 mi below the surface near Denver caused a series of shocks as described in Section 3.2.7.

Underground Mine Collapses

The collapse of salt and coal mines has been reportedly monitored as low-level earthquakes (Section 2.3).

Nuclear Explosions

Underground nuclear blasts cause readily detectable seismic tremors.

Focus and Epicenter

Nomenclature

The position of the earthquake source is described by the focus, or hypocenter, which is the location of the source within the Earth, as shown in Figure 3.2, and the epicenter, which is the location on the surface directly above the focus. Depending upon geologic conditions, the epicenter may or may not be the location where surface effects are most severe.

Earthquakes are classified on the basis of depth of focus as follows:

- Normal or shallow: 0 to 70 km (generally within the Earth's crust).
- Intermediate: 70 to 300 km.
- Deep: Greater than 300 km (None has been recorded greater than about 720 km, and no magnitude greater than 8.6 has been recorded below 300 km.).

Some Depth Relationships

Southern California events generally occur at about 5 km depth, whereas Japanese events generally occur at less than 60 km with more than half at less than 30 km.

Foci depths for a number of events apparently define the edge of a subducting plate in some locations, as shown in Figure 3.2. Focus depth can be significantly related to surface damage. The Agadir event (1960) of magnitude 5.8 had a very shallow focal depth of about 3 km, but since it was essentially beneath the city, its effects were disastrous. A magnitude 5.8 event is usually considered to be moderate, but the shallow focus combined with the very weak construction of the city resulted in extensive damage, although the total area of influence was small. The Chilean event of 1960 with a magnitude of 8.4, however, had a focal depth of about 65 km, and although it was felt over a very large area, there was no extreme damage.

3.2.2 Seismic Waves

Origin

Earthquake occurrence causes an energy release which moves as a shock front or strain pulse through the Earth, which is considered as an elastic medium. The pulse becomes an oscillatory wave in which particles along the travel paths are "excited" and move in orbits repeating cyclically. In a simple two-dimensional diagram, the oscillation is shown as a wave shape with a crest and a trough as given in Figure 3.4.

If the wave energy produces displacements in the geologic materials within their elastic limits, the materials return to their original volume and shape after the energy wave has passed. The waves are then termed elastic waves. They propagate through the Earth and along its surface as various types of seismic waves as shown in Figure 3.5.

Wave Types

Body Waves

Primary or compression waves (P waves) are generated by the initial shock, which applies a compressive force to the materials, causing a wave motion in which the particles move back

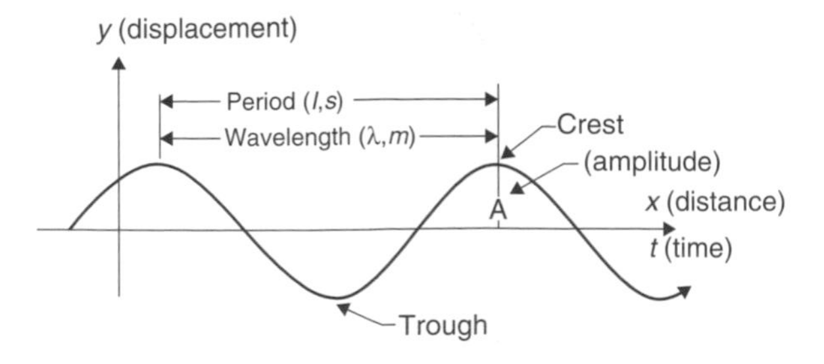

FIGURE 3.4
Characteristics of elastic waves. (From Hunt, R. E., *Geotechnical Engineering Investigation Manual*, McGraw-Hill Book Co., New York, 1984, p. 983.)

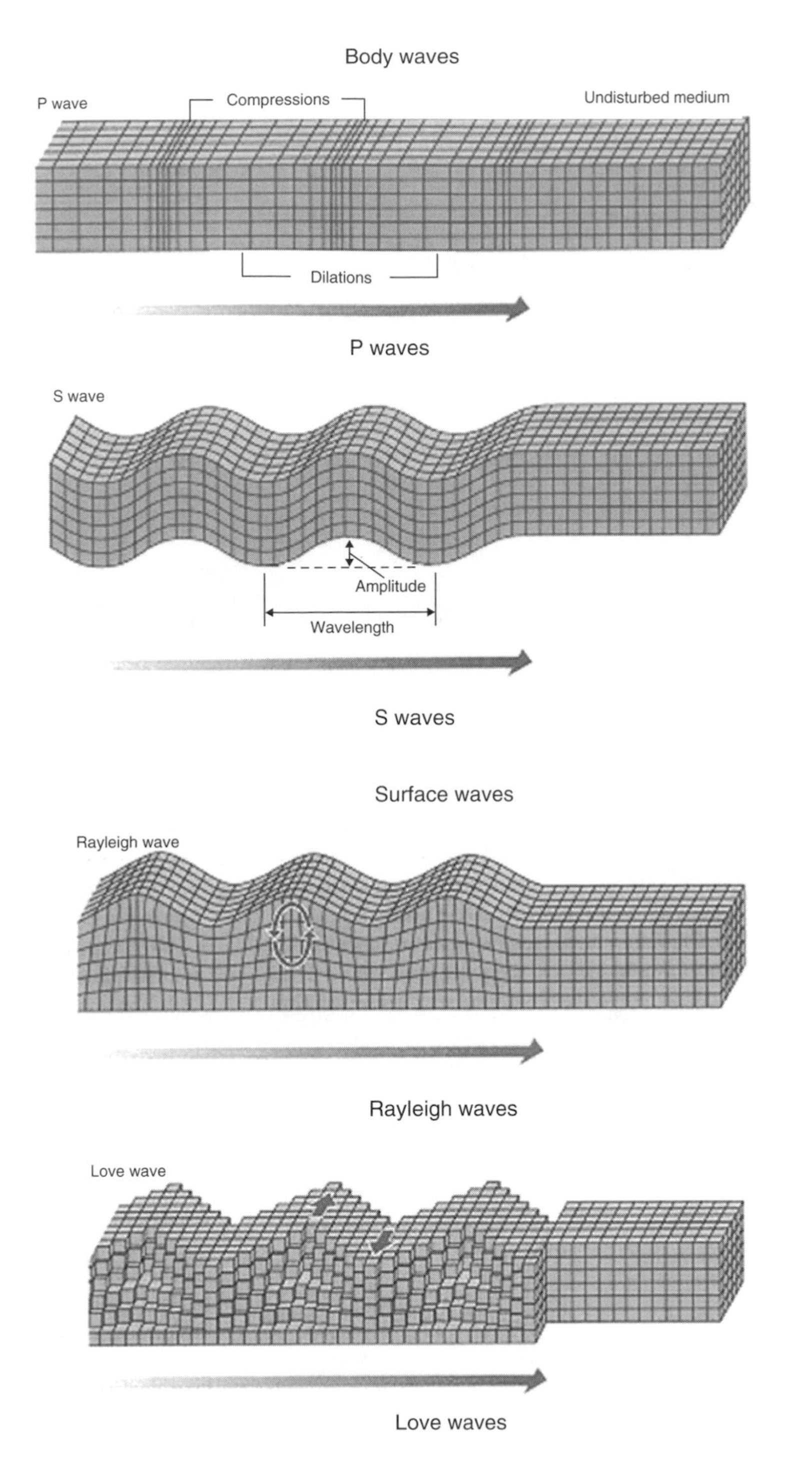

FIGURE 3.5
Types of seismic waves. (Courtesy of USGS.)

and forth in the direction of propagation. They are termed as longitudinal, compressional, or primary waves. Normally they are identified as primary waves (P waves) because they travel faster than any elastic waves and are the first to arrive at a distant point.

An indication of the compressive effect that seismic waves can have, even at great distances, has been reported by Rainer (1974). He made a comparison between the incidence of rock bursts occurring in a mine in Bleiberg, Germany, with earthquakes on a worldwide basis, and found a strong correlation between rock bursts at Bleiberg and the large shocks that occurred in Agadir (1960), Skopje (1969), San Fernando (1971), and Nicaragua (1972).

Shear, transverse, or secondary waves (S waves) are generated where the initial pressure pulse, or the P wave that it generates, strikes a free surface or a change in material in a direction other than normal. The shape of the transmitting material is then changed by shear rather than compression. S waves can travel only in a solid because their existence depends on the ability of the transmitting medium to resist changes in shape (the shear modulus). P waves can travel in any matter that resists compression or volume change to solid, liquid, or gas. The S waves move at slower velocities than P waves and arrive later at a distant point even though they are both generated at the same instant. Both P and S waves travel through the Earth in direct, refracted, or reflected paths, depending upon the material through which they are traveling.

Surface Waves or Long Waves (L Waves)

Long waves travel along the free surface of an elastic solid bounded by air or water. They are defined by the motion through which a particle in its path moves as the wave passes.

Rayleigh (R) waves cause the particles to move vertically in an elliptical orbit, or to "push up, pull down" in the direction of propagation.

Love (Q) waves cause the particles to vibrate transverse to the direction of wave advance, with no vertical displacement.

Both Rayleigh and Love waves move at slower velocities than P or S waves, and as they travel they disperse into rather long wave trains (long periods). (The comparative arrivals of P, S, and L waves are shown on the seismogram given in Figure 3.6.)

Propagation Velocity

The velocity with which seismic waves travel through the Earth is termed the propagation velocity, which can be expressed in terms of elastic moduli and material density for P waves (V_p) and S waves (V_s) as follows:

$$V_p = \sqrt{[K + (4/3)G]/\rho} \quad \text{m/sec}$$

$$V_s = \sqrt{G/\rho} \quad \text{m/s}$$

where K is the dynamic bulk modulus, G the dynamic shear modulus and ρ the material bulk density.

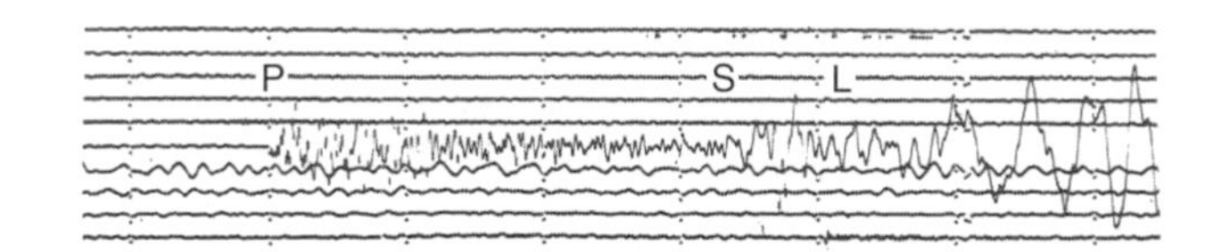

FIGURE 3.6
Seismograph from a long-period, vertical instrument for an event of September 26, 1959, located off the coast of Oregon and recorded in Tucson, Arizona. The long-period waves are well represented on the record and the P, S, and L waves are noted. The P and S waves have been damped (attenuated) as they traveled through the Earth. (From Neuman, 1966.)

Note: When these velocities are propagated synthetically in the field or in the laboratory, values for E_d, G_d, and K can be computed.

Characteristics

Seismic waves (Figure 3.4) may be described by the quantities of vibratory motion, i.e., amplitude, wavelength, period, and frequency. Amplitude and frequency are the two parameters commonly used to define vibratory motion in Earthquake studies. *Amplitude* A is the displacement from the mean position or onehalf the maximum displacement. *Wavelength* λ is the distance between crests. *Period* T is the time of a complete vibration, or the time a wave travels distance (λ), expressed as

$$T=1/f=2\pi/\omega \tag{3.1}$$

where f is the frequency and ω the circular frequency.

Frequency f is the number of vibrations per second (or oscillation in terms of cycles per unit of time), given normally in hertz (Hz) with units of cycles per second, expressed as

$$f=\omega/2\pi \quad \text{(Hz)} \tag{3.2}$$

Ground shaking is felt generally in the ranges from 20 Hz (high frequency) to less than 1 Hz (low frequency of long waves).

Circular frequency ω defines the rate of oscillation in terms of radians per unit of time; 2π rad is equal to one complete cycle of oscillations.

$$\omega=2\pi f=2\pi/T \tag{3.3}$$

3.2.3 Ground Motion

Elements

Ground motion occurs as the seismic waves reach the surface, and is described in terms of several elements derived from the characteristics of seismic waves, including displacement, velocity, and acceleration.

Displacement y at a given time t is a function of position x and time t in Figure 3.4, expressed as

$$y=A \sin 2\pi/\lambda\,(x+vt) \quad \text{cm} \tag{3.4}$$

Velocity v, termed the particle or vibrational velocity, is expressed as

$$v=\lambda/T=f\lambda, \text{ or } v=dy/dt=\dot{y} \quad \text{cm/sec} \tag{3.5}$$

Acceleration a is

$$\alpha=d^2y/dt^2=\ddot{y} \quad \text{cm/sý} \tag{3.6}$$

If Equation 3.5 is substituted for v and f is expressed in terms of the circular frequency ω from Equation 3.3, y, v, and a can be expressed as

$$\textit{ground displacement:} \quad y=A \sin \omega t \tag{3.7}$$

$$\text{ground velocity} \quad v = \dot{y} = \omega A \cos \omega t \tag{3.8}$$

$$\text{ground acceleration:} \quad \alpha = \ddot{y} = -\omega^2 A \sin \omega t = -\omega^2 y \tag{3.9}$$

Acceleration is given in terms of the acceleration due to gravity g: 1 g=32 ft/sec^2 or 980 cm/sec^2=980 gal where 1 gal=1 cm/sec^2.

Strong Ground Motion

Strong ground motion refers to the degree of ground shaking produced as the seismic waves reach the surface. It has an effect on structures, and is applied in both the horizontal and vertical modes.

Characteristics of ground motion are those of the waveform plus the duration of shaking.

- *Amplitude* is the differential along the wavelength and causes differential displacement of structures.
- *Wavelength*, when much larger (long period) than the length of the structure, will cause tall structures to sway, but differential displacement will be negligible.
- *Frequency* causes the shaking of structures as the wave crests pass beneath, and contributes to the acceleration magnitude.
- *Acceleration* is a measure of the force applied to the structure.
- *Duration* is the time of effective strong ground motion, and induces fatigue in structures and pore pressures in soils (Section 3.2.7).

Detecting and Recording

Seismographs

Seismic wave amplitudes are detected and recorded on *seismographs*. A seismometer, the detection portion of the instrument, is founded on rock and includes a "steady mass" in the form of a pendulum, which is damped. Seismic waves cause movement of the instrument relative to the pendulum, which remains stationary. In modern instruments, the movements are recorded electromechanically and stored on magnetic tape. Operation and recording are continuous.

Recording stations have "sets" of instruments, each set having three seismographs. Complete description of ground motion amplitude requires measurements of three components at right angles: the vertical component, and the north–south and east–west components. Instruments designed for different period ranges are also necessary, since no one instrument can cover all of the sensitivity ranges required. In North America it is common to have one set sensitive to periods of 0.2 to 2.0 sec, and another set sensitive to periods of 15 to 100 sec.

Seismograms are the records obtained of ground motion amplitudes; an example is given in Figure 3.6. The Richter magnitude (see Section 3.2.4) is assigned from the maximum amplitude recorded. The distance between the epicenter and the recording seismograph is determined from the arrival times of the P, S, and L waves. By comparing records from several stations, the source of the waves can be located in terms of direction and distance. Epicenters are calculated by National Oceanic Atmospheric Administration (NOAA) from information received from the worldwide network.

Seismographs are too sensitive to provide information of direct use in seismic design, and a strong earthquake near the normal seismograph will displace the reading off scale or even damage the instrument. Instruments are normally located on sound bedrock to eliminate local effects of soils or weakened rock structure, and therefore do not provide information on these materials.

Accelerographs (Strong-Motion Seismographs) (Accelerometers)

The *purpose* of the strong-motion seismograph is to provide ground-response criteria in an area of interest for the dynamic design of structures. Modern accelerographs measure and record the three components of absolute ground acceleration (one vertical, two horizontal) and are able to resolve peak accelerations to 0.1 cm/sec^2 or smaller. People at rest are able to feel motions as small as 1 cm/sec^2. In moderate magnitude earthquakes, damage to poorly designed structures occurs at accelerations of about 100 cm/sec^2 (10% g).

The *instrument* does not operate continuously, but is rather designed to begin operating and recording when affected by a small horizontal movement. The sensor is typically a damped spring–mass system. Through the 1970s the recording medium was usually photographic film and recording was slow. Modern designs convert an electrical signal into a digital format, which is recorded in a digital memory within the unit providing for rapid recording of accelerations. Accurate time recording is achieved by connection with satellites.

Location of most accelerographs is on the ground surface, and not necessarily on rock: therefore, data correlations between sites are difficult unless subsurface conditions are known for each. Many accelerographs are also located in buildings.

Accelerograms are the records obtained of ground accelerations g as illustrated in Figure 3.7. Ground motion displacements and velocities are then computed from the acceleration records by the integration of Equations 3.7 and 3.8.

The *peak horizontal ground acceleration* (PHGA) is the most common index of the intensity of strong ground motion at a site (Munfakh et al., 1998). The PHGA is directly related to

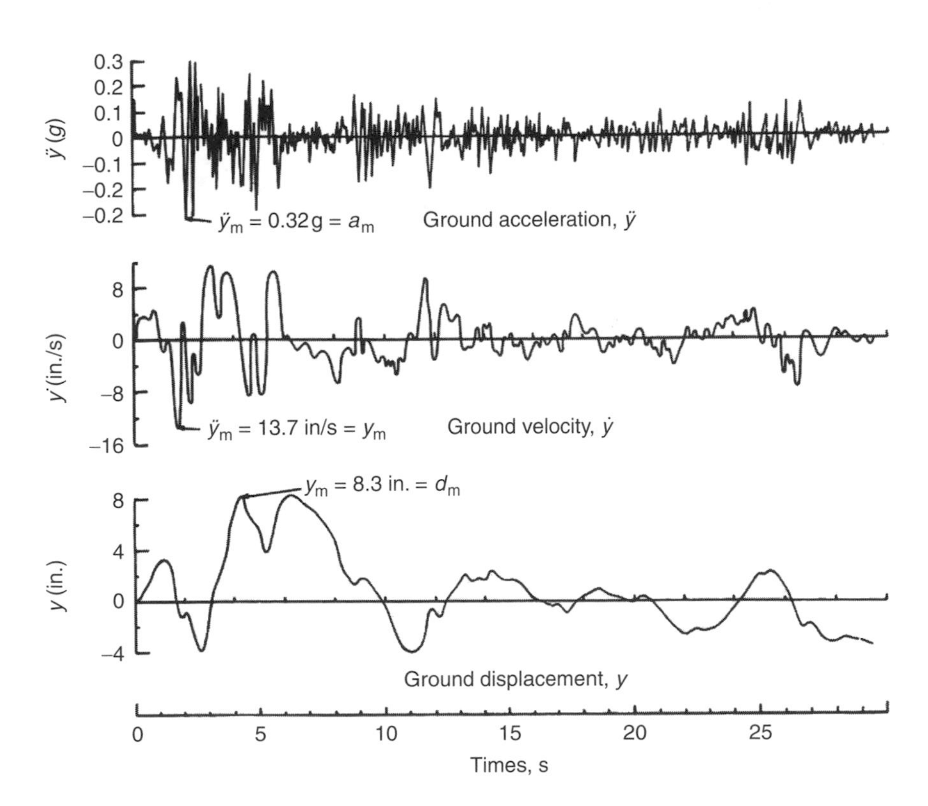

FIGURE 3.7
Strong ground-motion record of the N-S component from El Centro, California, earthquake of May 18, 1940. Ground acceleration is integrated to obtain velocity and displacement. (From USAEC, *Soil Behavior under Earthquake Loading Conditions*, National Technical Information Service TID-25953, U.S. Department of Commerce, Oak Ridge National Laboratory, Oak Ridge, Tennessee, January 1972. With permission.)

the peak inertial force imparted by strong shaking to a structure founded on the ground surface and to the peak shear stress induced within the ground itself. *Peak vertical ground acceleration* (PVGA) and *peak horizontal ground velocity* (PHGV) are used in some engineering analyses to characterize the damage potential of ground motions to buildings. The *peak horizontal ground displacement* (PHGD) may be used in the analysis of retaining walls, tunnels, and buried pipelines.

It is noted, however, that the recorded peak ground acceleration (PGA) is for an accelerograph placed at a particular location with particular conditions. Since accelerographs are now placed in arrays in many locations, isobars of recorded "*g*s" can be prepared.

Earthquakes with strong ground motions, recorded as of 1975, are summarized in Table 3.3. In the United States, El Centro (1940) remained for many years the severest ground motion recorded (0.32 g). The recording station was located 4 mi from the fault break. During the Parkfield event (1966) of $M=5.6$, accelerations of 0.5 g were recorded, but the acceleration was a single strong pulse. During the San Fernando quake (1971, $M=6.5$), an accelerograph registered the highest acceleration ever recorded at that time: in the 0.5 to 0.75 g range with peaks over 1.0. The instrument was located on the abutment of the Pacoima Dam, about 6 mi south of the epicenter, which was undamaged. Local topography and the location are believed to have influenced the very high peak acceleration (Seed et al., 1975). PGA $\approx$ 0.8 g was recorded for the Northridge event (1994, $M_w=6.7$) and the Kobe event (1995, $M_w=6.9$).

Network installations in the United States as of 2002 are given in Figure 3.8. Improvements in the instrument and costs has resulted in installations worldwide, and as of the year 2000 over 10,000 instruments were operating worldwide. The U.S. Geological Survey (USGS) is installing new networks under the "Advanced National Seismic

TABLE 3.3

Earthquakes with Strong Recorded Ground Acceleration[a]

Recording Station	Horizontal Distance to Epicenter (E) or Fault (F), km	Component	Maximum Acceleration (% gravity)	Remarks
May 16,1968, Japan; magnitude=7.9				
Hachinohe	ca. 200 (E)	N–S	24	Port area. Small shed
		E–W	24	Soft soil
July 21, 1952, Kern County, California; magnitude=7.7				
Taft	40 (E)	N21°E	15	In service tunnel between
		S69°E	18	buildings. Alluvium
October 17, 1966. Peru; magnitude=7.5				
Lima	200 (E)	N08°E	42	Small building. Coarse
		N82°W	27	dense gravel and boulders
April 13, 1949, Puget Sound, Washington; magnitude=7.1				
Olympia	16 (E)	S04°E	16	Small building. Filled land
		S86°W	27	at edge of Sound. Focal depth $h=50$ km
December 11, 1967, India; magnitude=6.5				
Koyna Dam	8 (E)	Along dam axis	63	Dam gallery
		Normal dam axis	49	

(Continued)

TABLE 3.3
(Continued)

Recording Station	Horizontal Distance to Epicenter (E) or Fault (F), km	Component	Maximum Acceleration (% gravity)	Remarks
January 21, 1970, Japan; magnitude=6.8				
Hiroo	18 (E)	E–W	44	Focal depth h=60 km
		N–S	≈30	
December 21, 1964, Eureka, California; magnitude=6.6				
Eureka	24 (E)	N79°E	27	Two-story building
		N11°W	17	Alluvium
August 6, 1968, Japan; magnitude=6.6				
Uwajima	11 (E)	Transverse	44	Itashima Bridge site
		Longitudinal	36	Soft alluvium
				Focal depth h=≈40 km
May 18, 1940, El Centro, California; magnitude=6.5				
El Centro	6 (F)	N–S	32	Two-story heavy reinforced concrete building with massive concrete engine pier. Alluvium
		E–W	21	
February 9, 1971, San Fernando, California; magnitude=6.5[b]				
Pacoima Dam Abutment	3 (F)	S14°W	115	Small building on rocky spine adjacent to dam abutment. Highly jointed diorite gneiss
		N76°W	105	
Lake Hughes Station No. 12	25 (E)	N21°E	37	Small building. 3 m layer of alluvium over sandstone
		N69°W	28	
Castaic Dam Abutment	29 (E)	N21°E	39	Small building. Sandstone
		N69°W	32	
March 10, 1933, Long Beach, California; magnitude=6.3				
Vernon	16 (E)	N08°E	13	Basement of six-story building. Alluvium
		S82°E	15	
December 23, 1972, Nicaragua; magnitude=6.2				
Managua	5 (F)	E–W	39	Esso Refinery. Alluvium
		N–S	34	
June 30, 1941, Santa Barbara, California; magnitude=5.9				
Santa Barbara	16 km (E)	N45°E	24	Two-story building
		S45°E	23	Alluvium
June 27, 1966, Parkfield, California; magnitude=5.6				
C–H No. 2	0.08 (F)	N65°E	48	Small building. Alluvium
		N25°W	Failed	
September 4, 1972, Bear Valley, California; magnitude=4.7				
Melendy Ranch	8.5 km (E)	N29°W	69	≈19 m from San Andreas fault. Small building Alluvium (no damage)
		N61°E	47	
June 21, 1972, Italy; magnitude=4.5				
Ancon[a]	Ca. 5 (E)	N–S	61	Rock
		E–W	45	

[a] From Bolt, B. A. et al., *Geological Hazards*, Springer, New York, 1975. Reprinted with permission of Springer.
[b] Maximum acceleration ≧ 0.15 g on 31 records within 42 km of the faulted zone.

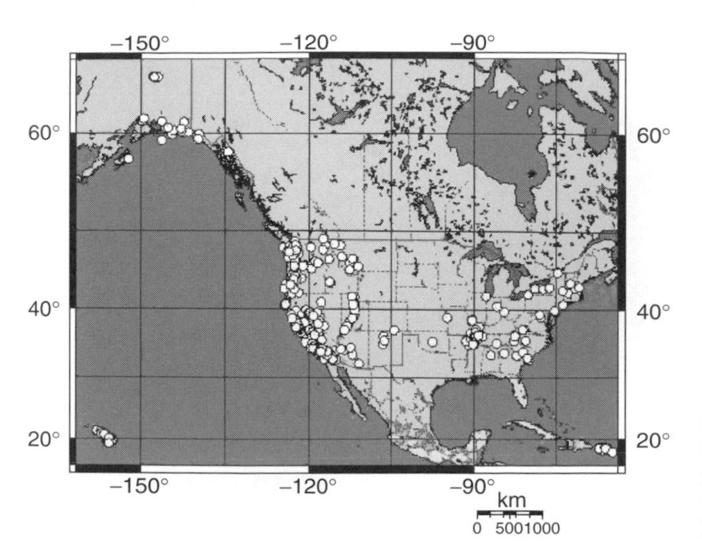

FIGURE 3.8
National Strong Motion Program recording stations as of April 10, 2002. (Courtesy National Strong Motion Program, USGS.)

System" (ANSS) with emphasis on locations with a history of high-intensity earthquake activity. A list of international sources of strong motion networks is available from the *National Center for Earthquake Engineering Research* (NCEER), in Buffalo, New York.

3.2.4 Intensity and Magnitude

Earthquake Strength Measurements

Two different scales are commonly used to provide a measure of Earthquake strength as related to ground motion forces at the surface: intensity and magnitude.

- Intensity is a qualitative value based on the response of people and objects on the Earth's surface. Given as "felt" reports, values and their geographic distribution very much reflect population density.
- Magnitude is a quantitative value computed from seismogram data. Presently, a number of different forms are recognized by seismologists.

Seismic moment is a parameter suggested in recent years to rate the strength of an Earthquake. It includes the rigidity of the rock in which the rupture occurs, times the length of fault face which moves, times the amount of slip. The San Fernando event (1971) has been computed to have a seismic moment of nearly 1026 erg.

Intensity (I or MM)

Modified Mercalli Scale of Intensity (MM)

Intensity scales were developed as a basis for cataloging the force of an event for comparison with others, and the change in force with distance (attenuation) from the epicenter. The first intensity scale was developed by DeRossi of Italy and Forel of Switzerland in 1883 (in the literature referred to as RF, followed by a Roman numeral representing the intensity).

The DeRossi–Forel scale was improved by Mercalli in 1902 and modified further in 1931. In 1956, Richter produced the version given in Table 3.4. It correlates ground motion with damage to structures having various degrees of structural quality. (The author has added the columns for approximate comparative peak ground velocity, acceleration, and magnitude. The magnitude is the Richter magnitude.)

TABLE 3.4
Modified Mercalli Scale, 1956 Version[a]

	Intensity	Effects	V[b](cm/sec.)	g[c]
M[d]	I	Not felt. Marginal and long-period effects of large earthquakes (for details see text)		
3	II	Felt by persons at rest, on upper floors, or favorably placed		
	III	Felt indoors. Hanging objects swing. Vibration like passing of light trucks. Duration estimated. May not be recognized as an earthquake		0.0035–0.007
4	IV	Hanging objects swing. Vibration like passing of heavy trucks; or sensation of a jolt like a heavy ball striking the walls. Standing motor cars rock. Windows, dishes, doors rattle. Glasses clink. Crockery clashes. In the upper range of IV wooden walls and frame creak		0.007–0.015
	V	Felt outdoors; direction estimated. Sleepers wakened. Liquids disturbed, some spilled. Small unstable objects displaced or upset. Doors swing, close, open. Shutters, pictures move. Pendulum clocks stop, start, change rate	1–3	0.015–0.035
	VI	Felt by all. Many frightened and run outdoors. Persons walk unsteadily. Windows, dishes, glassware broken. Knickknacks, books, etc., off shelves. Pictures off walls. Furniture moved or overturned. Weak plaster and masonry D cracked. Small bells ring (church, school). Trees, bushes shaken (visibly, or heard to rustle — CFR)	3–7	0.035–0.07
6	VII	Difficult to stand. Noticed by drivers of motor cars. Hanging objects quiver. Furniture broken. Damage to masonry D, including cracks. Weak chimneys broken at roof line. Fall of plaster, loose bricks, stones, tiles, cornices (also unbraced parapets and architectural ornaments — CFR). Some cracks in masonry C. Waves on ponds; water turbid with mud. Small slides and caving in along sand or gravel banks. Large bells ring. Concrete irrigation ditches damaged	7–20	0.07–0.15
	VIII	Steering of motor cars affected. Damage to masonry C; partial collapse. Some damage to masonry B; none to masonry A. Fall of stucco and some masonry walls. Twisting, fall of chimneys, factory stacks, monuments, towers, elevated tanks. Frame houses moved on foundations if not bolted down: loose panel walls thrown out. Decayed piling broken off. Branches broken from trees. Changes in flow or temperature of springs and wells. Cracks in wet ground and on steep slopes	20–60	0.15–0.35
7	IX	General panic. Masonry D destroyed; masonry C heavily damaged, sometimes with complete collapse; masonry B seriously damaged. (General damage to foundations — CFR.) Frame structures, if not bolted, shifted off foundations. Frames racked. Serious damage to reservoirs. Underground pipes broken. Conspicuous cracks in ground. In alluviated areas sand and mud ejected, earthquake fountains, sand craters	60–200	0.35–0.7
	X	Most masonry and frame structures destroyed with their foundations. Some well-built wooden structures and bridges destroyed. Serious damage to dam, dikes, embankments. Large landslides. Water thrown on banks of canals, rivers, lakes, etc. Sand and mud shifted horizontally on beaches and flat land. Rails bent slightly	200–500	0.7–1.2

(*Continued*)

TABLE 3.4
(Continued)

	Intensity	Effects	V^b(cm/sec.)	g^c
8	XI	Rails bent greatly. Underground pipelines completely out of service		>1.2
	XII	Damage nearly total. Large rock masses displaced. Lines of sight and level distorted. Objects thrown into the air	From Figure	3.12

Note: Masonry A, B, C, D. To avoid ambiguity of language, the quality of masonry, brick or otherwise, is specified by the following lettering (which has no connection with the conventional Class A, B, C construction).

- *Masonry A:* Good workmanship, mortar, and design: reinforced, especially laterally, and bound together by using steel, concrete, etc.; designed to resist lateral forces.
- *Masonry B:* Good workmanship and mortar; reinforced, but not designed to resist lateral forces.
- *Masonry C:* Ordinary workmanship and mortar; no extreme weaknesses such as nontied-in corners, but masonry is neither reinforced nor designed against horizontal forces.
- *Masonry D:* Weak materials, such as adobe; poor mortar; low standards of workmanship; weak horizontally.

[a] From Richter, C. F., *Elementary Seismology*, W.H. Freeman & Co., San Francisco, 1958. Adapted with permission of W. H. Freeman and Company.
[b] Average peak ground velocity (cm/sec).
[c] Average peak acceleration (away from source).
[d] Magnitude correlation.

Data Presentation

Isoseismal maps are prepared for affected areas showing zones of equal intensities. The intensity distribution for the Kern County shock of July 21, 1952, has been overlaid on a physiographic diagram of southern California in an attempt to show some relationship with geologic conditions in Figure 3.9.

Regional seismicity maps are also prepared on the basis of intensities (Figure 3.3 and Figure 3.10) and in some cases developed into seismic risk or seismic hazard maps (Figure 3.13). In recent years these maps have been prepared in terms of either the Richter magnitude or effective peak acceleration (Figure 3.14).

Magnitude (M)

The Richter Scale

The concept of magnitude was developed in 1935 by C. F. Richter for defining the total energy of seismic waves radiated from the focus based on instrumental data for shallow earthquakes in southern California. He defined the magnitude of local earthquakes M_L as "the logarithm to the base 10 of the maximum seismic wave amplitude (in thousands of a millimeter) recorded with a standard seismograph at a distance of 100 km from the earthquake epicenter." A nomograph, based on Richter's equation to determine magnitude, is given in Figure 3.11. It is considered applicable to moderate-size earthquakes, $3 < M_L < 7$ (USGS, 2003).

Other Magnitude Scales

Because of the limitations of the Richter Magnitude (M_L), a number of other magnitudes have been defined in recent years (Scawthorn, 2003). All of the currently used methods for measuring Earthquake magnitude yield results consistent with M_L (USGS, 2003):

- *Body wave magnitude* (M_b) used to measure small, local events up to about M_L=6.5 and is based on the amplitude of the P body waves.

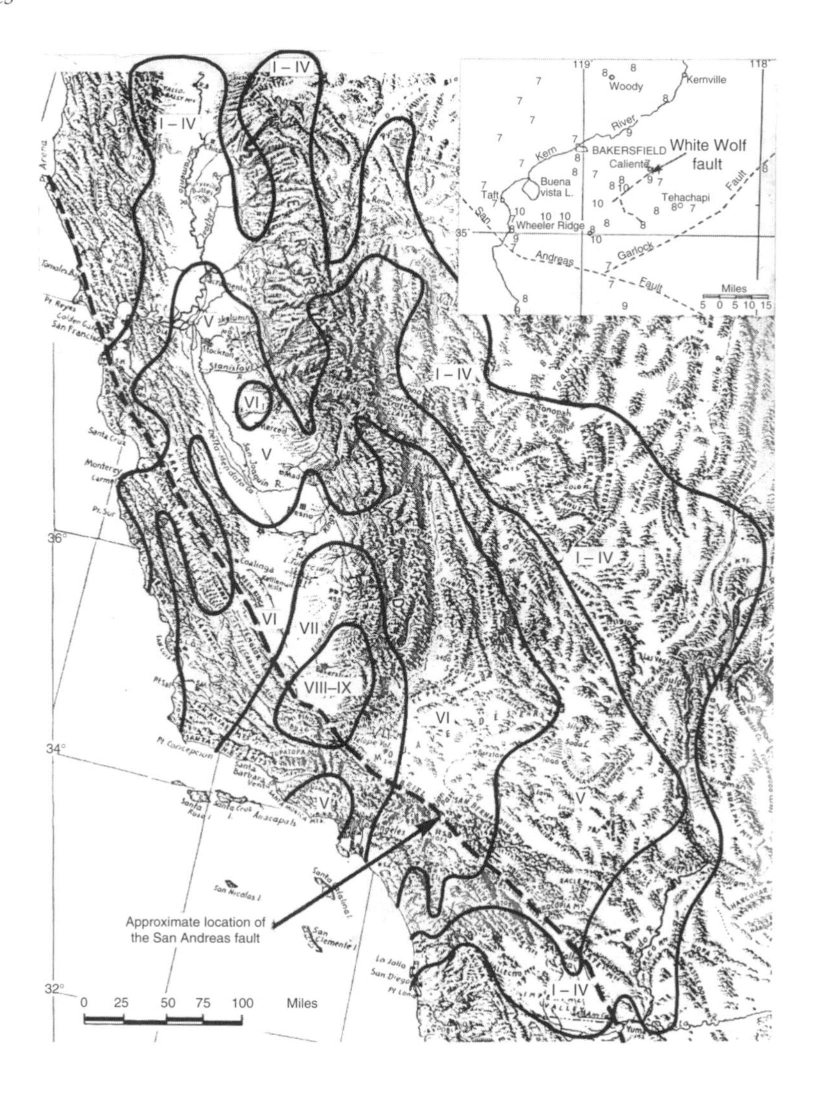

FIGURE 3.9
Intensity distribution of the July 21, 1952, earthquake in Kern Country, California, which occurred along the "inactive" White Wolf fault. (After Murphy, L.M. and Cloud, W.K., *Coast and Geodetic Survey, Serial No. 773*, U.S. Govt. Printing Office, 1954.) The map has been overlaid onto the physiographic diagram of southern California for comparison with geologic conditions as revealed by physiography. (Physiographic diagram from Raisz, E., *Map of the Landforms of the United States*, 4th ed., Institute of Geographical Exploration, Harvard University, Cambridge, Massachusetts, 1946.)

- *Surface wave magnitude* (M_s) used to measure larger quakes ($>M_L$ =6.5 to ~ 8.5).
- *Moment magnitude* (M_W) which is not based on seismometer readings, but on the energy released by the earthquake, termed the *seismic moment* (M_o). M_o=rock rigidity (shear modulus) × average physical area of the fault × the distance of fault slip.

Significance of Magnitude

Amplitudes vary enormously among different Earthquakes. An increase in one magnitude step has been found to correlate with an increase of 30 times the energy released as seismic waves (Bolt et al., 1975). An earthquake of magnitude 8.0, for example, releases almost 1 million times the energy of one of magnitude 4.0, hence the necessity for a logarithmic scale. The largest quakes have had a magnitude of 8.9 (Table 3.1). In general, magnitudes

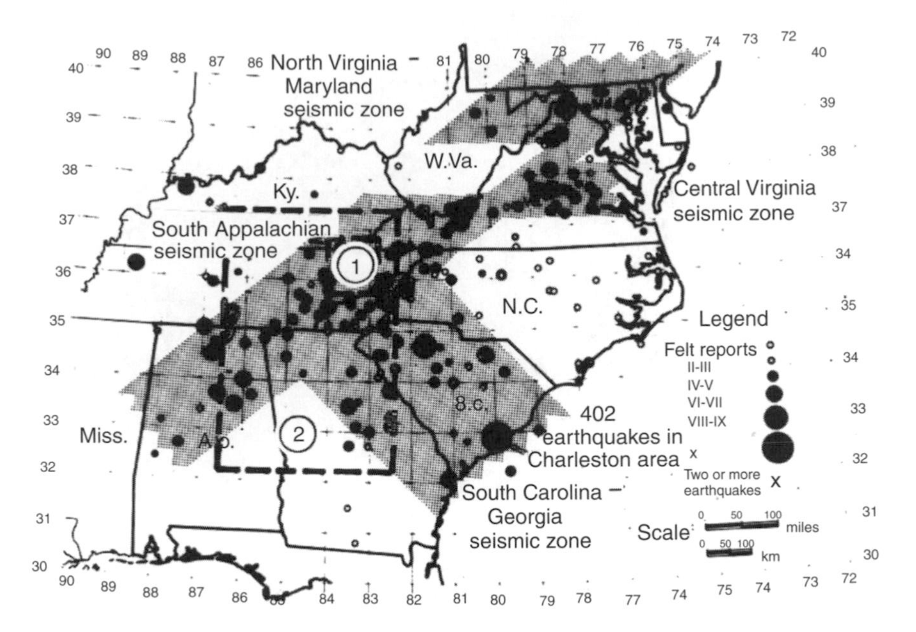

FIGURE 3.10
Seismicity (1754–1971) and earthquake zone map for the southeastern United States. (From Bollinger, G. A., *Am. J. Sci.*, 273A, 396–408, 1973. With permission.) Epicenters shown by open and solid circles; zones shown by stippling.

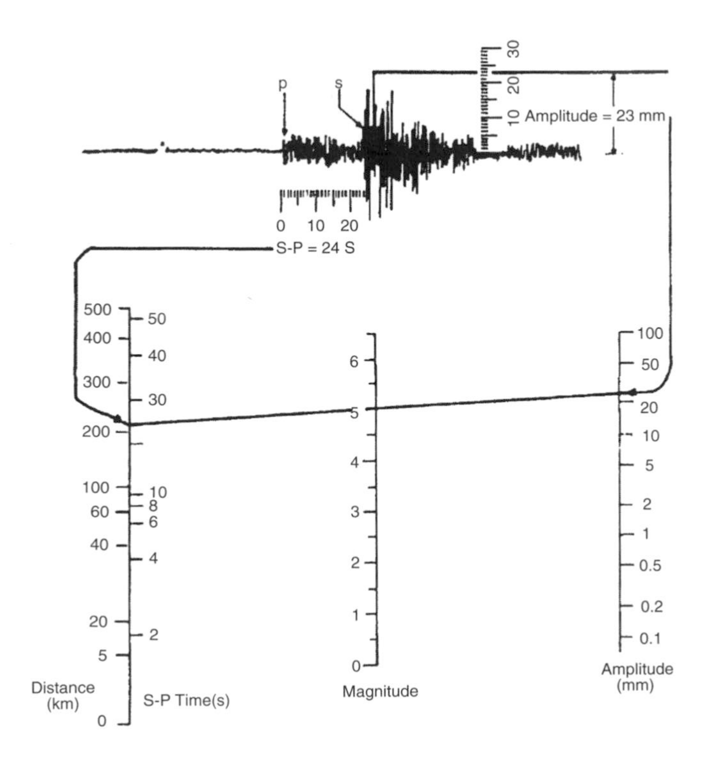

FIGURE 3.11
A nomogram illustrating the relationship between the amplitude in a Wood–Anderson seismograph and the Richter magnitude. The equation behind the nomogram was developed by Richter from Southern California earthquakes. (After Richter, C. F., *Elementary Seismology*, W. H. Freeman & Co., San Francisco, 1958. Reprinted with permission of W. H. Freeman.)

greater than 5.0 generate ground motions sufficiently severe to cause significant damage to poorly designed and constructed structures, and $M=4$ is generally considered as the damage threshold.

Magnitude is not, however, a measure of the damage that may be caused by an earthquake, the effect of which is influenced by many variables. These include natural conditions of geology (soil type, rock depth and structure, water-table depth), focal depth, epicentral distance, shaking duration, population density, and construction quality.

Correlations with Magnitude

Empirical correlations between M, I_o, and g are given in Figure 3.12, where I_o signifies epicentral intensity. Near the source there is no strong correlation with g (Bolt, 1978). Acceleration has also been related to magnitude by the expression given by Esteva and Rosenblueth (1969) as follows:

$$g = 2000\,e^{0.8M} R^{-2} \tag{3.10}$$

where R is the focal distance in kilometers and g is in cm/sec^2 per 1000 cm/sec^2.

Seismic Risk Maps

An early seismic risk map is given in Figure 3.3 with correlations between zones, maximum accelerations, and magnitude M. Zone 3 was considered as running a high risk of damaging earthquakes; zone 2, moderate risk; zone 1, low risk; and zone 0, essentially no risk. It was subsequently updated by Algermissen and others in 1969 and the boundary

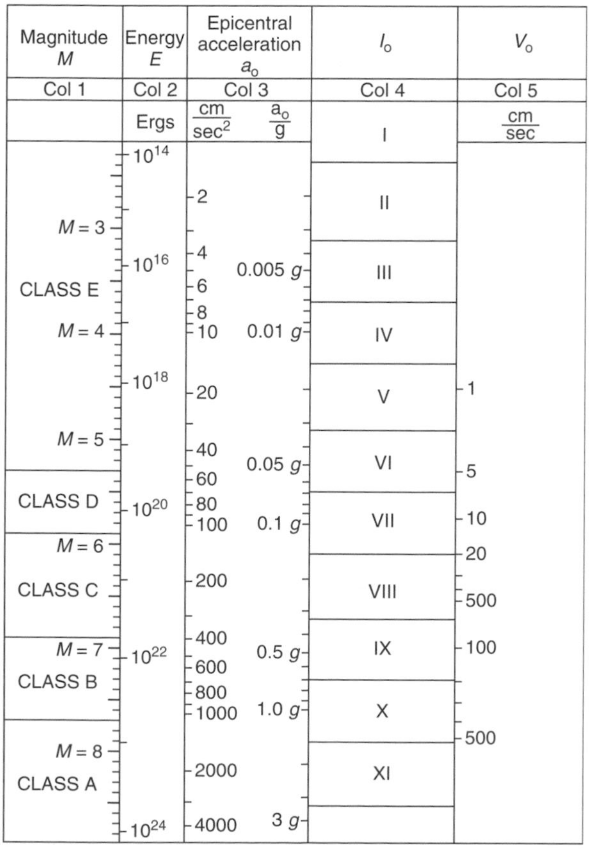

Magnitude M	Energy E	Epicentral acceleration a_o		I_o	V_o
Col 1	Col 2	Col 3		Col 4	Col 5
	Ergs	cm/sec^2	a_o/g	I	cm/sec
	10^{14}				
		2		II	
M = 3					
	10^{16}	4	0.005 g	III	
CLASS E		6			
		8			
M = 4		10	0.01 g	IV	
	10^{18}	20		V	1
M = 5		40	0.05 g	VI	5
		60			
CLASS D	10^{20}	80			
		100	0.1 g	VII	10
M = 6					20
CLASS C		200		VIII	500
M = 7	10^{22}	400	0.5 g	IX	100
CLASS B		600			
		800			
		1000	1.0 g	X	
M = 8					500
CLASS A		2000		XI	
	10^{24}	4000	3 g		

FIGURE 3.12
A summary of rough relationships between magnitude, energy, and epicentral acceleration and between acceleration, intensity, and ground velocity. Approximations are for an order of magnitude. (From Faccioli, E. and Resendiz, D., *Seismic Risk and Engineering Decisions*, Lomnitz, C. and Rosenblueth, Eds., Elsevier, New York, 1976, pp. 71–140, Chap. 4.)

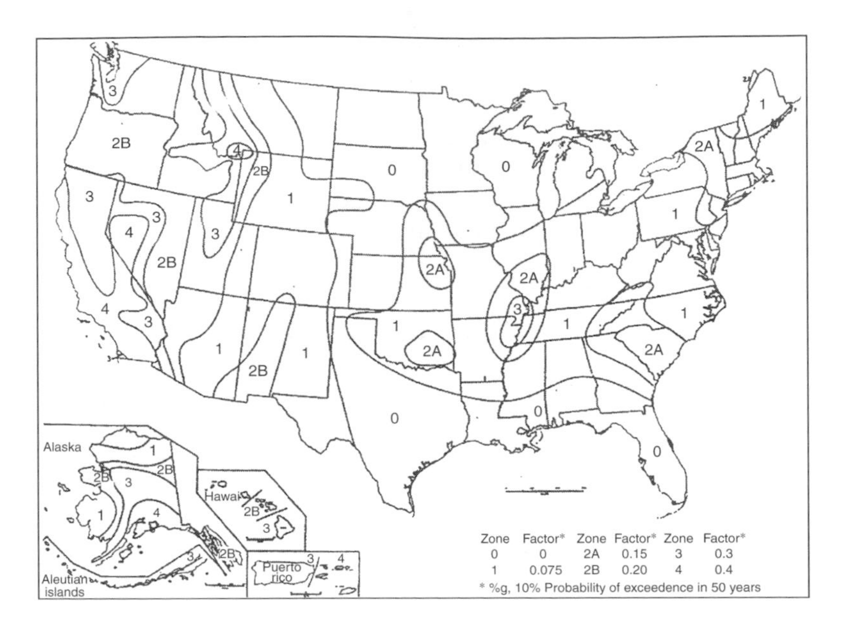

FIGURE 3.13
Seismic zone map of the United States. (From the International Conference of Building Officials, 1988.)

lines changed to incorporate new data. A recent seismic zone map of the Continental United States is given in Figure 3.13.

A recent seismic risk map for the United States is given in Figure 3.14. It presents contours of effective peak rock acceleration with a 90% probability of not being exceeded in a 50 year period.

3.2.5 Attenuation

Description

Attenuation is the decay or dissipation of energy or intensity of shaking with distance from the source, occurring as the seismic waves travel through the Earth, and results in the *site intensity of rock excitation*. In both deterministic and probabilistic seismic hazard analyses calculations are made of the ground motion parameter of interest at a given site from an earthquake of a given magnitude and site-to-source distance.

The epicentral area extends for some distance about the epicenter, in which there is no attenuation, then with increasing distance there is wide regional variation in intensity distribution. It is affected by geology, topography, and length of fault rupture. Variations are illustrated by isoseismal maps such as in Figure 3.9, Figure 3.15, or Figure 3.31; or by seismicity maps such in Figure 3.10, which are used to develop attenuation relationships. Figure 3.15 presents a comparison of intensity distributions from two earthquakes of different magnitudes but with fairly close epicenters.

Estimations

Theoretical Relationships

Previously, attenuation relationships were given in terms of intensity; at present most attenuation "models" are based on strong ground motion. Theoretical relationships are used to develop attenuation relations in areas where there are an insufficient number of

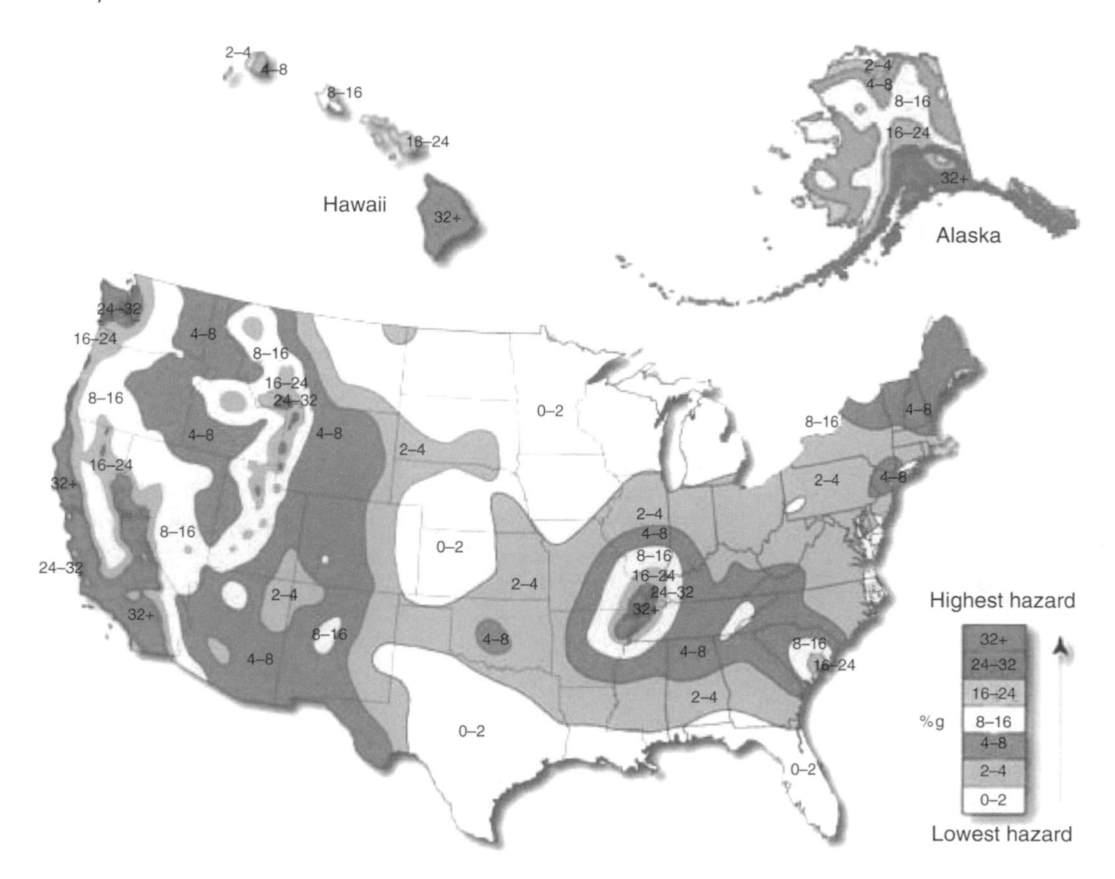

FIGURE 3.14
Peak acceleration (%*g*) with 10% probability of exceedence in 50 years. (Courtesy of USGS, National Seismic Hazard Mapping Project.)

strong motion recordings. The most common is the stochastic method in which the Fourier amplitude spectrum (FAS) of the average component of ground motion is described by the general relation (Campbell, 2003)

$$A(f) = \mathrm{Src}(f)\ \mathrm{Attn}(f,R)\ \mathrm{Amp}(f) \tag{3.11}$$

where Src(f) describes the earthquake source (magnitude), Attn(f,R) describes the attenuation caused by wave propagation through the crust (site-to-source distance), and Amp(f) describes the response of materials (geological conditions) beneath the site. Relatively complex equations describe each of the elements of Equation 3.11.

Engineering Models

A large number of attenuation relations have been developed to prepare engineering estimates of strong ground motion throughout the world. Campbell (2003) describes four models for "shallow active crust" in Western North America, three for "shallow stable crust" in Eastern North America, and several others for Europe, Japan, and worldwide. The various parameters in these models include ground motion, magnitude, distance, depth, faulting mechanisms, the site geologic conditions, fault hanging and foot-wall locations, source (fault) directivity (conditions), plus a number of miscellaneous parameters, described in detail by Campbell (2003). He notes that "all engineering models have limitations that results from the availability of recordings, the theoretical assumptions

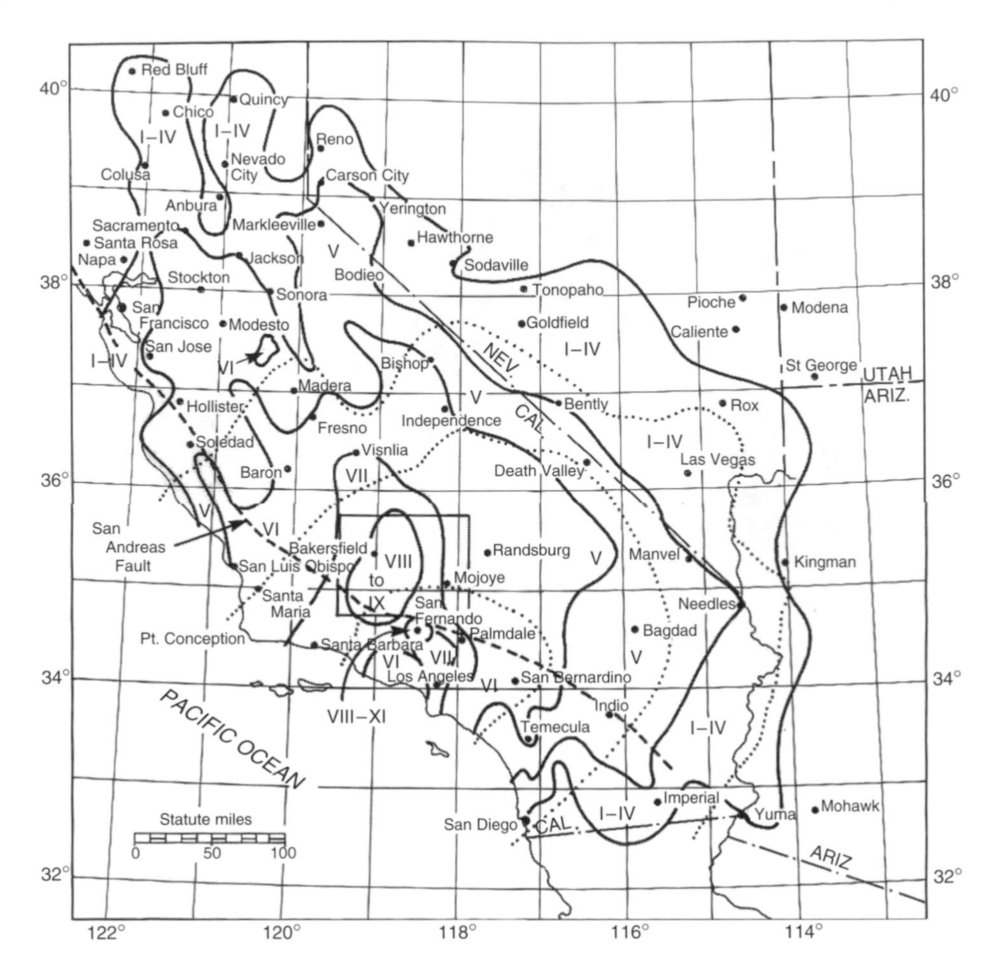

FIGURE 3.15
Isoseismal map showing the intensity distribution of the Kern County event of 1952 ($M = 7.7$) and that of the San Fernando event of 1971 ($M = 6.6$) (dashed lines). The differences in energy attenuation are clearly related to the differences in magnitude between the two earthquakes. (From United States Geological Survey.)

used to develop the models, and the seismological parameters used to define the source, path, and site effects" (see also Munfakh et al., 1998).

Graphs and Charts

A family of attenuation curves giving magnitude and acceleration as a function of distance from the source for the western, central, and eastern United States is given in Figure 3.16. Recent publications (Munfakh et al., 1998) provide similar relationships. A comparison of attenuation relationships for a strike-slip fault in the western United States by various investigators are given in Figure 3.17. The "models" used by the investigators can be found in Campbell (2003). Attenuation as a function of distance from the rupture of a causative fault is given in Figure 3.28 and Figure 3.29.

Comments

Attenuation does relate in many instances to focal depth; very shallow focus events will be felt over relatively small areas. However, geology and topography are also significant factors. In a comparison of earthquakes of similar intensities in the eastern United States

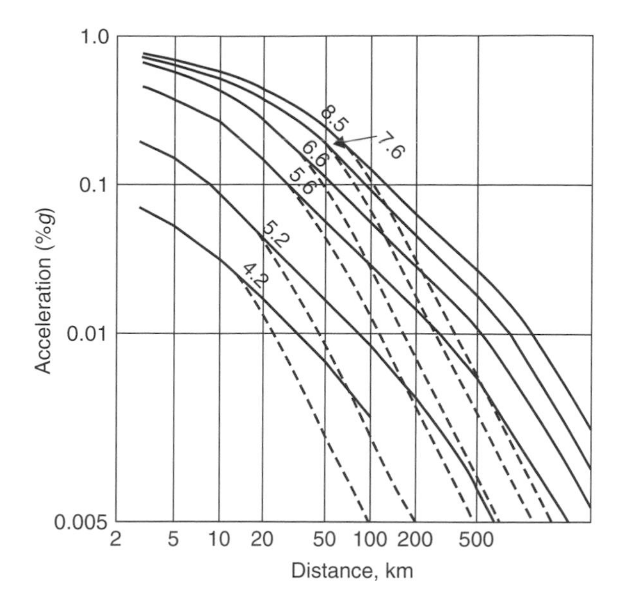

FIGURE 3.16
Acceleration attenuation curves for the United States. The solid lines are curves for the eastern region (east of longitude 105°). The dashed lines together with solid lines at close distances are the attenuation curves used for the western United States and are taken from Schnable and Seed (1973). It is to be noted that under certain conditions the area of shaking in the eastern United States is very much larger than in the western regions under similar earthquake conditions (see Figure 3.18). (From Algermissen, S. T. and Perkins, D. M., U.S. Geological Survey, Open File Report 76–416, 1976. With permission.)

with those in the western states, it can be shown that those in the east often affect areas 100 times greater than those in the west as illustrated in Figure 3.18. In addition, those in the east are not often associated with evidence of surface faulting (Nuttli, 1979).

3.2.6 Amplification

Description

Ground Amplification Factor

Site intensity is often amplified by soil conditions. An increase in ground acceleration with respect to base rock excitation is termed the ground amplification factor.

Stable Soil Conditions

Under conditions where the soils are stable (nonliquefiable), the influence of local soil conditions on ground motions can take the form of dynamic amplification, which can result in an increase in peak amplitudes at the surface or within a specific layer. The duration of shaking may also be increased. The factors influencing the occurrence are not well understood, although, in general, amplification is a function of soil type (densities and dynamic properties are most significant), depth, and extent. Attenuation may occur under certain conditions.

Unstable (liquefiable) soils are described in Section 3.3.3.

Influencing Factors

Soil Type and Thickness

Zeevaert (1972) concluded that in the valley of Mexico City, acceleration in the lacustrine soils is approximately two times larger than in the compact sand and gravel surrounding

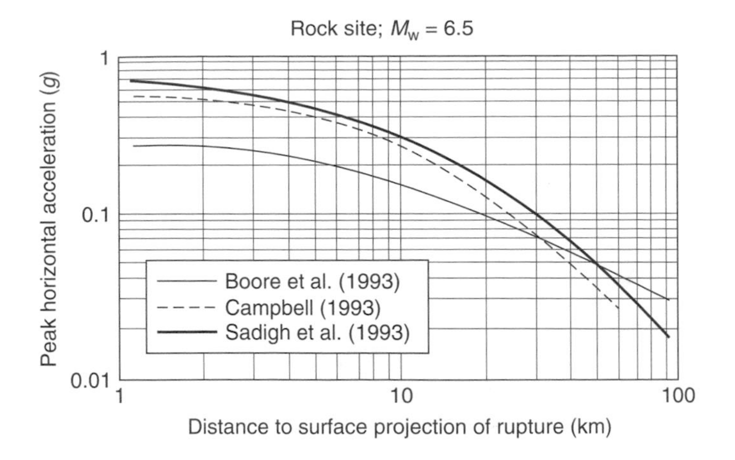

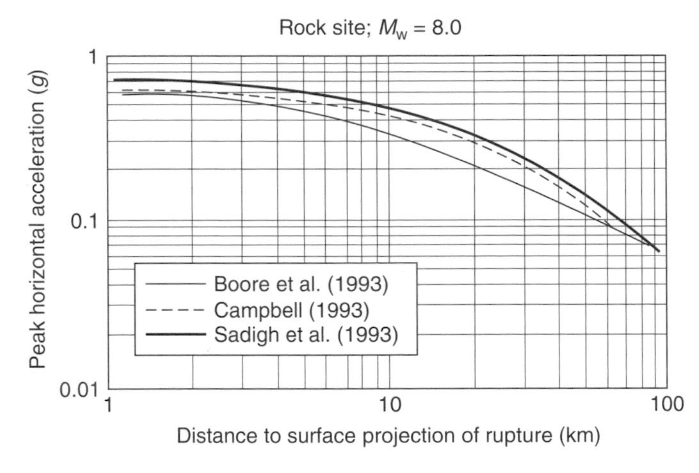

FIGURE 3.17
Comparison of mean value PHGA attenuation curves for M_w=6.5 and 8.0 events on a strike-slip fault calculated by three commonly used attenuation relationships for western United States. (From Munfakh, G. et al., *Geotechnical Earthquake Engineering Reference Manual*, Report No. FHWA-HI 99–012, FHA, Arlington, Virginia, 1998, Chap. 4.)

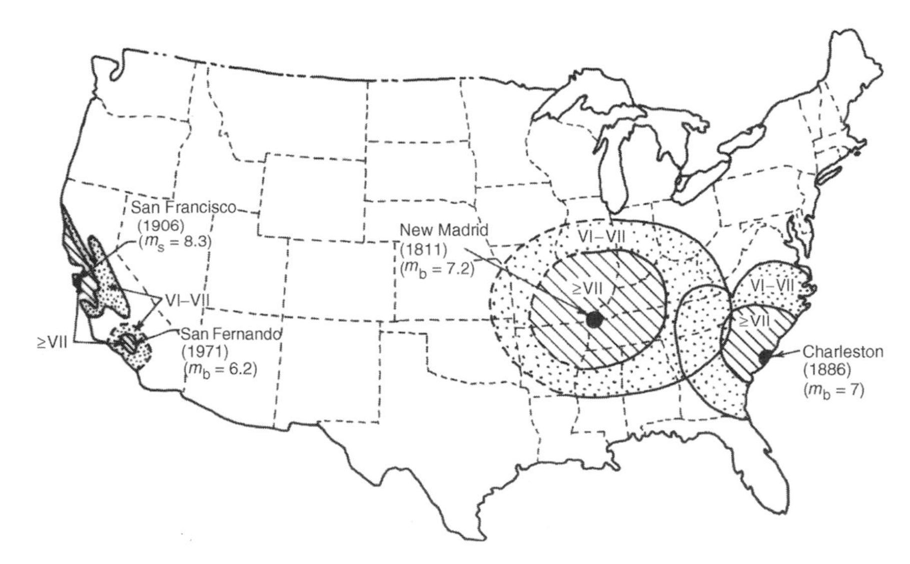

FIGURE 3.18
Comparison of areas of minor (I=VI – VII) and major (I>VII) damage for four major U.S. earthquakes. The damage area for the western half of the New Madrid event is inferred because there were no settlements in the area at that time (m_b=body wave magnitude; m_s=surface wave magnitude). (After Nuttli, O. W., *Reviews in Engineering Geology*, Vol. IV, Geological Society of America, Boulder, Colorado, 1979, pp. 67–93.)

the valley. This difference would probably result in intensities of VII in the lacustrine soils and VI in the sand and gravel. The spectral accelerations (PHGA) recorded during the September 19, 1985 Mexico City event for various subsurface conditions are given in Figure 3.19 (Response Spectra, Section 3.4.4). Since the earthquake epicenter (M= 8.0) was approximately 400 km distant from Mexico City, the rock and hard soil spectral accelerations were less than 0.1 g. At sites SCT and CAO, underlain by deep deposits of typical Mexico City soft clays and sands, it is shown that soil amplification was three to six times the rock response. The buildings at the CAO site generally had fundamental periods outside of the recorded periods and suffered minor damage. The buildings at the SCT site, however, had fundamental periods similar to the recorded periods, of the order of 1 to 2 s, and suffered major damage and even collapse (Bray, 1995).

Ground-motion records from a peat layer in Seattle showed that motion was amplified for a distant event (150 mi) but attenuated for one nearby (Seed, 1970). A possible explanation may be that the seismic waves from the distant event pass parallel to the layering in the peat, whereas those from nearby areas pass upward through the peat (the effect of different period characteristics).

Foundation Depth

Ground motions are usually given for the surface but design requires ground motion at the foundation level. Seed et al. (1975) present data showing that during the Tokyo–Higashi–Matsuyana earthquake of July 1, 1968, accelerations recorded for structures near the ground surface were on average about four times larger than for buildings founded at a depth of about 24 m.

Source Distance

The amplifications at close distances from the source appear to be more influenced by topographic expression and geologic structure than by local soil conditions. For large epicentral distances from the source, however, local amplification can be considerable, as it can be for motions of smaller intensities and stratigraphies characterized by sharp contrasts of seismic impedance (Faccioli and Rese'ndiz, 1976).

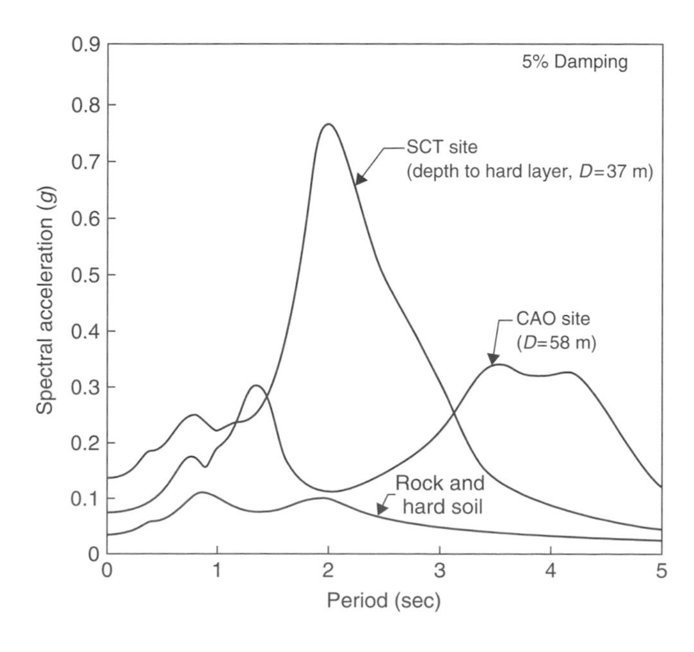

FIGURE 3.19
Acceleration response spectra for motions recorded in Mexico City during the 1985 event. (After Seed, H. B. et al., Earthquake Engineering Research Center Report No. UCB/EERC–87/15, University of California, Berkeley, 1987.)

Ground Amplification Factors

The Uniform Building Code (UBC, 1994) provides the normalized response spectrum for various soil types given in Figure 3.20. Such relationships should be considered as guides. Seismic response during the Loma Preita earthquake (1989) yielded spectral amplification factors for deep stiff clay sites in the order of 3 to 8, indicating that the seismic hazard at these sites may be underestimated (Bray, 1995).

Idriss (1991) studied the amplification of the peak ground acceleration recorded for the Mexico City and Loma Preita earthquakes. The results are given in Figure 3.21. Idriss concluded that the large accelerations for soft soil should tend to decrease rapidly as rock accelerations increase above about 0.1*g*.

Ground amplification factors are most reliably obtained from accelerograms from the site or from sites having similar conditions, and even these must be evaluated in the light of source distance and foundation depths. The database will improve with increasing strong motion data from accelerographs.

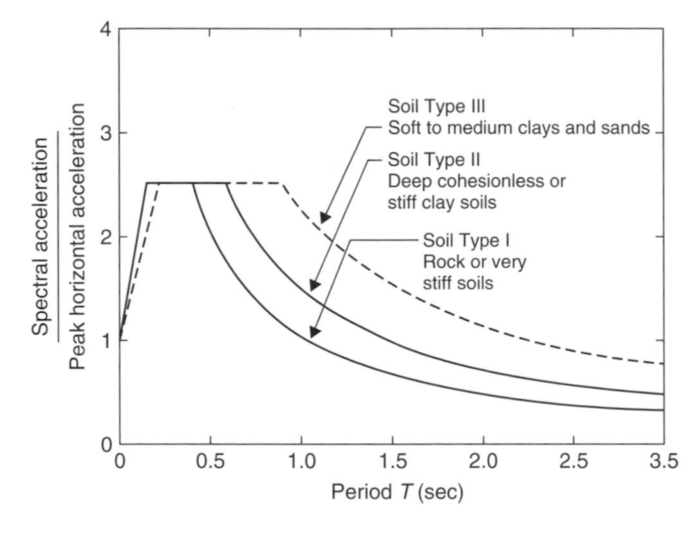

FIGURE 3.20
UBC (1994) normalized acceleration response spectra. (After Bray, J. D., *The Civil Engineering Handbook*, CRC Press, Boca Raton, Florida, 1995, Chap. 24, Figure 24.3.)

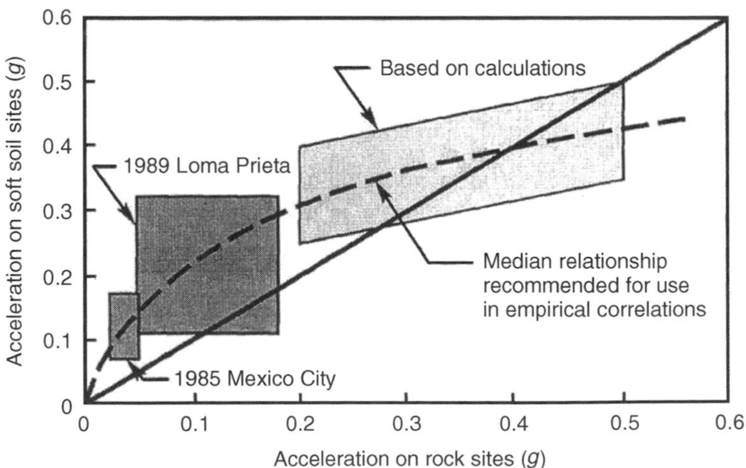

FIGURE 3.21
Variations in peak horizontal accelerations (PHGA) at soft soil sites with accelerations at rock sites. (After Idriss, I. M., *Proceedings of the 2nd International Conference on Recent Advances in Geotechnical Engineering and Soil Dynamics*, St. Louis, Missouri, 1991, pp. 2265–2273.)

Many cities, particularly in California, have prepared maps of the city giving ground amplification factors based on soil and rock conditions. They are available on the Internet.

3.2.7 Duration

The duration of strong ground motion plays a direct role in the destruction caused by an earthquake. It is a function of the size of fault rupture and fault type, path from the source to the site, and site geology.

Some examples illustrate the variability of duration:

- The San Francisco event (1906) started with a relatively small ground motion, which increased to a maximum amplitude at the end of about 40 sec, stopped for 10 sec, then began again more violently for another 25 sec.
- Agadir, Morocco (1960) was essentially destroyed in 15 sec and 12,000 of a population of 33,000 were killed.
- Guatemala City (1976) was first struck by an event of magnitude 6.5 which lasted for 20 sec. The first quake was followed by a number of smaller shocks (aftershocks). Two days later, two more shocks occurred, one with a magnitude of 7.5, and in the following week more than 500 shocks were registered.
- Anchorage, Alaska (1964) experienced a duration of the order of 3 min, which resulted in widespread slope failures where they had not occurred before, even though the area had been subjected to strong ground motion on a number of occasions (see Section 1.2.6 and Section 3.3.4).

Strong ground-motion duration is currently generally defined by (1) bracketed duration and (2) significant duration. An example is given in Figure 3.22. The *bracketed duration* is

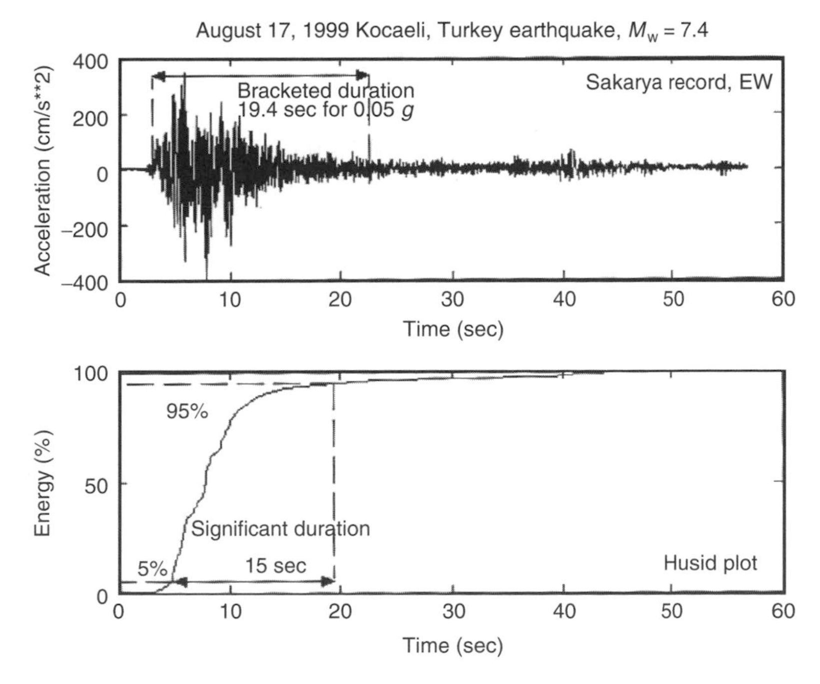

FIGURE 3.22
Bracketed duration and significant duration illustrated for the Sakarya record of August 17, 1999 Kocaeli, Turkey (M_w=7.4) earthquake. (From Erdik, M. and Durukal, E., *Earthquake Engineering Handbook,* Chan, W. and Scawthorn, C., Eds., CRC Press, Boca Raton, Florida, 2003. With permission.)

the interval between the two points in time where the acceleration amplitude is between the record start to 0.05 g. The *significant duration* is defined as the time required to build up from 5 to 95% of the energy in the ground-motion acceleration. A measure of the energy is the integral of $\int a^2\, dt$ for the total duration of the record, where a is the ground motion acceleration (Erdik and Durukal, 2003). Bolt (1973) provided relationships between bracketed duration, source to site distance, and magnitude given in Table 3.5.

3.2.8 Recurrence and Forecasting

General

Prediction Basis

Forecasting the location, magnitude, and time of occurrence of an Earthquake is the role of the seismologist, and is necessary for seismic design and for the early warning of an impending event.

A number of factors are considered in forecasting events:

- Statistical analysis of historical data (recurrence analysis).
- Measurements of fault movements, crustal warping, and stress increases.
- Changes in seismic wave velocities (dilatancy theory).
- Changes in the Earth's magnetic field and other geophysical properties.

Seismic Risk Analysis

Seismic risk analysis is based on probabilistic and statistical procedures to assess the probable location, magnitude, occurrence, and frequency of earthquake occurrence. Procedures require evaluation of historical records and of the regional and local geology, particularly with respect to faults and their activity. The recurrence of events of various magnitudes is examined, and then the attenuation relationships are evaluated to allow the development of the probability of ground motion at the site for various magnitudes in terms of the geologic conditions (Donovan and Bornstein, 1978).

In recent years, emphasis has been placed on "probabilistic seismic hazard analysis" (PSHA) (Figure 3.50). The methodology quantifies the hazard at a site from all earthquakes

TABLE 3.5

Bracketed Duration in Seconds[a]

	Magnitude						
Distance (km)	**5.5**	**6.0**	**6.5**	**7.0**	**7.5**	**8.0**	**8.5**
10	8	12	19	26	31	34	35
25	4	9	15	24	28	30	32
50	2	3	10	22	26	28	29
75	1	1	5	10	14	16	17
100	0	0	1	4	5	6	7
125	0	0	1	2	2	3	3
150	0	0	0	1	2	2	3
175	0	0	0	0	1	2	2
200	0	0	0	0	0	1	2

[a] From Bolt, B. A., *5th World Conference on Earthquake Engineering*, Rome, 1973. With permission.

Note: 1. *Bracketed duration*: The elasped time, for a particular frequency range, between the first and last acceleration excursions on an accelerogram record greater than a given amplitude level.

2. Acceleration >0.05 g, frequency >2 Hz.

of all possible magnitudes, at all significant distances from the site of interest, as a probability by taking into account their frequency of occurrence (Thenhaus and Campbell, 2003).

Statistical Analysis and Recurrence Equations

Limiting Factors

Prediction of an event for a given location during investigations is usually based on statistical analysis of recorded historical events, but the limitations in the accuracy of such predictions must be recognized. It is known where earthquakes are likely to occur from recorded history, but it must be considered that major events can occur in areas where they would be totally unexpected. New Madrid and Charleston, for example, are essentially singular events.

Comparing the span of modern history and its recorded events with the span of even recent geologic history results in the realization that data are meager as a basis for accurate prediction. For example, activity can be cyclic. A region can apparently go through several centuries without seismic activity and then enter a period with numerous events. The Anatolian zone of Turkey, with 2000 years of recorded events, now is an active seismic area, although it has had periods of inactivity for as long as 250 years (Bollinger, 1976). By comparison, the history of significant population in the United States is scarcely 250 years, not adequately long for effective predictions.

On a historical basis, records of events are closely related to population density and area development, especially for events of moderate to low magnitudes felt over limited areas.

General Recurrence Relationships

Occurrence frequency of shocks of any given magnitude for the world in general and most of the limited areas that have been studied is roughly about 8 to 10 times that for shocks about one magnitude higher. The relationship (Richter, 1958; Lomnitz, 1974) can be represented by

$$\log_{10} N = a - bM \tag{3.12}$$

where N is the number of shocks of magnitude M or greater per unit of time, a, b are constants for a given area based on statistical analysis of recorded data,

$a = \log_{10} N(0)$, or the logarithm of the number of earthquakes greater than $M=0$ for a given time period, given in units of earthquakes per year and

$b = \log_{10}\{[1 - F(M)]/M\}$ where $F(M)$ is the cumulative probability distributionof earthquake magnitudes.

On a semi-log plot for $5.5 < M < 8.5$, Equation 3.12 plots as a straight line for number of events against M. Studies by the Japan Meteorological Society found a drop off in the number of events above $M=7.5$. Data were for the period of 1885 to 1990 (Scawthorn, 2003).

Recurrence relation is also found expressed in terms of I_o as

$$\log N = \alpha - \beta I_o \tag{3.13}$$

where N is the annual number of earthquakes with epicentral intensities equal to or greater than I_o and α, β are the empirical constants which describe the decay rate of occurrence with increasing epicentral intensity in a manner similar to a and b in Equation 3.12 (Christian et al., 1978).

Some Regional Relationships

Recurrence equations have been developed for various regions expressing the number of earthquakes per year (N) in terms of the maximum magnitude M or MM intensity I, for the magnitude range of interest. The data are taken from seismicity maps such as Figure 3.10. Statistically, the computed number of events per year for a given magnitude is usually presented as a return time once in so many years. The general relationships vary from region to region with geologic conditions, stress level, and perhaps magnitude. The variation in stress level is important in that both short-term and secular changes in earthquake frequency are thereby permitted (Bollinger, 1976).

In the *southeastern United States* (Bollinger, 1976), from Figure 3.10,

$$\log N = 3.01 - 0.59 I_o \text{ (for } V \le I_o \le \text{VIII)} \tag{3.14}$$

The frequency of occurrence for events of various intensities expressed by Equation 3.14 is given in Table 3.6. Several interpretations are possible from these data: the region is overdue for the occurrence of a damaging shock (I_o=VII or VIII), there is a change toward a lower level of activity, or the maximum intensity in some of the historical data has been overestimated.

For the *Ramapo Fault* in New Jersey (period 1937–1977) (from Aggarwal and Sykes, 1978),

$$\log N = 1.70 \pm 0.13 - 0.73M \tag{3.15}$$

Equation 3.15 gives a recurrence of shocks of M=7.0 of once every 97 years.

Early Warning Indicators

General

There is no certain way of predicting where or when an earthquake may occur, although a number of tentative methods are under long-term study, most of which are related to subtle geologic changes with time.

Geologic changes occurring with time include:

- Fault displacement, tilting, or warping of the surface.
- Stress increase in fault zone or in surface rocks.
- Fluctuation of gravitational or magnetic fields above normal levels (Before the Hollister, California event of November 1974, of M=5.2, the magnetic field rose above the normal level.).
- Change in arrival times of transient P waves (dilatancy theory).
- Change in radon emissions from soils and subsurface waters.

Animal reactions are considered significant by the Chinese. It appears that domestic animals can sense microseisms, and shortly before an earthquake they become highly nervous. The Chinese have even evacuated cities in recent years on the basis of animal reactions preceding an event.

TABLE 3.6

Frequency of Earthquake Occurrence in the Southeastern United States[a]

I_o	Return Period (years)	Years Since Last Occurrence	Number Expected per 100 Years
V	0.9	0	115
VI	3.4	1	30
VII	13.0	48	8
VIII	51.0	63	2
(IX, X)	(200, 780)[b]	(90)	(0.5, 0.1)

[a] From Bollinger, G. A., *Proc. ASCE, Numer. Methods Geomech.*, 2, 917–937, 1976. With permission.

[b] Extrapolated values in parentheses.

Dilatancy Theory

The *dilatancy theory* (or V_p/V_s anomaly, or seismic velocity ratio method) was based on the observation that the arrival times of transient P waves traveling through the Earth's crust undergo a gradual decrease when compared with the arrival times of S waves, until just before an earthquake. Then the arrival time difference returns to normal relatively quickly and is followed by the shock (Scholtz et al., 1973; Whitcomb et al., 1973). The *time period* from the return to normal of the P wave velocity until the actual event has been found by examination of pre-earthquake records to be roughly onetenth the time interval during which the decreasing velocities occurred. If the time of the initial decrease and the return to normal are known, predictions as to when the earthquake is likely to occur and what the anticipated magnitude is likely to be can be made.

Rock-mass behavior under stress provides an explanation for the dilatancy theory. As the crustal pressures preceding a quake approach the failure point in rock masses under high stress, a myriad of tiny cracks open. This causes the decrease in the velocity of the P waves, since their velocity is reduced when they travel across air-filled openings. As groundwater seeps into the cracks, the velocity increases until all of the cracks are filled and velocity returns to normal. The presence of the water "lubricates" the cracks, reducing rock strength, permitting failure, and producing the earthquake. Some investigators now believe that the stress level on faults is quite low and dilatancy, if it occurs, is likely to be confined to small areas of stress concentration and not spread over significant volumes of rock where it can be readily measured.

Deep-well injection has apparently verified the "lubrication" effect (which is in reality probably a pore-pressure effect). A series of shocks occurred between 1962 and 1967 near Denver, Colorado, following the pumping of liquid wastes down a borehole into rock at a depth of 3 mi below the Rocky Mountain arsenal, a region where earthquakes were almost unknown. After the waste pumping was suspended, the number of events declined sharply. A similar experiment was carried out by the USGS at Chevron Oil Company's oil field in Rangely, Colorado, in 1972. Water was forced under high pressure into a number of deep wells and a series of minor earthquakes occurred. Activity ended immediately when the water was pumped from the wells (Raleigh et al., 1972).

Surface Warping

Overstresses in the Earth's crust cause surface warping, which may predate an earthquake. Records from the literature regarding the phenomenon appear meager. Data are available on ground surface elevation changes for many locations in the United States from the Vertical Division Network, National Geodetic Survey, Silver Springs, Maryland.

Niigata, Japan: Japanese geologists reported that a land area near Niigata had risen 13 cm in 10 years before the 1964 event (M=7.5).

Palmdale, California: Measurements by the USGS determined that an area of about 4500 km^2 around Palmdale rose by as much as 45 cm between 1959 and 1974 as shown in Figure 3.23. The area, now known as the Palmdale bulge, is centered on the San Andreas Fault. Subsequent data indicated that between 1974 and 1977, Palmdale had dropped by 18 cm (Hamilton, 1978). Two earthquakes of M=5.7 and 5.2, which centered on the bulge, occurred on March 15, 1979 (ENR, 1979).

Research and Monitoring Networks

Space-based instruments image Earth movements to millimeters, measuring the slow buildup of deformation along faults, and mapping ground deformation after an earthquake. The primary methods are the global positioning system (GPS) navigation system and Interferometric Synthetic Aperture Radar. Numerous GPS systems have been, and are

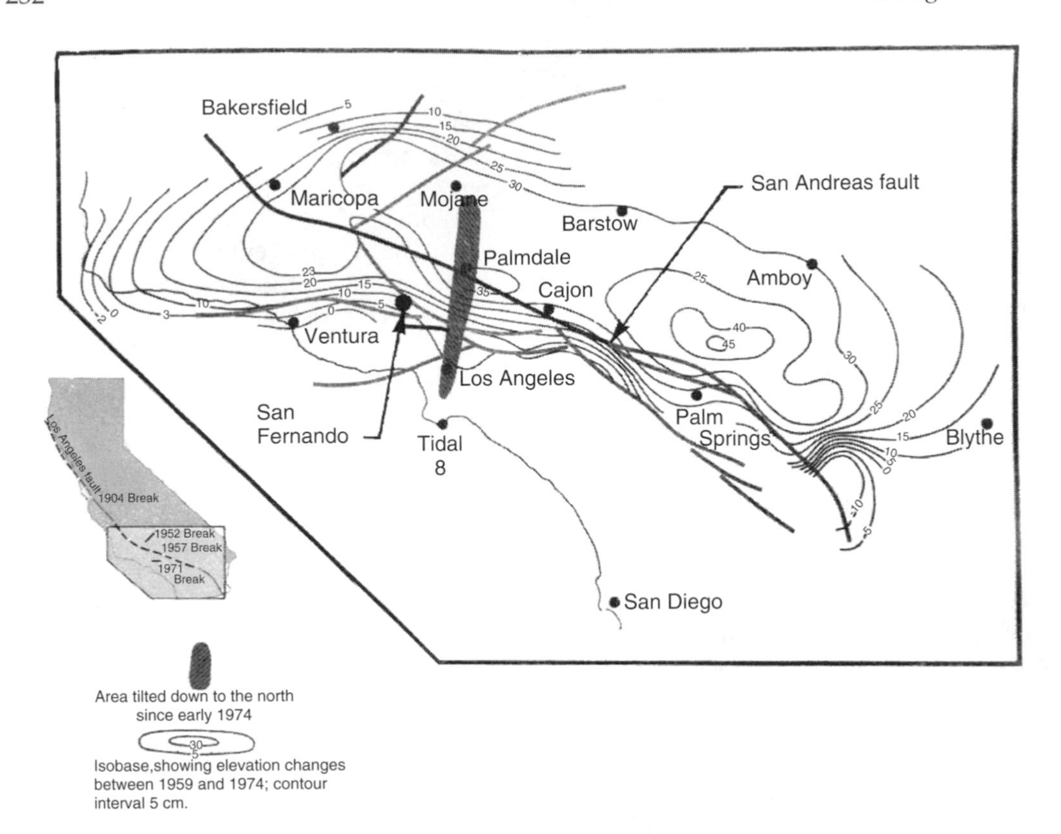

FIGURE 3.23
Contours of surface warping (cm), area of Palmdale, California. Area uplifted 45 cm between 1959 and 1974, but between 1974 and 1977 an area at Palmdale had dropped 18 cm. Recent surveys show that the uplifted area is larger than previously thought, and that the shape and size of uplift change with time. (From Hamilton, R. M., Geological Survey Circular 780, U.S. Department of the Interior, 1978. With permission.)

being installed, around historically seismically active areas. They continuously monitor and record horizontal and vertical movements. InSAR complements GPS data by providing an overview of whole regions on a periodic basis. In 2000, NASA began the Global Earthquake Satellite System (GESS) study, which advocates the installation of a number of satellites to transmit InSAR imagery.

National Center for Earthquake Research (USGS) Menlo Park, California, was established in 1966. An array of monitoring instruments were installed along the Hayward Fault in the San Francisco Bay area and the San Andreas Fault in Parkfield (1985). Included for continuously monitoring movements are creepmeters for near-surface movement, tiltmeters for ground rotation and tilting, dilatometers for volumetric contraction and extension, borehole tensor strainmeters for directional contraction and extension, and GPS systems.

The San Andreas Fault Observatory at Depth (SAFOD) is a deep borehole observatory proposed to measure the physical conditions under which plate boundary earthquakes occur. Planned for 2004, it is designed to directly sample fault zone materials, and monitor the San Andreas Fault zone near Parkfield. Moderate-size earthquakes of about $M=6$ have occurred on the Parkfield section of the fault at fairly regular intervals: 1857, 1881, 1901, 1922, 1934, and 1966. The location of Parkfield is given in Figure 3.24a; an intensity map of the 1966 event is given in Figure 3.24b.

A 3.2-km-deep hole will be drilled through the fault zone close to the hypocenter of the $M=6$ Parkfield 1966 quake, where the fault slips through a combination of

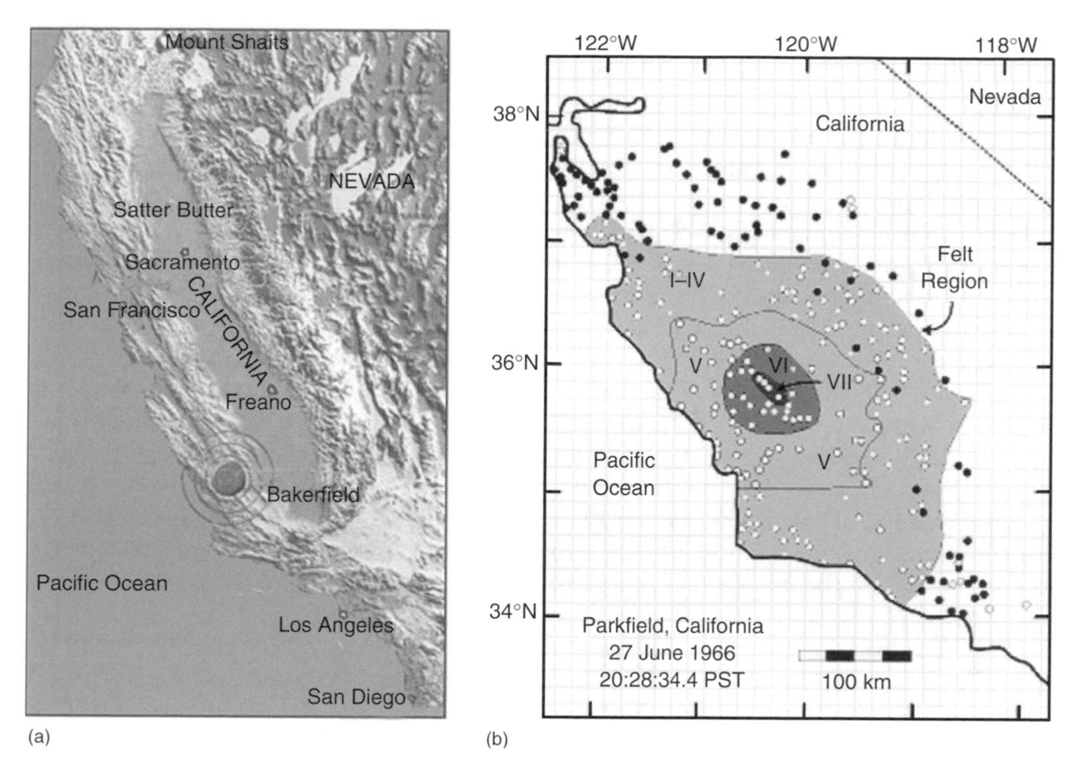

FIGURE 3.24
Parkfield, California: (a) location map, (b) intensity distribution of 1966 earthquake. (Courtesy of USGS.)

small-to-moderate magnitude earthquakes. The location will be sufficiently far from the fault to allow for drilling and coring deviated holes through the fault zone starting at a depth of about 3 km, continuing through the fault zone until relatively undisturbed rock is reached on the other side. In addition to sampling the rock and fluids, continuous monitoring will be made of pore pressure, temperature, and strains within and adjacent to the fault.

In addition to the USGS, NASA, and the University of California at Berkeley, earthquake research is being conducted by MCEER (Multidisciplinary Center for Earthquake Engineering Research), headquartered at the University of Buffalo.

3.3 Surface Effects on the Geologic Environment

3.3.1 Faulting

General

Importance in Earthquake Engineering

Shallow-focus earthquakes, usually the most destructive, are frequently associated with faulting, which can consist of a single main fracture, or of a system including subsidiary fractures. Fault identification is an important element in studies that aim to evaluate the probability of earthquake occurrence and magnitude.

Modern earthquake engineering places a lot of emphasis on evaluations of fault criteria, particularly to arrive at estimated values of magnitude and ground motion. There is much to learn, however, about faults. Some faults considered as dead or inactive have been the location of quakes, such as the White Wolf Fault in California and the Ramapo Fault in New Jersey. Some faults are buried and undetected such as the fault beneath the Northridge event.

The North Antolian Fault in Turkey has been the subject of a number of studies because of its unusual history (Okumura et al., 1993). Similar to the San Andreas in length, its strike-slip form, and its long-term movement rates, the fault has been the location of eight events of M=6.8 to 7.8 between 1939 and 1999 along a 900 km length (Figure 3.25). Of particular interest is the fact that rather than being located in a few areas, the earthquakes have been migrating westward in sequence from Erzincan (1939, M=7.8) to Izmit (1999, M=7.4). The Izmit event is reported to have resulted in at least 13,000 deaths and estimates of up to 35,000 missing (Associated Press release). Radiocarbon dating in deep trenches dug at several locations by a joint Japan–Turkey research team recognized eight events dating from about 30 B.C. with a recurrence interval estimated between 200 and 300 years (Okumura et al., 1993).

Correlations have been made from earthquake data in some geographic regions (principally in the United States) to develop a number of relationships:

- Length of fault rupture vs. earthquake magnitude.
- Distance from the causative fault vs. the acceleration due to gravity.
- Fault displacement vs. magnitude.

Fault Study Elements

During engineering studies for seismic design, the following aspects related to faulting are considered:

- Positive identification that a fault (or faults) is present.
- Fault activity: establish the "capable fault" by judging if it is potentially active or inactive.
- Displacement amount and form (dip-slip, strike-slip, etc.) that might be expected.
- Earthquake magnitude that might be generated by rupture (related generally to length).
- Estimated site acceleration after attenuation from the capable fault.

Fault Activity (The Capable Fault)

Significance

In recognition that shallow-focus events are associated with faulting, but that many ancient faults are not under stress and therefore are "dead" or inactive and not likely to be the source of a shock, it becomes necessary, in order to predict a possible earthquake, to identify a fault as active, potentially active, or inactive (dead). Seismic design criteria are often based on the identification of active or potentially active faults (capable faults) and their characteristics.

The Capable Fault

U.S. Nuclear Regulatory Commission (NRC, 2003) defines the *Capable Tectonic Source* as a tectonic structure that can generate both vibratory ground motion and tectonic surface deformation, such as faulting or folding. It is considered as capable if there is the presence of

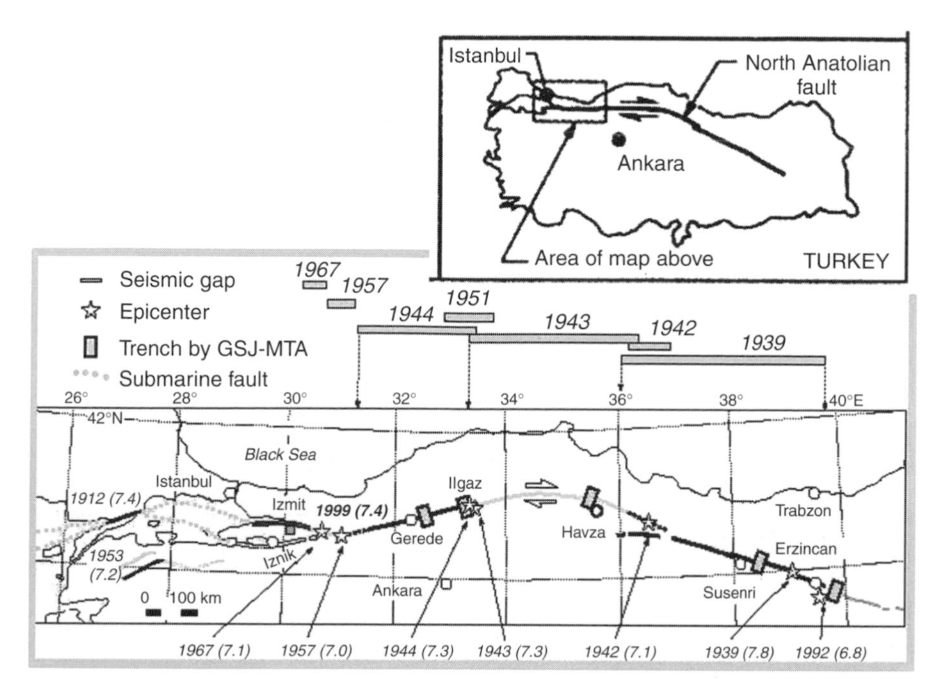

FIGURE 3.25
Rupture history of the North Anatolian Fault 1939 to 1967 and location of the 1999; $M=7.4$, Izmit earthquake. (After Okumura, K. et al., U.S.G.S. Open File Report, 94–568, 1993, pp. 143–144.)

surface or near-surface deformation of a recurring nature within the last approximately 500,000 years, or at least once in approximately the last 35,000 years; if there is a reasonable association with sustained earthquake activity; or if there is a connection to a capable tectonic source.

International Atomic Energy Commission (*IAEC*) considers a fault capable if it has undergone movement in late Quaternary, if there is topographic evidence of surface rupture, if there are instrumentally recorded and located events, if there is creep along the fault, or if it is connected to a capable fault.

Japan grades active faults according to the amount of displacement per unit of time, creep along the fault, movement during the Quaternary, and expected future movement.

Identification and Classification

The general criteria for recognition of an active fault are given in Table 3.7, the methods of dating the minimum age of the last fault displacement are listed in Table 3.8, and a system for the classification of fault activity based on available data is given in Table 3.9.

Limitations in Identification

Stratigraphy: In glaciated areas, it may be difficult to find a well-stratified Holocene section older than 50,000 years; therefore, the favored approach of trenching to permit examination for bedding ruptures in Holocene strata may not be applicable. In addition, fault displacement does not always extend to the surface; weathered rock and soil near the surface can sometimes adsorb the slip. After the Alaska event (1964), locations were found where 2 m of displacement were adsorbed by 20 m of weathered rock (Bolt et al.,1975). In such cases, trenches may prove inconclusive.

TABLE 3.7

Criteria Used for Recognizing an Active Fault[a]

General Criteria	Specific Criteria
Geological	Active fault indicated by the following features: *Young geomorphic features*: Fault scarps, triangular facets, fault scarplets, fault rifts, fault slice ridges, shutter ridges, offset streams, enclosed depressions, fault valleys, fault troughs, side-hill ridges, fault saddles *Ground features*: Open fissures, "mole tracks" and furrows, rejuvenated streams *Subsurface features*: Stratigraphic offset of Quaternary deposits, folding or warping of young deposits, en echelon faults in alluvium, groundwater barriers in recent alluvium
Historical	Description of past earthquakes, surface faulting, landsliding, fissuring, and other phenomena from historical manuscripts, news accounts, and other publications. Indications of fault creep or geodetic monument movements may be indicated in recent reports
Seismological	High-magnitude earthquakes and microearthquakes, when instrumentally well-located, may indicate an active fault. A lack of known earthquakes cannot be used to indicate that a fault is inactive

[a] After Cluff et al. (1972).

TABLE 3.8

Some Methods of Dating the Minimum Age of Last Displacements on Faults[a]

- Determining the age of undisplaced strata overlying the fault through the use of fossils, radiometric dating, or paleomagnetic studies
- Determining the age of cross-cutting undisturbed dikes, sills, or other intrusions
- Determining the rate of development of undisturbed soil profiles across a fault
- Radiometric dating of minerals caused by the fault movement or of undeformed minerals in the fault zone
- Dating of geomorphic features along or across the fault
- Dating techniques in fault investigations — see Appendix A.4

[a] From Adair, M. J., *Reviews in Engineering Geology*, Vol. IV, Geological Society of America, 1979, pp. 27–39. With permission.

Present "dead" faults: Many faults that have not been carefully studied may be considered to be dead or inactive because they have not been the locus of recorded events or activity, but may be potentially active. The Kern County event (1952) occurred along the White Wolf fault (Figure 3.9), which was little known and considered to be dead fault, although it was approximately 64 km in length. It may be connected to either the San Andreas or the Garlock Fault, the two largest in California, which are located only about 24 km apart. In August 1975, an earthquake of $M=5.7$ had its epicenter near the Oroville Dam on one of the faults of the Foothills System of the Sierra Nevada range of California, which was considered to be a dead fault. In 1952, the Ramapo Fault in northeastern New Jersey was considered to be long dead. In the intervening years, the installation of a seismograph station at Lamont, in addition increased area development and habitation, have revealed that there is a substantial amount of activity along the fault and today it even has its own recurrence equation (see Equation 3.15).

Fault Displacements

Importance

Correlations have been made among the amount of displacement, the fault length along which displacement occurs, and the magnitude of the event. Displacement can vary substantially and does not occur in uniform amounts along the fault, and some sections may not displace at all. In general, displacement is related to the magnitude of an event.

The maximum displacement (MD) along a fault rupture length for the western United States can be estimated (NRC, 1997) by the relationships developed by Wells and Coppersmith (1994):

$$\log(\mathrm{MD}) = a + b\,M_w \tag{3.16}$$

where
- strike-slip: $a=-7.03$m, $b=1.03$m, $s=0.34$
- normal: $a=-5.90$m, $b=0.89$m, $s=0.39$
- all: $a=-5.46$m

with M_w being the moment magnitude, and s the standard deviation.

The work of Wells and Coppersmith (1994) is widely referenced in the literature, including the NRC Regulatory Guide 1.165 (NRC, 1997). They studied 167 events and developed regressions of rupture length, rupture width, rupture area, and displacement, in terms of moment magnitude.

Creep (Slip Rate)

Before or after an earthquake, slow movement can occur along a fault (tectonic creep), which can range from a few millimeters to a centimeter or more every year. This fault slippage apparently occurs in faults filled with gouge from previous rupture as strain energy accumulates in the rock below the gouge zone. Creep does not usually occur along an extensive line, but rather is limited to certain areas.

In recent years, increased attention is being paid to slip rates as an expression of the long-term activity of a fault. They reflect the rate of strain energy release on a fault, which can be expressed as the seismic moment. Because of this they are being used to estimate earthquake recurrence, especially in probabilistic seismic hazard analysis (Schwartz and Coppersmith, 1986).

The Southern California Earthquake Data Center (Internet 2004) gives some slip rates for California faults as follows: San Andreas, 20 to 35 mm/year; San Jacinto, 7 to 17 mm/year; Garlock, 2 to 11 mm/year; Elismore, 4 mm/year; and Owens Valley, 0.1 to 2.0 mm/year.

Strike-Slip Displacement

One of the largest movements on record is the 20 ft of horizontal displacement that occurred during the San Francisco quake of 1906; vertical movement did not exceed 3 ft. Horizontal movement of the Imperial Valley event (El Centro) of 1940 reached 10 ft, East of El Centro, at a location along Highway 40, displacement across the roadway was 18 in. By 1966, displacement was 25 in. because of fault creep and slips over the 26 year interval.

Dip-Slip Displacement

Surface tilting and warping often result in dip-slip displacement. California events normally have about a few feet of displacement. The largest recorded vertical displacement appears to be the 30 ft that may have occurred in Assam in 1897, or possibly the 45 ft that may have occurred in the Yakutat Bay, Alaska, quake of 1899. The problem in determining displacements on old scarps is that of erosional changes as discussed by Wallace (1980).

Strong ground motions are strongly influenced by fault geometry. Now referred to as *directivity*, it should be considered in selection of the design earthquake. Two sites located at similar distances but on opposite sides of a fault may experience significantly different ground motions; larger, short-period motions may occur on the hanging wall side. For

TABLE 3.9
System for Classification of Fault Activity Based on Available Data[a]

Activity Classification and Definition	Criteria			
	Historical	**Geological**	**Seismological**	**Studies to Define Further Activity**
Active — a tectonic fault which has a history of strong earthquakes and surface rupture, or a fault which can be demonstrated to have an interval of recurrence short enough to be significant during the life of the particular project. The recurrence time period considered significant for individual projects will vary with the consequence of activity	1. Surface faulting and associated strong earthquakes 2. Tectonic fault creep, or geodetic indications of movement	1. Geologically young[b] deposits have been displaced or cut by faulting 2. Fresh geomorphic features characteristic of active fault zones present along fault trace 3. Physical groundwater barriers produced in geologically young[b] deposits	Earthquake epicenters are assigned to individual faults with a high degree of confidence	Additional investigations and explorations are needed to define: 1. The exact location of individual fault traces 2. The recurrence interval 3. The projected magnitude of future events 4. The type of surface deformation associated with the surface faulting 5. The probable source of energy release with respect to the site
Potentially active — a tectonic fault which has not ruptured in historic time, but available evidence indicates that rupture has occurred in the past and the recurrence period could be short enough to be of significance to particular projects	No reliable report of historic surface faulting	1. Geomorphic features characteristic of active fault zones subdued, eroded and discontinuous 2. Faults are not known to cut or displace the most recent alluvial deposits, but may be found in older alluvial deposits 3. Water barrier may be found in older materials 4. Geological setting in which the geometric relationship to active or potentially active faults suggests similar levels of activity	Alignment of some earthquake epicenters along fault tract, but locations are assigned with a low degree of confidence	Additional investigations are needed to resolve: 1. The time interval of past activity 2. The recurrence of activity 3. The possible locations of the individual fault traces The classification becomes less important if the fault does not cross the project site and a known active fault which is capable of producing frequent high-magnitude earthquakes is located closer to the structure or project under study, and would therefore be the more significant fault

Activity uncertain — a reported fault for which insufficient evidence is available to define its past activity or its recurrent interval. The following classifications can be used until the results of additional studies provide definitive evidence	Available information is insufficient to provide criteria that are definitive enough to establish fault activity. This lack of information may be due to the inactivity of the fault or due to a lack of investigations needed to provide definitive criteria			This classification indicates that additional studies are necessary if the fault is found to be critical to the project. The importance of a fault with this classification depends upon the type of structure involved, the location of the fault with respect to the structure, and the consequences of movement
Tentatively active — predominant evidence suggests that the fault may be active even though its recurrence interval is very long or poorly defined	Available information suggests evidence of fault activity, but evidence is insufficient to be definitive			
Tentatively inactive — predominant evidence suggests that fault is not active	Available information suggests evidence of fault inactivity, but evidence is insufficient to be definitive			
Inactive — a fault along which it can be demonstrated that surface faulting has not occurred in the recent past, and that the recurrence interval is long enough not to be of significance to the particular project	No historic activity	Geomorphic features characteristic of active fault zones are not present and geological evidence is available to indicate that the fault has not moved in the recent past and recurrence is not likely during a time period considered significant to the site. Should indicate age of last movement: Holocene, Pleistocene, Quaternary, Tertiary, etc.	Not recognized as a source of earthquakes	No additional investigations are necessary to define activity

[a] From Cluff et al. (1972).

[b] The exact age of the deposits will vary with each project and depends upon the acceptable level of risk and the time interval which is considered significant for that project.

vertical or dipping faults, the direction of rupture propagation in the near field can cause substantial differences in the level of shaking for different orientations relative to the fault's strike. In the Kobe (1996) and Northridge (1994) events it was found that sites close to the source experienced strike-normal peak velocities substantially larger than strike-parallel velocities, particularly during periods longer than about 0.5 sec (Bray, 1995).

Displacement vs. Magnitude

Relationships between the maximum surface displacement for various types of faults and earthquake magnitude based on recorded events world wide, as of 1977, is given in Figure 3.26.

Wells and Coppersmith (1994) have proposed a relationship for the western United States. The magnitude vs. the MD along a fault rupture length for the western United States can be estimated (NRC, 1997) from the following:

$$M_w = a + b \log(\text{MD}) \tag{3.17}$$

where

strike-slip: a=6.81m, b=0.78m, s=0.29
normal: a=6.61m, b=0.89m, s=0.39
all: a=6.69m, b=0.74m, s=0.40

with MD being the maximum displacement.

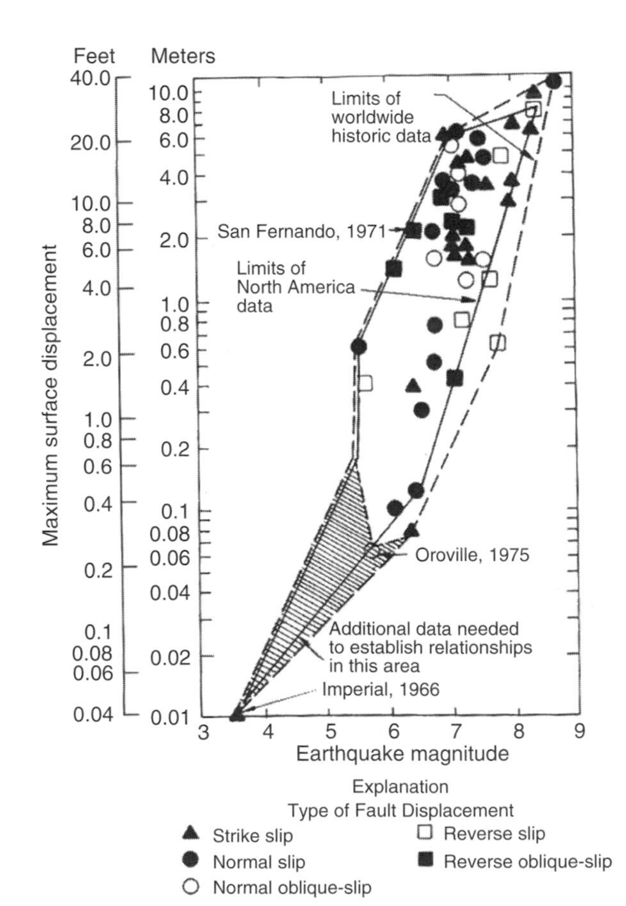

FIGURE 3.26
Relationship between maximum surface displacement and earthquake magnitude reported for historic events of surface faulting throughout the world. (From Taylor, C. L. and Cluff, L. S., Proceedings of ASCE, The Current State of Knowledge of Lifeline Earthquake Engineering, Specialty Conference, University of California, Los Angeles, 1977, pp. 338–353. With permission.)

Rupture Length

General

Surface rupture length along a fault varies greatly; in California earthquakes it has been generally in the range of 1 to 40 mi. The crustal deformation that occurred during the Alaskan event of 1964 was the most extensive yet studied in a single earthquake (Bolt et al., 1975). Vertical displacements occurred along the Alaskan coastline for a distance of almost 600 mi, including a broad zone of subsidence of as much as 6 ft along the Kodiak–Kenai–Chugach mountain ranges, and a major zone of uplift of as much as 35 ft along the coastline. The large extent is perhaps the reason for the unusual duration of 3 min. During the 1906 San Francisco event, the San Andreas Fault is estimated to have ruptured for a length of 260 mi.

Rupture Length vs. Magnitude

The energy released by a shallow-focus earthquake has been related to the surface length of fault rupture as shown in Figure 3.27. The majority of events plotted are from the western United States, Alaska, and northern Mexico; and may not apply elsewhere, although good agreement has been found at the higher magnitudes with quakes in Turkey and Chile. Such relationships as given in the figure have been used to estimate the potential magnitude of an earthquake by assuming that a fault will rupture along its entire identified length, or perhaps only one half to one third of its length, depending on its activity and the degree of risk involved.

For a specific fault, the M_w of a potential earthquake can be estimated by relating it to the potential rupture length of the fault in the western United States (NRC, 1997) using the Wells and Coppersmith (1994) relationship

$$M_w = a + b \log(\text{SRL}) \tag{3.18}$$

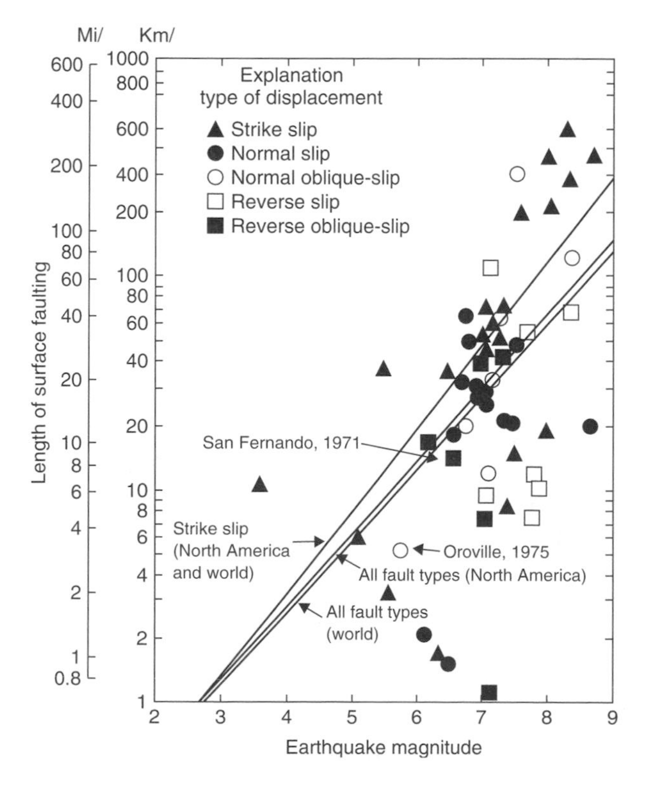

FIGURE 3.27
Scatter diagram of length of surface faulting related to earthquake magnitude from historical events of surface faulting throughout the world. Lines are least squares fits. (From Taylor, C. L. and Cluff, L. S., Proceedings of ASCE, The Current State of Knowledge of Lifeline Earthquake Engineering, Specialty Conference, University of California, Los Angeles, 1977, pp. 338–353. With permission.)

where

strike-slip:	a=5.16 m,	b=1.12 m,	s=0.28
reverse:	a=5.00 m,	b=1.22 m,	s=0.28
normal:	a=4.86 m,	b=1.32 m,	s=0.34
all:	a=5.08 m,	b=1.16 m,	s=0.28

with SRL being the surface rupture length in km.

Various investigators have proposed relationships to predict ground-motion relationships with fault length. In two recent events (Izmit, Turkey, and Chi-Chi, Taiwan, both in 1999 with M_w=7.6) investigators found that peak accelerations were significantly below predictions when several of the relationships were used. An explanation may be in terms of fault characteristics. Faults with a large total slip may be smooth and devoid of asperities, thus radiating less high-frequency energy than a fault with less total slip and a rough interface.

Rupture Width and Area

For a specific fault, the M_w of a potential earthquake can be estimated by relating it to the *potential rupture width* of the fault in the western United States (NRC, 1997) using the Wells and Coppersmith (1994) relationship

$$M_w = a + b \log(\mathrm{RW}) \tag{3.19}$$

where

strike-slip:	a=3.80 m,	b=2.59 m,	s=0.45
reverse:	a=4.37 m,	b=1.95 m,	s=0.32
normal:	a=4.04 m,	b=2.11 m,	s=0.31
all:	a=4.06 m,	b=2.25 m,	s=0.41

with RW being surface rupture width in km.

For a specific fault, the M_w of a potential earthquake can be estimated by relating it to the *potential rupture area* of the fault in the western United States (NRC, 1997) using the Wells and Coppersmith (1994) relationship

$$M_w = a + b \log(\mathrm{RA}) \tag{3.20}$$

where

strike-slip:	a=−3.42 m,	b=0.90 m,	s=0.22
reverse:	a=−3.99 m,	b=0.98 m,	s=0.26
normal:	a=−2.87 m,	b=0.82 m,	s=0.22
all:	a=−3.49 m,	b=0.91 m,	s=0.24

with RA being surface rupture area in km^2.

Duration vs. Length

Shaking duration in large earthquakes depends heavily on the length of faulting. The longer the length of fault rupture, the greater is the duration of the time in which the seismic waves reach a given site. A fault can, however, rupture progressively over a long duration such as in Alaska (1964), or it can rupture in a sequence of breaks resulting in a duration such as in San Francisco in 1906 (see Section 3.2.3).

Attenuation from the Fault

Close Proximity to Rupture

Peak intensities or magnitudes are not necessarily located at the surface expression of the fault; the depth of focus and fault inclination will affect surface response. Housner (1970a) suggests that the rate of decrease in magnitude is relatively small over a distance from the fault equal to the vertical distance to the focus, and beyond this point the drop-off increases rapidly. Many investigators report the lack of markedly greater shaking damage to structures adjacent to the fault as compared with the damage some distance beyond (Bonilla, 1970).

Moderate Distances and Beyond

In most geological environments high accelerations are severely reduced at even moderate distances. Wave attenuation from frictional resistance occurs exponentially, and if shaking is caused by shear waves in the surface of the crust, then even at short distances the exponential decay becomes very effective.

A relationship between peak bedrock acceleration, magnitude, and distance from the causative fault for focal depths of 0 to 20 km is given in Figure 3.28.

Attenuation of maximum acceleration with distance from fault rupture for California earthquakes, prepared from strong motion records, is given in Figure 3.29. Curve I applies to high-intensity sources, associated with effective modes of dislocation, rock types, rupture depths, etc., as, for example, San Fernando, 1971. Peak acceleration near a high-intensity fault would appear to be close to 0.6*g* on average. The expectation ranges indicate that 90% of the time the peak acceleration would be less than 0.4 *g*. Curve II applies to medium-intensity sources, produced by fault dislocations of a less efficient kind, such as El Centro, 1940. In either case, for distances over 100 km from the source, peak accelerations on rock are unlikely to exceed 0.1 *g*.

Relationships among predominant periods of maximum acceleration, magnitude, and distance from the causative fault are given in Figure 3.30.

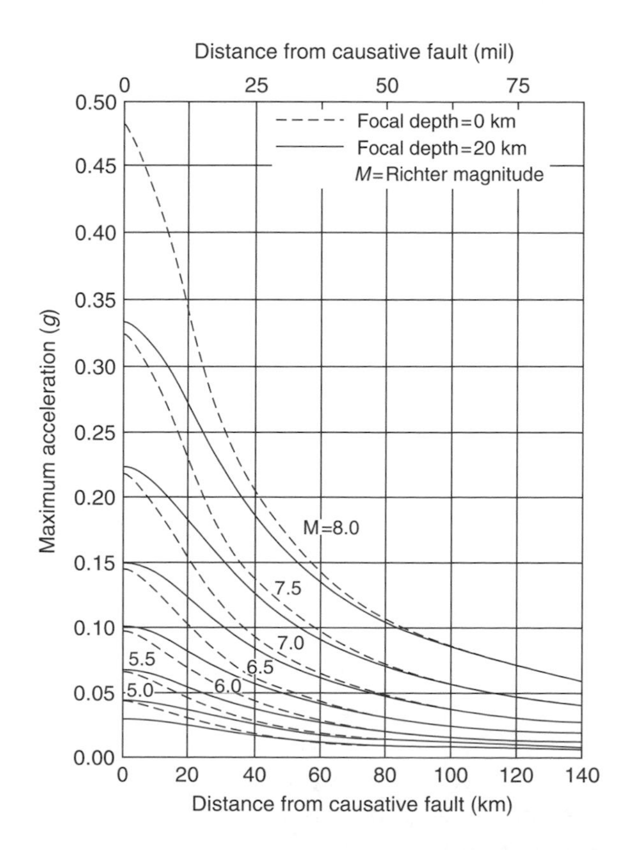

FIGURE 3.28
Relationship between peak bedrock acceleration, earthquake magnitude, and distance from the causative fault for focal depths of 0 and 20 km. (From Leeds, D. J., *Geology, Seismicity and Environmental Impact*, Special Publication Association of Engineering Geology, Los Angeles, California, 1973. With permission.)

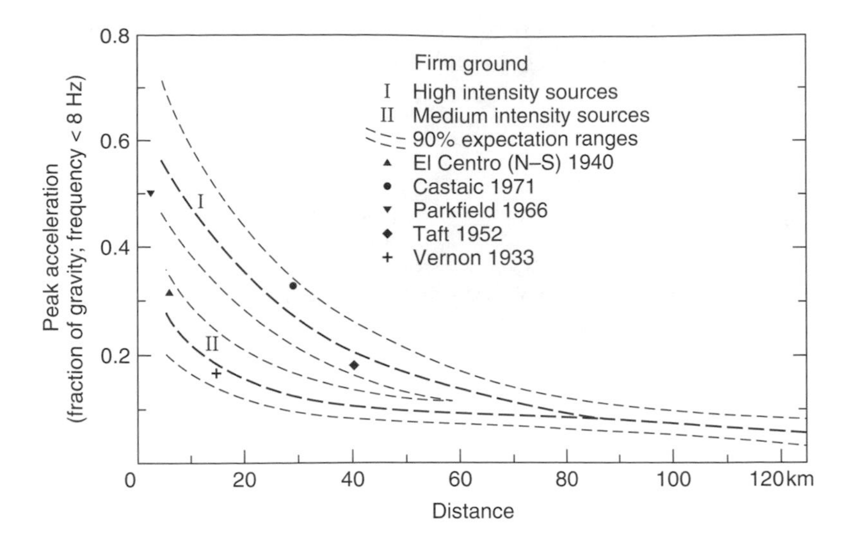

FIGURE 3.29
Attenuation of maximum acceleration with distance from fault rupture for California earthquake prepared from strong-motion records. Frequencies are less than 8 Hz, and generally in the low to intermediate range. Acceleration is for rock or strong soils. (After Bolt, *5th World Conference on Earthquake Engineering*, Rome, 1973.)

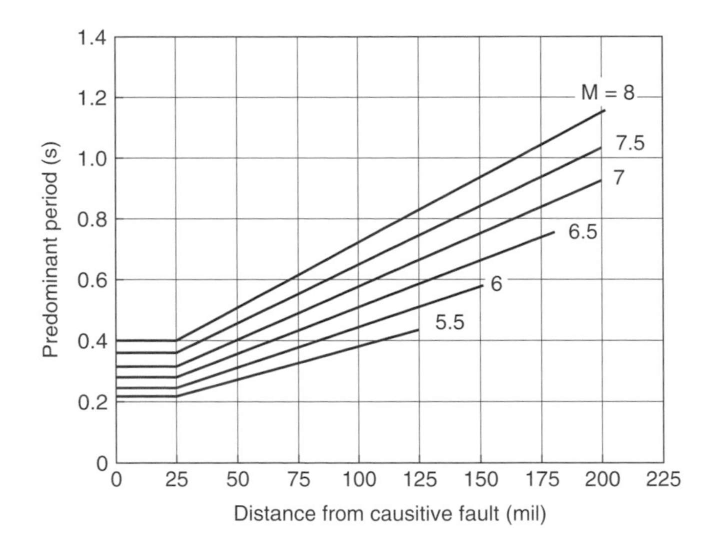

FIGURE 3.30
Relationship between predominant periods of maximum acceleration, magnitude, and distance from the causative fault. (From Seed, H. B. et al., *Proc. ASCE, J. Soil Mech. and Found. Eng. Div.*, 95, 1969. With permission. As presented in Atomic Energy Commission, 1972.)

Isoseismal Maps

The strong relationship between intensity distribution and fault rupture, where rupture length is long, is illustrated in Figure 3.31, the intensity distribution for the San Francisco 1906 earthquake. Intensities generally of VIII to IX are given along the fault, but intensities of IX are given also in isolated areas as far as 64 km from the fault. Two branches of the San Andreas Fault are also shown, the Hayward and the Calaveras faults. The map suggests that movement may have also occurred along these faults during the 1906 quake.

Seismicity Maps and Tectonic Structures

In studies for seismic design, seismicity maps are overlain with geologic maps to obtain correlations with tectonic structures. For example, a comparison of the seismicity map of

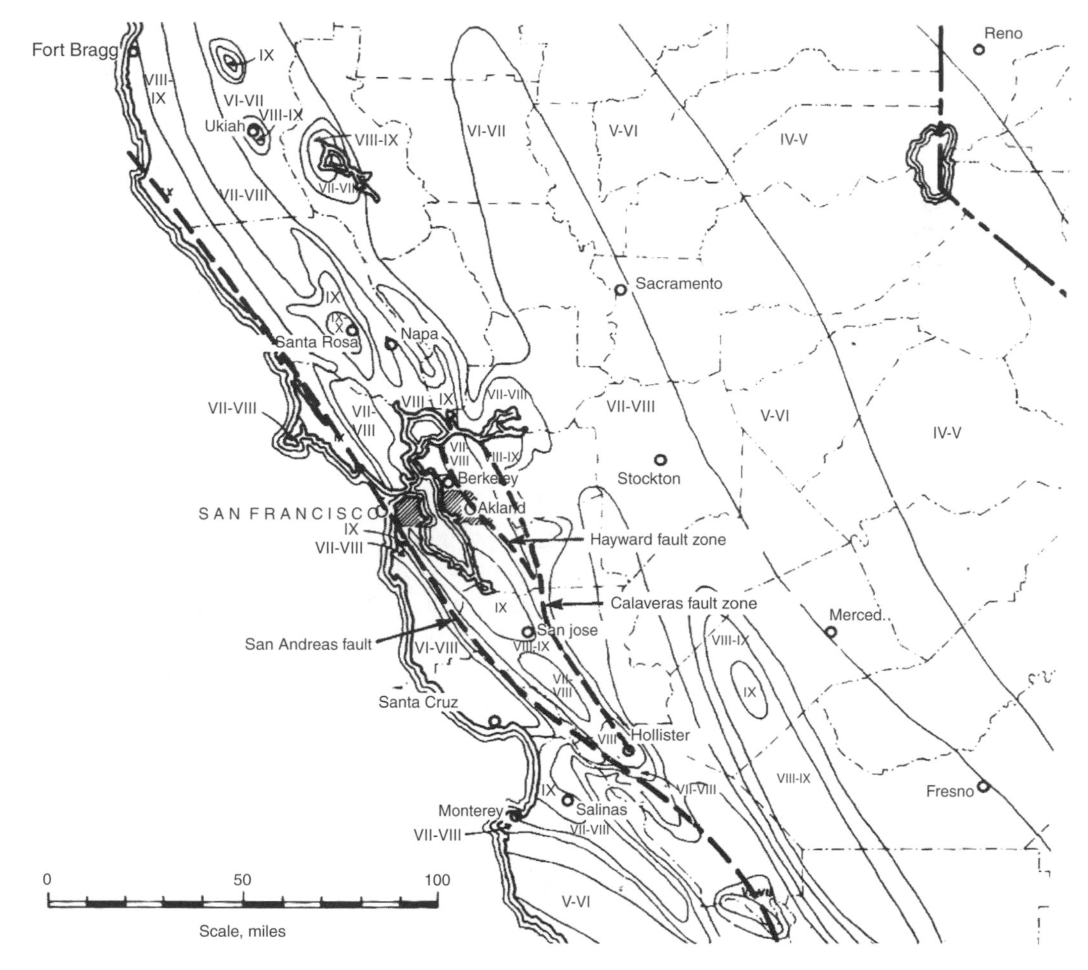

FIGURE 3.31
Isoseismal map of the 1906 San Francisco earthquake given in the Rossi-Forel scale which roughly parallels the modified Meracalli through I=X. (After Environmental Science Services Administration, 1969; and Lawson et al., 1908.)

the southeastern United States (see Figure 3.10) with the geologic map of the area shows a very strong correlation between the epicenters and faults in the zone trending SW–NE through Tennessee, Alabama, and Georgia. The recurrence equation for the area is given in Equation 3.14.

3.3.2 Soil Behavior

General

Excitation emanating as a stress wave from an underlying bedrock surface applies a cyclic shearing stress to soils. On the basis of response to bedrock motions, soils are divided into two general classes:

- *Stable soils* undergo elastic and plastic deformations but serve to dampen seismic motion and still maintain some strength level.
- *Unstable soils* are subject to sudden compaction or a complete loss of strength by cyclic liquefaction (see Section 3.3.3).

Characteristic Properties of Soils under Cyclic Strain

Shear modulus G is the relation between shear stress and shear strain, which occurs under small amplitudes, such as earthquake loads.

Internal damping ratio D or λ pertains to the dissipation of energy during cyclic loading. Shear modulus and damping are the most important characteristics required for the analysis of most situations.

Strength and *stress–strain relationships* in general must be considered for large deformations such as are produced by sea-wave forces on pile-supported structures.

Poisson's ratio ν is required for the description of dynamic soil response, but varies within relatively close limits and affects seismic response only slightly. It is independent of frequency in the range of interest in earthquake engineering, and, in contrast to E and G, is insensitive to thixotropic effects. General ranges are $\nu=0.25$ to 0.35 for cohesionless soils and $\nu=0.4$ to 0.5 for cohesive soils.

Soil Reaction to Dynamic Loads

Initially, cyclic loading causes partially irreversible deformations, irrespective of strain amplitude, and load–unload stress–strain curves do not coincide.

Subsequently, after a few cycles of similar small-strain amplitudes, differences between successive reloading curves tend to disappear and the stress–strain curve becomes a closed loop. This can be described by two parameters: shear modulus, defined by the average slope, and damping ratio, defined by the ratio of the specific enclosed areas as shown in the figure. It reflects the energy that must be fed into the soil to maintain a steady state of free vibration.

Cyclic Shear Related to Earthquake Characteristics

Simple shear stress–strain characteristics at low strains are important in site response analysis because the significant earthquake strain amplitudes normally do not exceed 10^{-4} or 10^{-5}, and are usually in the range of 10^{-1} to 10^{-3}. Higher strains might occur during site response to a large earthquake, but the number of cycles at high strain amplitude are likely to be few.

The effect of the *number of cycles* at low strain amplitudes is not great.

The effect of *loading frequency* is negligible within the range encountered in most earthquakes, i.e., 0.1 to 20 Hz.

Strain amplitude is the most significant characteristic. Shear modulus decreases markedly with an increase in strain amplitude (Taylor and Larkin, 1978).

Shear Modulus and Damping Ratio

Factors Affecting Values

Main factors affecting values for shear modulus and damping ratio in all soils are shear strain amplitude, initial effective mean principal stress, void ratio, shear stress level, and the number of loading cycles.

Cohesive soil values are affected also by stress history (OCR), saturation degree, effective strength parameters, thixotropy, and temperature.

Shear strain amplitude affects the shear modulus as follows:

- *Cohesionless soils*: Shear modulus G decreases appreciably for amplitudes greater than 10^{-4}, below which G is nearly constant.
- *Cohesive soils*: G decreases with increase in amplitude at all levels (Faccioli and Resendiz, 1976).

Measurement of Values

Laboratory tests are used to measure the variation in shear modulus and damping as a function of stress–strain amplitude up to levels of strong motion interest.

In situ or field tests take the form of direct-wave seismic surveys, which provide compression and shear-wave velocities from which G and other dynamic properties are computed. Because the moduli are obtained at lower amplitudes than those imposed by earthquakes, they are likely to be somewhat higher than reality. Values obtained from *in situ* testing are scaled down by comparing the results with those obtained for the same soils from laboratory testing. Approximate strain ranges for earthquake laboratory and field tests are compared in Figure 3.32.

A recent discussion of dynamic soil behavior and testing is found in Brandes (2003).

Evaluation of Data

The evaluation of shear modulus and damping ratio data is described in USAEC (1972) and Hardin and Drenvich (1972a, 1972b).

Applications to soil–structure interaction (SSI) problems are discussed in Section 3.4.5.

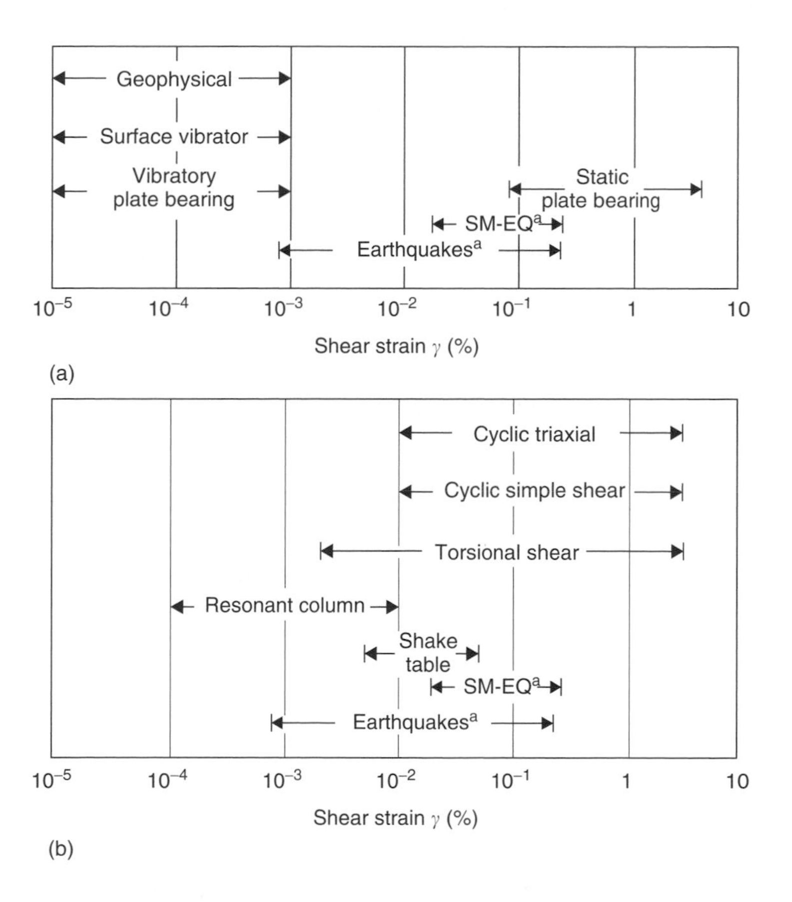

FIGURE 3.32
Comparison of approximate strain ranges for earthquakes, field and laboratory testing: (a) field tests and (b) laboratory tests. (From USAEC, *Soil Behavior Under Earthquake Loading Conditions,* National Technical Information Service TID-25953, U.S. Department of Commerce, Oak Ridge National Laboratory, Oak Ridge, Tennessee, January 1972.)

[a] Range of shear strain denoted as "earthquake" represents an extreme range for most earthquakes. "SM-EQ" denotes strains induced by strong-motion earthquakes.

3.3.3 Subsidence and Liquefaction

General

Earthquake-induced vibrations can be the cause of several significant phenomena in soil deposits, including:

- *Compaction* of granular soils resulting in surface subsidence, which at times occurs over very large areas.
- *Liquefaction* of fine sands and silty sands, which results in a complete loss of strength and causes structures to settle or even overturn and slopes to fail.
- *Reduction in strength* in soft, cohesive soils (strain softening), which results in the settlement of structures that can continue for years and also results from a form of liquefaction.

Subsidence from Compaction

Causes

Cyclic shear strains densify granular soils, resulting in subsidence. *Horizontal motions* induced by shocks cause compaction as long as the cycles are relatively close together, even if the cyclic shear strains are relatively small. Vertical accelerations in excess of 1*g* are required to cause significant densification of sands, which is far greater than most surface accelerations during earthquakes. This has been demonstrated by laboratory tests (Whitman and DePablo, 1969).

Susceptibility Factors

As noted in the discussion of liquefaction below, the susceptibility of soils to compaction during ground shaking depends on soil gradation, relative density or void ratio, confining pressure, amplitude of cyclic shear stress or shear strain, and number of stress cycles or duration.

Compaction subsidence and liquefaction are closely related; the major difference in occurrence is the ability of the material to drain during cyclic loading. Compaction occurs with good soil drainage.

Occurrence

New Madrid Events of 1811 and 1812: Ground subsidence extended over enormous areas, and was reported to be as high as 15 to 23 ft in the Mississippi valley.

Homer, Alaska (1964): A deposit of alluvium 450 ft in original thickness subsided 4 ft (Seed, 1970, 1975).

Niigata, Japan (1964): Many structures underlain by sand settled more than 1 m (Seed, 1970, 1975).

The Liquefaction Phenomenon

Cyclic Liquefaction in Granular Soils

Defined: Cyclic liquefaction refers to the response of a soil, subjected to dynamic loads or excitation by transient shear waves, which terminates in a complete loss of strength and entry into a liquefied state. (Cyclic liquefaction differs from the liquefaction that occurs during the upward flow of water under static conditions.)

Described: If a saturated sand is subjected to ground vibrations it tends to compact and decrease in volume; if the sand cannot drain rapidly enough, the decrease in volume results in an increase in pore pressure. When the pore pressure increases until it is equal to the overburden confining pressures, the effective stress between soil particles becomes zero, the sand completely loses its shear strength, and enters a liquefied state.

Origin: Wylie and Streeter (1976) hypothesized that the shearing motion of the soil causes a slippage or sliding of soil grains, which weakens the soil skeleton temporarily and causes the constrained modulus to be reduced. At the time of shear reversal, the particles do not slide, so the skeleton recovers much of its original strength, but in a slightly consolidated form. The consolidation reduces the pore volume, thereby tending to increase pore pressure and to reduce the effective stress in the soil skeleton. Since shear modulus and maximum shear stress depend on effective stress, the horizontal shaking causes a trend towards zero effective stress, and hence liquefaction. Drainage by percolation tends to reduce pore-pressure rise and cause stabilization.

Ground response: The phenomenon can occur in a surface deposit or in a buried stratum. If it develops at depth, the excess hydrostatic pressures in the liquefied zone will dissipate by upward water flow. A sufficiently large hydraulic gradient will induce a "quick" or liquefied condition in the upper layers of the deposit. The result is manifested on the surface by the formation of boils and mud spouts and the development of "quicksand" conditions. As the ground surface liquefies and settles in an area with a high groundwater table, the water will often flow from the fissures of the boils and flood the surface. Even if surface liquefaction does not occur, subsurface liquefaction can result in a substantial reduction in the bearing capacity of the overlying layers.

Surface effects can be significant as shown by occurrences in Alaska and Japan in 1964 and Chile in 1960. Buildings settled and tilted (Figure 3.33), islands submerged, dry land became large lakes, roads and other filled areas settled, differential movement occurred between bridges and their approach fills, and trucks and other vehicles even sank into the ground.

Soft Cohesive Soils

Partial liquefaction can be said to occur in soft cohesive soils. Longitudinal waves, because of their characteristics of compression and dilation, induce pore-water pressures in saturated clays. The seismically induced pore pressures reduce the shear strength of the soil, and subsequently the bearing capacity, resulting in partial or total failure. Deformation in

FIGURE 3.33
Overturning of buildings during 1964 Niigata earthquake. (Photo from Internet: www.nd.edu/~quake/education/liquefaction/.)

soft to medium consistency clays from the horizontal excitation will be essentially pure shear (Zeevaert, 1972).

Increased settlements of existing structures may result. An example of partial liquefaction is a building in Mexico City founded on soft silty clays that was not undergoing significant settlements until the July 28, 1957 earthquake. During the quake the building settled 5 cm and continued to settle for years afterward at rates of 3 to 5 cm/year. The shear forces from the earthquake-induced seismic waves reduced the shear strength of the clay and significantly increased compressibility by the phenomenon of "strain softening".

Rupture of foundation members may result as the seismic shear forces cause buildings to translate horizontally, imposing high earth pressures on walls and bending forces on piles and piers, especially where soft clays are penetrated. Softening of the clay due to the cyclic strains may be a factor in inducing rupture.

Occurrence of Liquefaction

Geographic Distribution

Incidence of liquefaction is not great in comparison with the large number of earthquakes that occur annually. Studies of earthquake records have produced relatively few cases where liquefaction was reported, even though the records extended back to 1802 (Seed, 1975; Christian and Swiger, 1975). Known cases of liquefaction were reported for 13 locations of which two were earth dams. Magnitudes were generally greater than 6.3.

- *Japan*: Mino Qwari (1891), Tohnankai (1944), Fukui (1948), Niigata (1964), and Tokachioki (1968).
- *United States*: Santa Barbara (1925, the Sheffield Dam), El Centro (1940), San Francisco (1957), San Fernando (1971), Van Norman Reservoir Dam, San Francisco (1971).
- *Others*: Chile (1960), Alaska (1964), and Caracas (1967).

Since the 1975 reports, liquefaction has been reported to have occurred in Loma Preita (1989), California, Kobe, Japan (1994), and Izmit, Turkey (1999). In recent years, the potential for liquefaction and analytical procedures have received considerable attention from investigators. Many municipalities have prepared liquefaction zonal maps.

Geologic Factors and Susceptibility

Geologic factors influencing the susceptibility to liquefaction include sedimentation processes, age of deposition, geologic history, water table depth, gradation, burial depth, ground slope, and the nearness of a free face.

The potential *susceptibility* for soils of various geologic origins in terms of age are summarized in Table 3.10. Susceptibility is seen to decrease as the age of the deposit, which reflects prestressing by removal of overburden or densification by ancient earthquakes, increases. The greatest susceptibility is encountered in coastal areas where saturated fine-grained granular alluvium predominates, often with limited confinement, and where recent alluvium appears more susceptible than older alluvia. Offshore liquefaction must be considered since the seafloor can become unstable from earthquakes or wave forces during large storms (see Section 3.3.4).

Factors of Liquefaction Potential

General

Gradation: As shown in Figure 3.34, fine sands and silty sands are most susceptible, especially when they are poorly graded. Permeability is relatively low and drainage slow.

TABLE 3.10

Estimated Susceptibility of Sedimentary Deposits to Liquefaction during Strong Seismic Shaking[a]

Type of Deposit	General Distribution of Cohesionless Sediments in Deposits	Likelihood That Cohesionless Sediments, When Saturated, Would Be Susceptible to Liquefaction (by Age of Deposit)			
		<500 Year	Holocene	Pleistocene	Prepleistocene
Continental Deposits					
River channel	Locally variable	Very high	High	Low	Very Low
Floodplain	Locally variable	High	Moderate	Low	Very low
Alluvial fan and plain	Widespread	Moderate	Low	Low	Very low
Marine terraces and plains	Widespread		Low	Very low	Very low
Delta and fan-delta	Widespread	High	Moderate	Low	Very low
Lacustrine and playa	Variable	High	Moderate	Low	Very low
Colluvium	Variable	High	Moderate	Low	Very low
Talus	Widespread	Low	Low	Very low	Very low
Dunes	Widespread	High	Moderate	Low	Very low
Loess	Variable	High	High	High	Unknown
Glacial till	Variable	Low	Low	Very low	Very low
Tuff	Rare	Low	Low	Very low	Very low
Tephra[b]	Widespread	High	High	?	?
Residual soils	Rare	Low	Low	Very low	Very low
Sebka[c]	Locally variable	High	Moderate	Low	Very low
Coastal Zone					
Delta	Widespread	Very high	High	Low	Very low
Estuarine	Locally variable	High	Moderate	Low	Very low
Beach					
High wave energy	Widespread	Moderate	Low	Very low	Very low
Low wave energy	Widespread	High	Moderate	Low	Very low
Lagoonal	Locally variable	High	Moderate	Low	Very low
Fore shore	Locally variable	High	Moderate	Low	Very low
Artificial					
Uncompacted fill	Variable	Very high			
Compacted fill	Variable	Low			

[a] From Youd, T. L. and Perkins, D. M., *Proc. ASCE, J. Geotech. Eng. Div.*, 104, 433–446, 1978. With permission.
[b] Tephre — coastlines where slopes consist of unconsolidated volcanic ash or bombs.
[c] Sebkha — flat depression, close to water table, covered with salt crust, subject to periodic flooding and evaporation. Inland or coastal.

Groundwater conditions: To be susceptible, the stratum must be below the groundwater level and saturated, or nearly so, without the capacity to drain freely.

Relative density: D'Appolonia (1970) suggested that liquefaction might occur where D_R values were as high as 50% during ground accelerations > 0.1 *g*, but for sands with D_R in the range of 75% or greater, liquefaction was unlikely.

Boundary drainage conditions and soil stratigraphy: These factors affect the rate of pore pressure increase.

Initial effective overburden pressure: Also known as depth effect, this pressure influences susceptibility.

Duration, amplitude and period of induced vibrations: These factors influence liquefaction potential. It appears that liquefaction does not occur for 0.1 *g* or less.

Soil Conditions

Susceptible soils: From a study of four case histories, Seed (1975) concluded that liquefaction occurs in relatively uniform, cohesionless soils for which the 10% size is between 0.01

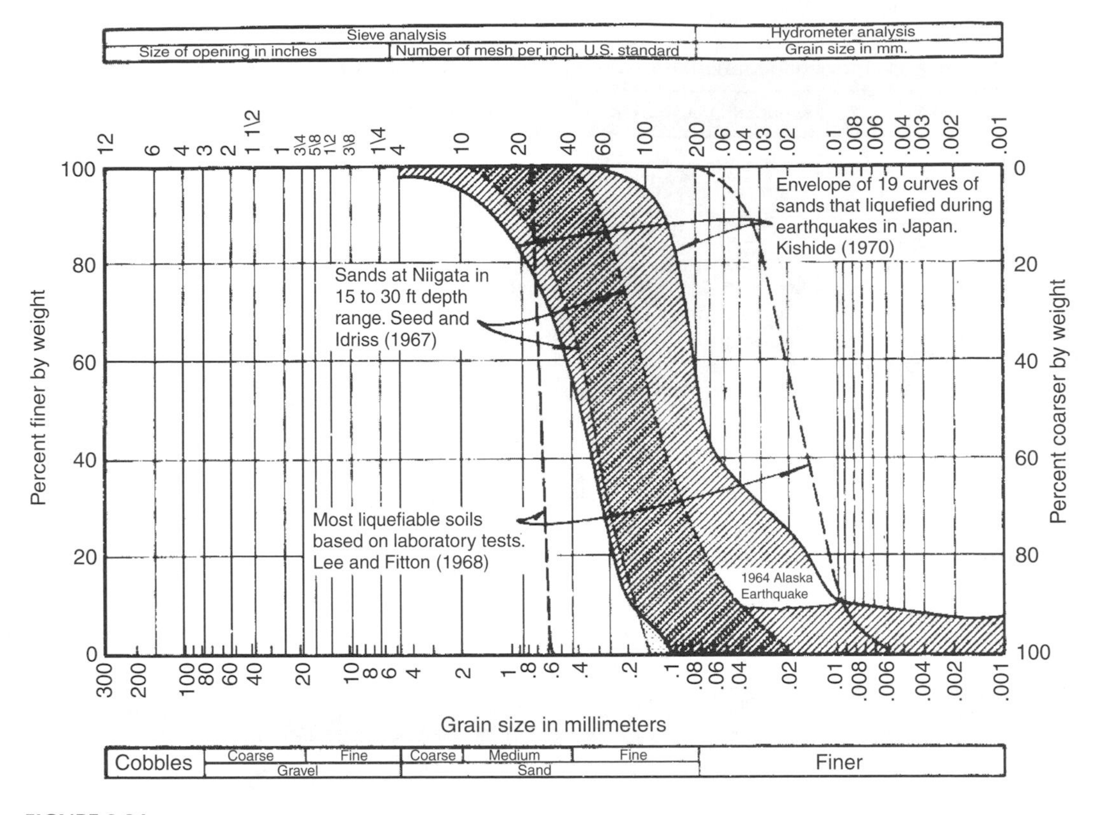

FIGURE 3.34
Effect of grain size distribution on liquefaction susceptibility (cyclic mobility). (After Shannon and Wilson, Inc. and Agbabian-Jacobsen Associates, 1971; from Finn, W. D., *Proceedings of the International Conference on Microzonation*, Seattle, Vol. I, November 1972, pp. 87–112. With permission.)

and 0.25 mm, and the uniformity coefficient is between 2 and 10. In general, the liquefiable soils had SPT N values < 25. Zeevaert (1972) considers that under certain conditions soft to medium clays undergo a partial liquefaction, although they do not become fluid on level ground. Many types of clays and clayey silts are susceptible to liquefaction on slopes as discussed in Section 3.3.4.
Nonsusceptible soils: Gravels and sandy gravels, regardless of N values (Seed, 1975), and stiff to hard clays or compact sands, regardless of being situated on level ground or in slopes, appear to be nonsusceptible to liquefaction.

Foundation Damage Susceptibility

Case Study: Niigata, Japan (Seed, 1975)

General: During the earthquake in Niigata, Japan, in 1964, liquefaction caused a great amount of damage but distribution was random. The city is underlain by sands up to depths of 30 m. Gradation is characterized generally by a 10% size ranging from about 0.07 to 0.25, with a uniformity coefficient between 2 and 5 (uniformly or poorly graded).

Damage zoning: Three zones were established relating damage to soil conditions:

- *Zone A*: Coastal dune area with dense granular soils and a relatively deep water table experienced very little damage to structures.
- *Zone B*: Relatively old alluvium of medium-compact to loose sands with a high groundwater table experienced relatively light damage to structures.

- *Zone C*: Recent alluvium of loose sands with high water table experienced heavy damage.

The primary difference between zones B and C appears to lie in soil density as revealed by SPT *N* values. Heavy damage implies that buildings suffered large settlements or tilted, such as illustrated in Figure 3.33, a building supported on spread footings. Reinforced concrete buildings in zone C were supported either on shallow spread footings or on piles.

Damage vs. SPT values: Japanese engineers found that buildings supported on shallow foundations suffered heavy damage where *N* values were less than 15, and light to no damage where *N* values were between 20 and 25. For buildings supported on pile foundations (lengths ranged from 5 to 18 m), damage was heavy if the *N* value at the tip was less than 15. A relationship among foundation depth, SPT value, and extent of damage, developed from Seed's (1975) study, is given in Table 3.11. Seed concluded that two important factors vary with depth: the SPT value as affected by soil confinement, and ground acceleration, which is usually considered to be larger at the ground surface and to decrease with depth.

Predicting the Liquefaction Potential

General

Various investigators have studied locations where liquefaction did not occur in attempts to obtain correlations for predicting liquefaction potential on the basis of material density, initial effective overburden stress, and earthquake-induced cyclic horizontal shear stress (Castro, 1975; Christian and Swiger, 1975; Seed et al., 1975; Seed, 1976).

Cyclic Stress Ratio

The cyclic stress ratio has been proposed as a basis for anticipating liquefaction potential (Seed et al., 1975; Seed, 1976; Seed et al., 1985). It is defined as the ratio of the average horizontal shear stress (λh) induced by an earthquake to the initial effective overburden pressure (σ'_o).

For sites where liquefaction occurred, a lower bound was plotted in terms of the cyclic stress ratio vs. corrected SPT values $(N1)_{60}$, where $(N1)_{60}$ is the measured *N* value corrected to an effective overburden pressure of approximately 100 kPa (1 tsf) and other factors. The correlation is given in Figure 3.35.

The cyclic stress ratio (CSR) at any depth in the ground causing liquefaction can be calculated with reasonable accuracy from the relationship given by Seed et al. (1985):

$$\mathrm{CSR}=\lambda_{hav}/\sigma'_{o=}0.65\,(\alpha_{max}/g)(\,\sigma_{vo}/\sigma'_{vo})\,\mathrm{r}_d \tag{3.21}$$

TABLE 3.11
Relationship between Foundation Depth, SPT, and Damage at Niigata[a]

Foundation Depth Range (m)	*N* Value at Foundation Base	Damage Relationship
0–5	14	Apparently adequate to prevent damage by settlement or overturning
5–8	14–28	Required to prevent heavy damage
8–16	28	Required to prevent heavy damage

[a] After Seed, H. B., *Foundation Engineering Handbook*, Winterkorn and Fang, Eds., Van Nostrand Reinhold, New York, 1975.

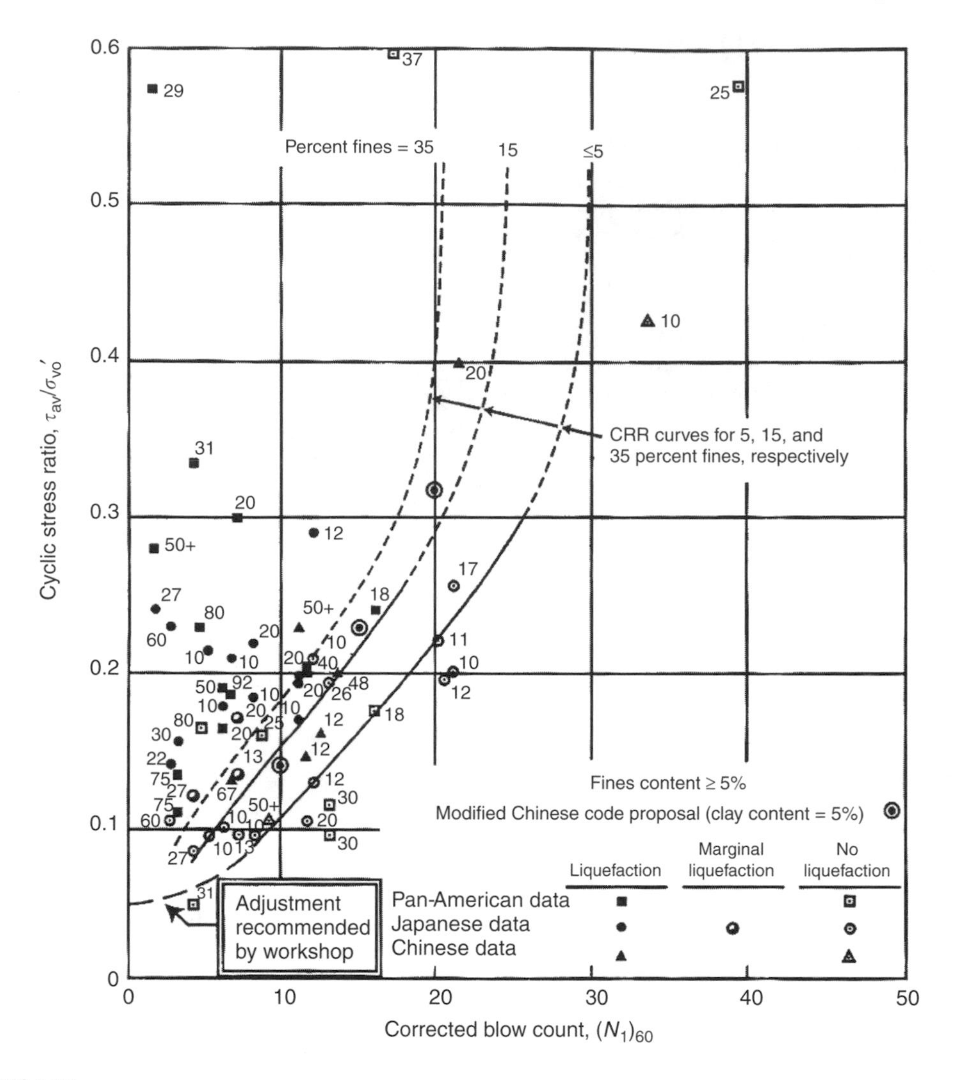

FIGURE 3.35
Liquefaction resistance based on the SPT. (From Seed, H. B. et al., *Proc. ASCE, J. Geotech. Eng. Div.*, 111, 1425–1445, 1985. With permission.)

where α_{max} is the peak horizontal acceleration at the ground surface, g the acceleration due to gravity, σ_{vo} the total vertical overburden stress on the stratum under consideration, σ'_{vo} the effective vertical overburden stress on the stratum under consideration, and r_d a stress reduction factor (from Liao and Whitman, 1986).

For $z \leq 9.15$ m, $r_d = 1.0 - 0.00765z$ and for $9.15 \text{ m} < z \leq 23$ m, $r_d = 1.174 - 0.0267z$

Using Equation 3.21 a value for CSR(1) is obtained. The value for $(N1)_{60}$ is entered on the curve representing the percent fines to obtain a value for CSR(2).

The factor of safety against liquefaction = CSR(2)/CSR(1). (The important factor of duration is not considered in the evaluation.)

CSR plotted vs. the corrected CPT tip resistance for $M = 7.5$ is given in Figure 3.36. Because data are limited compared with SPT data, the NCEER Workshop (Youd and Idriss, 1997) recommended that the CPT data be used with at least some correlative data. As the information obtained with the CPT is more detailed than the SPT it is expected that as more data become available in the future, the CPT correlations will find increased application.

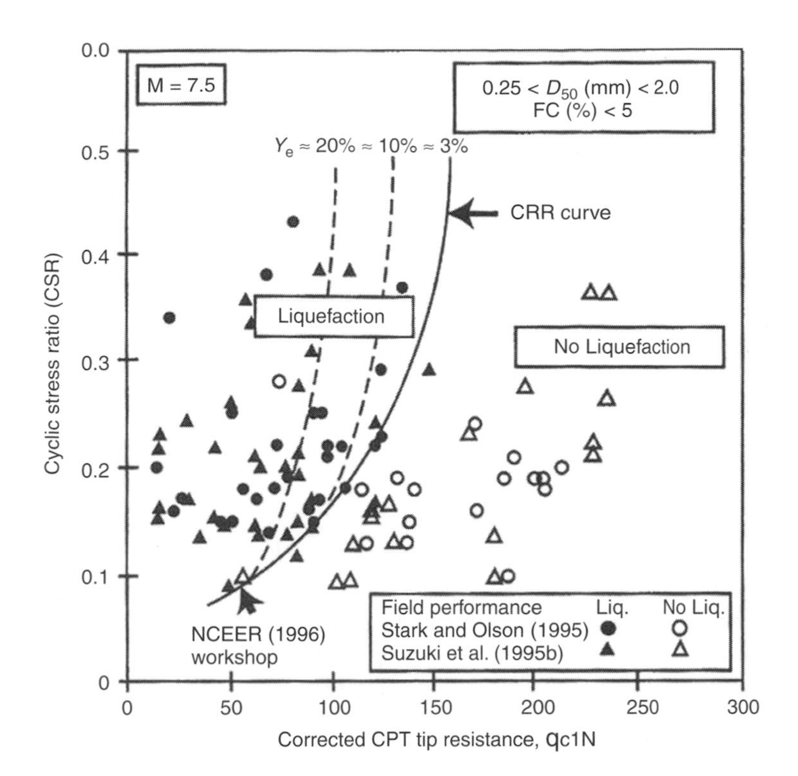

FIGURE 3.36
Liquefaction criteria based on CPT data. (From Robertson, P. K. and Wride, C. E., *Proceedings NCEER Workshop on Evaluation of Liquefaction Resistance of Soils*, Salt Lake City, Utah, 1997. Used with permission of the Multidisciplinary Center for Earthquake Engineering Research, Buffalo, New York.)

Treatment for Liquefaction Prevention

Avoid construction in seismically active areas underlain by loose fine-grained granular soils where the water table may rise to within about 30 ft (10 m) of the surface, especially adjacent to water bodies.

Relatively shallow deposits may be treated by excavation and replacement of the susceptible soils with engineered compacted fill, or by the support of structures on foundations bearing on nonsusceptible soils.

Moderately deep deposits may be treated by densification with vibroflotation or dynamic compaction, by strengthening with pressure grouting, or by improvement of internal drainage. The last may be accomplished with cylindrical, vertical gravel, or rock drains (stone columns). A series of charts is presented by Seed and Booker (1977) which provides a basis for the design and selection of a suitable drain system for the effective stabilization of potentially liquefiable sand deposits by relieving pore pressures generated by cyclic loading as rapidly as they are generated.

3.3.4 Slope Failures

Natural Slopes

General Occurrence

Seismic forces cause numerous slope failures during earthquakes, often as a result of the development of high pore pressures. Such pressures are most likely to be induced in heavily jointed or steeply dipping stratified rock on steep slopes, and in saturated fine-grained soils even on shallow slopes. Loess or other deposits of fine sands and silts, and clays with seams and lenses of fine sand or silt are all highly susceptible.

Debris Slides and Avalanches

Shallow debris slides are probably the most common form of slope failure during earthquakes and can be extremely numerous in hilly or mountainous terrain. Very large mass movements occur on high, steep slopes such as in the avalanche that buried most of the cities of Yungay and Ranrahirca during the 1970 Peruvian event (see Section 1.2.8).

An earthquake-induced debris avalanche in relatively strong materials occurred at Hebgen Lake, Montana, during the August 17, 1959 event (M=7.1). Approximately 43 million yd^3 of rock and soil debris broke loose and slid down the mountainside, attaining speeds estimated at 100 mi/h when it crossed the valley. Its momentum carried it 400 ft up the opposite side of the valley, and the material remaining in the valley formed a natural dam and new lake.

Lateral Spreading

Common in lowlands along water bodies, lateral spreading results in considerable damage, especially to bridges and pipelines. During the Alaska quake of 1964, 266 bridges were severely damaged as a consequence of lateral spreading of floodplain deposits toward stream channels. During the San Francisco event of 1906, every major pipeline break occurred where fills overlay the soft bay muds (Youd, 1978).

The Turnagain Heights failure that occurred during the 1964 Alaska quake is described in detail in Section 1.2.6 (see Figure 1.44). Of interest is the previous earthquake history for the area without the incidence of major sliding including M=7.3 (1943) with an epicentral distance of 60 km, M=6.3 (1951) with an epicentral distance of 80 km, and M=7 (1954) with an epicentral distance of 100 km. The 1964 event had M=8.3 with an epicentral distance 120 km, but a duration of about 3 min, which appears to have been the cause of many large slope failures.

Flows

Flows can be enormous in extent under certain conditions. During the 1920 earthquake in Kansu, China, formations of loess failed, burying entire cities. Apparently, the cause was the development of high pore-air pressures. The flow debris, which extends for a distance of 25 km down the valley of the Rio La Paz near La Paz, Bolivia, is considered to be the result of an ancient earthquake (see Section 1.2.11).

Offshore

Flows or "turbidity currents" offshore can also reach tremendous proportions. An earthquake during November 1929 is considered to be the cause of the enormous "turbidity current" off the coast of Newfoundland. It is speculated that a section of the Continental Shelf broke loose from the Grand Banks, mixed with seawater and formed a flow that moved downslope along the continental rise to the lower ocean floor for a distance of about 600 mi. Its movements were plotted from the sequential breaking of a dozen marine cables in about 13 h, which yielded an average velocity of 45 mi/h (Hodgson, 1964).

Large submarine flows occurred during the Alaskan event of 1964, carrying away much of the port facilities of Seward, Whittier, and Valdez. At Valdez, 75 million yd^3 of deltaic sediments moved by lateral spreading, resulting in displacements in the city behind the port as large as 20 ft (Youd, 1978).

Large submarine slides may also occur, such as the one depicted on the high-resolution seismic profile given in Figure 1.57, presumably caused by an earthquake affecting the Kayak Gulf of Alaska.

Earth Dams and Embankments

Occurrence

If *earth embankments*, such as those for roadway support, fail during earthquakes, it is usually by lateral spreading due to foundation failure such as that occurred in San Francisco and the coastal cities of Alaska as described in the previous chapter.

Earth dams, when well built, can withstand moderate shaking, of the order of 0.2 g or more, with no detrimental effects. Dams constructed of clay soils on clay or rock foundations have withstood extremely strong shaking ranging from 0.35 to 0.8 g (from an M=8.5 event) with no apparent damage. The greatest risk of damage or failure lies with dams constructed of saturated cohesionless materials that may be subjected to strong shaking. A review of the performance of a large number of earth dams during a number of earthquakes is presented by Seed et al. (1978). They list six dams in Alaska, California, Mexico, and Nevada that are known to have failed, three dams in California and Nevada known to have suffered heavy damage, and numerous dams in Japan that suffered embankment slides.

Foundation failure appears to have caused the collapse and total failure of the Sheffield Dam, near Santa Barbara, California, during the 1925 quake (Seed et al., 1969).

Case Study: The Upper and Lower San Fernando Dams (Seed et al., 1975)

Event: During the 1972 San Fernando earthquake (M=6.6), the two dams on the lower Van Norman Reservoir complex, located about 9 mi from the epicenter, suffered partial failures. If either dam had failed completely, a major disaster would have occurred, since some 80,000 people were living downstream.

Description: The Upper San Fernando Dam was 80 ft high at its maximum section. During the earthquake, the crest moved downstream about 5 ft and settled about 3 ft. Severe longitudinal cracking occurred on the upstream slope, but there was no overtopping or breaching. The Lower San Fernando Dam was 142 ft high at its maximum section, and it suffered a major slide in the upstream slope and part of the downstream slope, leaving about 5 ft of freeboard in a very precarious position as shown in the photo (Figure 3.37).

Both dams were constructed by a combination of compacted fill and semihydraulic fill placed during times when little was known about engineered compacted fill. The Lower Dam was completed in 1915, and raised twice, in 1924 and 1930. The Upper Dam was completed in 1922. The dams had withstood the not-so-distant Kern County event of 1952 (M=7.7). The locations of the events of 1971 and 1952 are shown on the isoseismal maps, (Figure 3.15).

Instrumentation had been installed in recent years, including piezometers in the Upper Dam and two seismoscopes on the Lower Dam, one on the crest and one on the rock of the east abutment.

Pseudostatic analysis performed before the earthquake had evaluated the stabilities against strong ground motion and found the dams to be safe.

Dynamic analysis of the response of the dams to earthquake loadings appeared to provide a satisfactory basis for assessing the stability and deformations of the embankments (Seed et al., 1975).

- *Lower Dam*: Dynamic analysis indicated the development of a zone of liquefaction along the base of the upstream shell, which led to failure. Evidence of liquefaction was provided by the seismoscopes, which indicated that the slide had developed

FIGURE 3.37
The Lower San Fernando Dam after the San Fernando earthquake of February 9, 1971. A major slide occurred along the upstream side by liquefaction of hydraulic fill in the dam body. (Photograph courtesy of the U.S. Geological Survey.)

after the earthquake had continued for some time, when ground motions had almost ceased following the period of strong ground shaking. The investigators concluded that since the slide did not occur when the induced stresses were high, but rather under essentially static load conditions, there was a major loss of strength of some of the soil in the embankment during ground shaking.

- *Upper Dam*: Dynamic analysis indicated that the dam would not undergo complete failure, but that the development of large shear strains would lead to substantial deformations in the embankment.

Analytical Methods

Dynamic Analysis

The procedure is essentially similar to that performed for structures as described in Section 3.4. Characteristics of the motion developed in the rock underlying the embankment and its soil foundation during the earthquake are estimated, the response of the foundation to the base rock excitation is evaluated, and the dynamic stresses induced in representative elements of the embankment are computed.

Representative samples of the embankment and foundation soils are subjected to laboratory testing under combinations of preearthquake stress conditions and superimposed dynamic stresses to permit assessments of the influence of earthquake-induced stresses on the potential for liquefaction and deformations. From the data, the overall deformations and stability of the dam sections are analyzed.

Pseudostatic Analysis (see also Section 1.3.2)

In the conventional approach, the stability of the potential sliding mass is determined for static loading conditions and the effects of an earthquake are accounted for by including equivalent vertical and horizontal forces acting on the mass. The horizontal and vertical forces (Figure 3.38) are a product of the weight and seismic coefficients k_v and k_h. The effects of pore pressure are not considered, and a decrease in soil strength is accounted for only indirectly.

Various applications of the coefficient k can be found in the literature. It seems reasonable to decrease the resisting force by adding the term $k_v W$, or considering that the vertical acceleration of gravity is less than the horizontal aceleration, $0.67k_v W$; and to increase the driving force by adding the term $W(k_h \cos\theta)$ to Janbu's equation (Equation 1.17) as follows:

$$FS = f_o\,(\Sigma\{[c'b + (W - 0.67k_v W - ub)\tan\phi']\,[1/\cos\theta\, M_i(\theta)]\}/\Sigma W\,(\tan\theta + k_h\cos\theta) + V) \qquad (3.22)$$

Where $\cos\theta\, M_i(\theta)$ is given in Equation 1.18, and f_o in Equation 1.19.

The selection of the seismic coefficient is empirical. After a review of various investigators, it is suggested that the values given in Table 3.12 are considered reasonable.

Earthquake Behavior Analysis of Earth Dams

Pseudostatic analysis is now generally recognized as being inadequate to predict earth dam behavior during earthquakes. The Committee on Earthquakes of the International

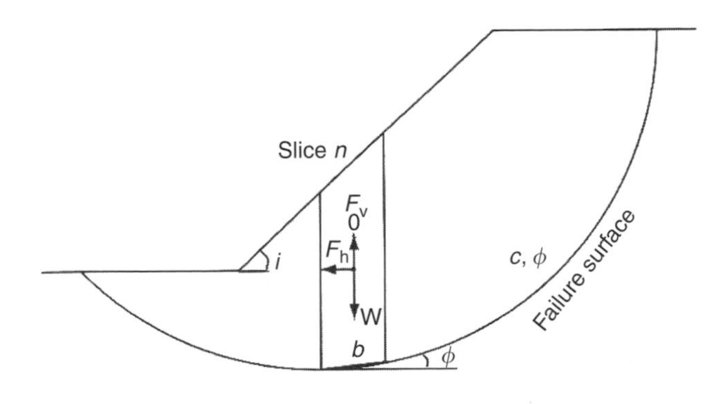

FIGURE 3.38
Pseudostatic seismic forces acting on a slice in Janbu Method.

TABLE 3.12
Seismic Coefficient *K* (Suggested Values)

K	Zone	Intensity I
0.0	0	I–IV
0.05	1	V–VI
0.10	2	VII
0.15	3	VIII–IX
0.25	3	X or greater
0.50	3	VIII or greater and high risk

Commission on Large Dams (ICOLD, 1975) has recommended the dynamic analysis approach for high embankment dams whose failure may cause loss of life or major damage.

Design proceeds first by conventional methods followed by dynamic analysis to investigate any deficiencies which may exist in the pseudostatic design. A simplified procedure for estimating dam and embankment earthquake-induced deformations is presented by Makdisi and Seed (1978). Dynamic analysis is also discussed by Brandes (2003).

3.3.5 Tsunamis and Seiches: Response of Large Water Bodies

Tsunamis

General

Tsunamis are long sea waves that can reach great heights when they encounter shorelines, where they represent a very substantial hazard. Generally associated with earthquakes, they also result from underwater landslides. Seismologists generally agree that they usually reflect some sudden change in seafloor topography such as up-thrusting or down-dropping along faults, or less frequently the sliding of unconsolidated material down continental shelves. They are potentially very damaging.

Occurrence: Synolakis (2003) lists 152 tsunamis that have occurred in the past 100 years.

Geographic Occurrence

Pacific Ocean regions: The regions of most frequent occurrence include the Celebes Sea, Java Sea, Sea of Japan, and the South China Sea, but in recent years, tsunamis have struck and caused damage in Crescent City, California and Alaska from the 1964 event; and in Hilo, Hawaii, Japan, and Chile from the 1960 Chilean quake. Japan probably has the greatest incidence of tsunamis of any country, and has been subjected to 15 destructive tsunamis since 1956, eight of them disastrous (Leggett, 1973). One of the worst tsunamis in history occurred along the northeast coast in June 1896 when a wave 76–100 ft (23–30 m) above sea level rushed inland, destroying entire villages and killing more than 27,000 people. The cause was considered to be a nearby earthquake. The Hawaiian Islands are subjected to a serious tsunami about once in every 25 years (Leggett, 1973).

Atlantic Ocean: Occurrence is very infrequent. Following the Lisbon earthquake of 1755, the sea level was reported to have risen to 6 m at many points along the Portuguese coast, and in some locations to 15 m. The Grand Banks event of 1929, with the epicenter located about 400 km offshore, caused a great tsunami that was very destructive along the Newfoundland coast and took 27 lives. It may have been caused by the turbidity current described in Section 3.3.4.

Indian Ocean (Bay of Bengal): On December 26, 2004, as this book was going to print, a Magnitude 9 earthquake occurred 155 miles off the coast of Aceh Province of Sumatra. Located about 6 miles below the seabed of the Indian Ocean, where the Indo-Australian Plate subducts beneath the Philippine Plate, it caused a tsunami that resulted in sea waves reaching heights of 35 ft or more, devastating shorelines in Sri Lanka, India, Indonesia, Malaysia, Thailand and Maldives. Deaths were reported to exceed at least 50,000, many of which occurred in Sri Lanka, 1000 miles distant, where the sea wave arrived 2 h after the earthquake. A number of tsunamis have occurred in recent years in the East Indian Ocean including Flores in Indonesia in 1992 resulting in more than 2000 deaths, as reported by Synolakis (2003).

Mediterranean Sea: Occasional occurrence.

Characteristics

At sea: Tsunamis can be caused by nearby earthquakes, or as often occurs, by earthquakes with epicenters thousands of kilometers distant from the land areas they finally affect. They are never observed by ships at sea because in the open sea, wave amplitudes are only a meter or so. They travel at great velocities, of the order of 700 km/h or more. The tsunami caused by the 1960 Chilean quake (M=8.4) reached Hawaii, a distance of 10,500 km, in a little less than 15 h, and Japan, 17,000 km distant, in 22 h. (The velocity of water waves is given approximately by the relationship $v=\sqrt{gD}$, where g is the acceleration due to gravity and D is the water depth. Wavelength λ is given by the relationship $\lambda=vT$, where T is the period. Tide gages around the Pacific showed the Chilean tsunami to have a period of about 1 h; therefore, its wavelength was about 700 km.)

Coastal areas: The magnitude of a tsunami at its source is related to the earthquake's magnitude. When it arrives at a coastline, the effect is influenced by offshore seafloor conditions, wave direction, and coastline configuration. Wavelengths are accentuated in bays, particularly where they have relatively shallow depths and topographic restrictions. The wave funnels into the bay and builds to great heights. Containing tremendous energy, the wavefront runs up onto the shore, at times reaching several kilometers inland. The crest is followed by the trough during which there is a substantial drawdown of sea level, exposing the seafloor well below the low tide level. After an interval of 30 min to an hour, depending upon the wave period, the water rises and the second wave crest, often higher than the first, strikes the beach. This sequence may continue for several hours, and the third or fourth wave may sometimes be the highest. At Hilo, Hawaii, after the Chilean event of 1960, the first wave reached 4 ft above mean sea level, the second 9 ft, and the third, 33 ft. On Honshu and Hokkaido, Japan, the water rose 10 ft along the coast during the Chilean tsunami.

Early Warning Services

After the very damaging 1946 tsunami, an early warning service was established by the USGS and centered in Hawaii. When seismograph stations in Hawaii show a Pacific Ocean focus earthquake, radio messages are sent to other Pacific seismograph stations requesting data from which to determine the epicenter. Adequate time is available to compute when tsunami waves might arrive and to so warn the public in coastal areas.

In general, a "watch" is initiated for magnitudes of 7.5 or greater. "Warnings" are issued if tide gages detect a tsunami (Kerr, 1978). Unless waves strike the shores near the epicenter, however, there is no way for people on distant shorelines to know if a tsunami has been generated. Even though the Chilean earthquake caused tsunamis along the Chilean coastline, many people in Hawaii chose to ignore the warning and not to move to higher ground. Japanese officials similarly ignored the warning, since a Chilean earthquake had never before caused a tsunami in Japan. The tsunami reached Hawaii within 1 min of the predicted arrival time.

Hazard Prediction

When important structures or new communities are located along shorelines, the potential for the occurrence of tsunamis should be evaluated. Some procedures for evaluating the tsunami hazard are given by Synolakis (2003).

Qualitatively there are several high-hazard conditions to evaluate:

- Regional tsunami history and recurrence.
- Near onshore earthquakes: recurrence and magnitude.
- Coastline configuration: Irregular coastlines with long and narrow bays and relatively shallow waters appear to be more susceptible than regular coastal plains when exposed to tsunami waves generated by distant earthquakes of large magnitudes.

Seiches

Description

Seiches are caused when ground motion starts water oscillating from one side to the other of a closed or partly closed water body, such as a lake, bay, or channel.

Occurrence

Large seiches are formed when the period of the arrivals of various shocks coincides with the natural period of the water body, which is a function of its depth, and sets up resonance. During the 1959 earthquake at Hebgen Lake, Montana, a witness standing on Hebgen Dam saw the water in the reservoir disappear from sight in the darkness, then return with a roar to flow over the dam. The fluctuation continued appreciably for 11 h with a period of about 17 min. The first four oscillations poured water over the dam. The Lisbon event of 1977 ($M=8.7$) set up seiches all over Europe with the most distant ones reported from Scandinavia, 3500 km away.

3.3.6 The Volcano Hazard

Eruptions

Eruptions, lava flows, and particles thrown into the atmosphere present the hazardous aspects of volcanic activity. In the last 2000 years, there have been relatively few tremendous and disastrous eruptions. Mt. Vesuvius erupted in 79 A.D. and destroyed the ancient city of Pompeii. Mont Pele in Martinique erupted violently in 1902, destroying the city of St. Pierre and leaving but two survivors. Krakatoa in Indonesia literally "blew up" in 1883 in what was probably the largest natural explosion in recorded history.

Two modern significant events were Mt. St. Helens in the United States and Mt. Pinatubo, in the Phillipines. On March 27, 1980, after a week of intermittent Earth shaking, Mt. St. Helens, 50 miles northeast of Portland, Oregon, started ejecting steam, ash, and gas. Finally on May 28, 1980, the mountain top exploded, sending ash and debris some 15 mi into the air. Avalanches, debris flows, and huge mudflows, in addition to the blast forces, caused widespread devastation and flooding, and about 60 deaths. A thick layer of ash was deposited over thousands of square miles. Mt. Pinatubo, in June 1991, was one of the largest eruptions of the 20th century. It destroyed Clark Airbase and displaced thousands of Filipino citizens. Enormous amounts of ash were released, which blanketed large areas.

Flowing Lava

Flowing lava is perhaps the most common cause of destruction. An example is Iceland's first geothermal power plant (ENR, 1976). As work on the $45 million plant was nearing completion, the site was shaken for about a year with tremors, and then fissures opened about 1.6 km away and lava was spewed, threatening the installation. Apparently, the lava was contained by dikes erected around the plant area. Many of the 11 deep wells drilled to capture the geothermal energy were severely damaged by subsurface movements. Between 1983 and 1990, Kilauea volcano on Hawaii erupted with a series of lava flows that eventually destroyed over 180 homes in Kalapana.

Lahars

Lahars have received much attention in recent years as an important geologic hazard. The term lahar describes a hot or cold mixture of water and rock fragments flowing down the slopes of a volcano. In its most destructive mode the mass is very large and moves at high velocities. The greatest danger exists where the volcano has a deep snow cover that can be melted by volcanic activity. Such was the case with Mt. St. Helens; at Mt. Pinatubo, a lahar was initiated by heavy rains. In 1985, a small eruption at a Columbian volcano, Nevado del Ruiz, caused a lahar that took more than 20,000 lives in Armero, located in a valley about 75 km from the volcano's summit.

In the United States there is concern over a potential lahar impacting on the urban area near Portland Oregon, around the base of Mt. Rainier. Evidence of lahars, about 500 years old, have been identified in the valleys near the base of the volcano. This has prompted the USGS to set up a network of stations to monitor and send out alarms when ground vibration thresholds are exceeded so that evacuation can be initiated (Internet: http;//volcanoes.usgs.gov/RainierPilot.html).

Volcanic Hazard Manuals have been prepared by the USGS and UNESCO.

3.4 Earthquake-Resistant Design: An Overview

3.4.1 Introduction

Ground Motion

Dynamic Forces

The large amounts of energy released during earthquakes travel through the Earth as various types of seismic waves with varying oscillation frequencies and amplitudes or displacements. The oscillating particles in the wave possess velocity and exert a force due to the acceleration of gravity (see Section 3.2.3). Earthquake ground motions for structural analysis are usually characterized by peak ground acceleration (or a fraction thereof), response spectra, and acceleration time histories.

In rock and other nearly elastic geologic materials, these dynamic forces result in transient deformations, which are recovered under the low strains of the seismic waves. Interest in the properties of these materials lies primarily in their ability to transmit the seismic waves.

In weaker deposits, such as alluvium, colluvium, and aeolian soils, however, the base-rock excitation transmitted to the soils is usually amplified. In addition, some materials respond to the cyclic shear forces by densifying, liquefying, or reducing in shear strength (unstable soils).

Significance

The effect of the dynamic forces, therefore, can be divided into two broad categories:

1. The effect on structures subjected to the forces transmitted through the ground, which result in ground shaking.
2. The effect on the geologic material itself, primarily in the form of response to cyclic shear forces.

Field Measurements

Ground displacement (amplitude), used in Richter's relationship to compute magnitude, is measured by seismographs.

The force imposed on structures, in terms of the acceleration due to gravity, which has both horizontal and vertical components, is measured by accelerographs.

Surface Damage Relationships

Factors

Earthquake destruction is related to a number of factors including magnitude, proximity to populated areas, duration of the event, the local geologic and topographic conditions, and the local construction practices.

Example

The effects of some of these factors are illustrated by a comparison of damage to two cities during earthquakes of similar magnitude: Managua, Nicaragua (1972), M=6.2, and San Fernando, California (1971), M=6.6. Both events affected an area with about 400,000 inhabitants. The Managua quake resulted in 6000 deaths compared with 60 in San Fernando; the difference is related to soil conditions and local building practices. Managua is located over relatively weak lacustrine soils, and relatively few structures have been constructed with consideration for seismic forces. The San Fernando valley is filled with relatively compact soils and most major structures have been constructed according to modern practices. Even with "modern" practices, however, a number of new structures were severely damaged as shown in Figure 3.39 and Figure 3.40.

FIGURE 3.39
Modern freeway structures at the interchange of Highways 5 and 210, San Fernando, California, damaged by the earthquake of February 9, 1971. (Photograph courtesy of the U.S. Geological Survey.)

FIGURE 3.40
Damage suffered by the Olive View Hospital during the San Fernando earthquake of February 9, 1971 (M=6.6; I_{max} =VIII–XI). Damage to the newly constructed reinforced buildings included the collapse and "pancaking" of the two-story structure in the upper left (1), the collapse of the garage and other structures in the foreground (2), and the toppling of three four-story stairwell wings (3). Vertical accelerations combined with horizontal accelerations were a significant cause in the collapse. (Photograph courtesy of the U.S. Geological Survey.)

3.4.2 Structural Response

General Characteristics

Reaction to Strong Ground Motion

Dynamic forces are imposed on structures by strong ground motion. Structural response is related to the interaction between the characteristics of the structure in terms of its mass, stiffness, and damping capability, and the characteristics of the ground motion in terms of the combined influence of the amplitude of ground accelerations, their frequency components, and the duration. *Damping* refers to a resistance that reduces or opposes vibrations by energy absorption.

The elements of ground motion are amplitude A, displacement y, frequency f (Hz), and period T (sec), and the derivatives velocity v and acceleration α.

Energy Transmitted to Structures

Maximum vibration velocity imposed by the elastic wave as it passes beneath the structure in terms of circular frequency w and amplitude is

$$V_{max}=2\pi fA \text{ cm/sec} \tag{3.23}$$

The greatest acceleration to which the structure is subjected is

$$\alpha_{max}=4\pi^2f^2A \text{ cm/sec}^2 \tag{3.24}$$

The greatest force applied to the structure is found when acceleration is expressed in terms of the mass. This allows for the development of an expression relating force to the acceleration due to gravity, frequency, and amplitude as follows:

$$F_{max}=m\ \alpha_{max}=(W/g)(\alpha_{max})=(W/g)4\pi^2f^2A \quad \text{dyn} \tag{3.25}$$

where W is the weight. It is seen that the force varies as the square of the frequency.

Kinetic energy is possessed by a body by virtue of its velocity. The energy of a body is the amount of work it can do against the force applied to it, and work is the product of the force required to displace a mass and the distance through which the mass is displaced. Kinetic energy is expressed by

$$KE_{max} = 1/2\, mv^2_{max} = 1/2\, (W/g)v^2_{max} \quad \text{erg} \tag{3.26}$$

where v is the velocity with which a structure moves back and forth as the seismic wave passes (Leet, 1960).

Energy transmitted to the structure may be represented in several ways:

- Motion amplitude (displacement), frequency or acceleration (results from combining amplitude and frequency).
- Force with which the energy moves the structure.
- Energy itself, defined in terms of the motion which it produces (kinetic energy).

Ground Shaking and Analysis

Base excitation of the structure from ground shaking results in horizontal and vertical deflections for some interval of time and imposes strains, stresses, and internal forces on the structural elements. The shaking intensity depends on the maximum ground acceleration, frequency characteristics, and duration.

Analysis requires definition of the system motion in terms of time-dependent functions, and determination of the forces imposed on the structural members.

Response Modes

Structures exhibit various modes of response to ground motion depending on their characteristics.

Peak horizontal ground acceleration relates closely to the lateral forces imposed on a structure and is the value used normally for approximating earthquake effects.

Vertical acceleration of ground motion can cause the crushing of columns. During the downward acceleration of a structure, the stress in the columns is less than static. When the movement reverses and becomes upward, an acceleration is produced, which causes an additional downward force that adds loads to the columns. This effect was a major factor in the collapse of the Olive View Hospital during the San Fernando event (see Figure 3.40). Vertical accelerations are also important input to the design of certain structures such as massive dams and surface structures such as pipelines.

Differential displacements beneath structures can cause distortion and failure of longitudinal members.

Frequency, in relation to the natural frequency of the structural elements, governs the response of the structure. Low frequencies (long periods) cause tall structures to sway.

Duration of shaking or *repeated application* of forces causes fatigue in structural members (and a continuous increase in soil pore pressures).

Dynamic Reaction of Structures

Source

The dynamic reaction of a structure to ground motion is governed by its characteristic period T or structural frequency of vibration ω which is related to structural mass, stiffness, and damping capability.

Characteristic Periods

- Very rigid, 1 story structure: 0.1 sec (very high frequency, $f=10$ Hz).
- Relatively stiff structures of 4 to 6 stories: 0.4–0.5 sec ($f=2$ Hz).
- Relatively flexible structures of 20 to 30 stories: 1.5 to 2.5 sec ($f=0.5$ Hz).
- Very flexible structures where deformation rather than strength governs design, and wind loads become important: 3 to 4 sec or higher (very low, $f=0.25$ Hz).

The ground vibrates at its natural period which in the United States varies from about 0.4 to 2 sec depending generally on ground hardness. When the ground period approaches the natural period of the structure, *resonance* may occur. This can result in a significant increase in the acceleration of the structures.

Forces on Structural Members

Static loads result in stresses and deflections.

Dynamic loads result in time-varying deflections, which involve accelerations. These engender inertia forces resisting the motion, which must be determined for the solution of structural dynamics problems. The complete system of inertia forces acting in a structure is determined by evaluating accelerations, and therefore displacements, acting at every point in the structure.

Deflected Shape of Structure: The deflected shape of a structure may be described in terms of either a lumped-mass idealization or generalized displacement coordinates.

In the *lumped-mass idealization*, it is assumed that the entire mass of the structure is concentrated at a number of discrete points, located judiciously to represent the characteristics of the structure, at which accelerations are evaluated to define the internal forces developing in the system.

Generalized displacement coordinates are provided by Fourier series representation.

In either case, the number of displacement components of coordinates required to specify the position of all significant mass particles is called the number of degrees of freedom of the structure (Clough, 1970).

Single-Degree-of-Freedom System

Two types of single-degree-of-freedom systems are shown in Figure 3.41. In both cases, the system consists of a single rigid (lumped) mass M so constrained that it can move with only one component of simple translation. The dynamic forces acting on a simple building frame founded on the surface may be represented by a simple mass–spring–dashpot system as shown in Figure 3.41a. (A dashpot is an energy adsorber.)

Translation motion is resisted by weightless elements having a total spring constant K (stiffness) and a damping device which adsorbs energy from the system. The damping force C is proportional to the velocity of the mass. The fundamental period T is expressed by

$$T=2\pi(M/K)^{1/2} \tag{3.27}$$

During earthquakes, the motion is excited by an external load $P(t)$ which is resisted by an inertia force F_I, a damping force F_D, and an elastic force F_s. The resisting forces are proportional to the acceleration, velocity, and displacement of the mass given in terms of the

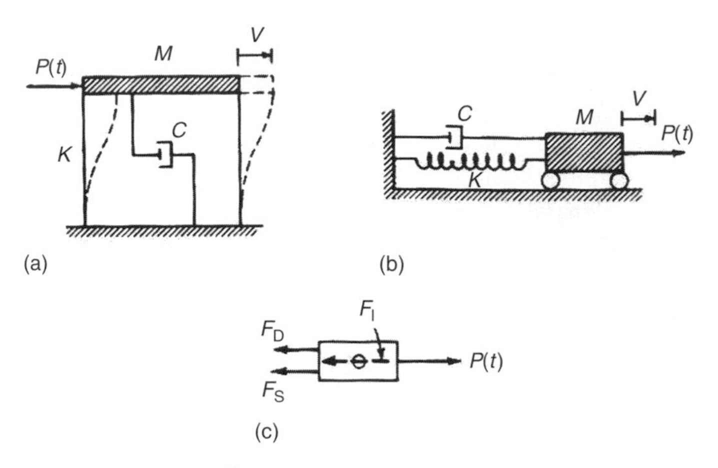

Inertia force $F_1 = M/\ddot{V}$
Damping force $F_D = C\dot{v}$ $P(t) = F_1 + F_D + F_S$
Elastic force $F_S = Kv$
where M = mass; C = damping coefficient; K = total spring constant; v = displacement; $P(t)$ = external load [period $T = 2\pi\ (M/K)^{1/2}$]

FIGURE 3.41
Single-degree-of-freedom systems: (a) simple frame; (b) spring–mass system; (c) forces acting on mass.

differential of displacement v with respect to time as follows: $F_I = M$v (second derivative), $F_D = C$v (first derivative), and $F_s = K$v. Equilibrium requires that

$$P(t) = F_I + F_D + F_s$$

where the force $P(t)$ is the ground acceleration input, equal to the product of the mass and the acceleration, or

$$P(t) = M\alpha_{max} \tag{3.28}$$

Multi-Degree-of-Freedom Systems

In *multistory buildings*, each story can be considered as a single-degree-of-freedom element with its own mass concentrated at floor levels and its own equations of equilibrium similar to the single-story building. One can proceed with analysis by assuming that displacements are of a specific form, for example, that they increase linearly with height.

Free-foundation bases, such as those for mechanical equipment, can move in a number of directions as illustrated in Figure 3.42.

Dynamic Response

The dynamic response of a structure is defined by its displacement history, i.e., by the time variation of the coordinates that represent its degrees of freedom, and by its period of vibration T or frequency ω. The displacements are determined from equations of motion, which are expressions of the dynamic equilibrium of all forces acting on the structure.

External Forces

General

The external forces acting on the structure can be assigned on the basis of the conventional or simple approach or of the comprehensive approach.

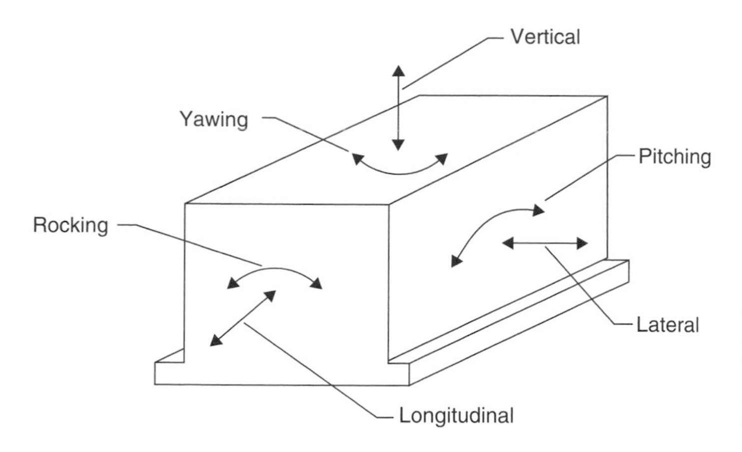

FIGURE 3.42
Six modes, or degrees of freedom, of foundation vibration. Translation modes: vertical, longitudinal, lateral. Rotational modes: rocking, pitching, yawing.

Conventional or Simple Approach

The conventional or simple approach considers only values for acceleration g, which are obtained from provisions in the International Building Code (IBC) or local building codes. (In 2000, the Uniform Building Code [UBC], the Building Officials Code Administrators [BOCA], and the Standard Building Code [SBC] were combined to become the International Building Code [IBC]). These codes often refer to seismic probability maps such as Figure 3.14, or simply give values for acceleration in terms of g for various locations. This approach, at times referred to as the *seismic coefficient method*, does not take into account all of the significant dynamic properties of either structures or earthquakes.

Comprehensive Approach

All ground response factors (see Section 3.4.3) are considered in the comprehensive approach. The design earthquake (see Section 3.4.6) is considered in terms of peak acceleration, frequency content, and duration by statistical analysis of recorded events; information obtained by strong motion seismographs; and geologic conditions. In terms of the structure, the design spectrum, damping, and allowable design stresses are specified. This approach is becoming standard practice for all high-risk structures such as nuclear power plants, 50-story buildings, large dams, long suspension bridges, and offshore drilling platforms, regardless of location.

3.4.3 Site Ground-Response Factors

Design Bases

Seismic design criteria are based on ground-motion characteristics, including acceleration, frequency content, and shaking duration, which are normally given for excitation of rock or strong soils. Soil conditions have an effect on these values.

Maximum Acceleration

Peak horizontal acceleration (PGHA) is considered to be closely related to the lateral forces imposed on a structure, and is the value normally used for approximating earthquake forces.

Peak vertical acceleration (PGVA) has been generally accepted as about one half of the mean horizontal acceleration, but close to dip-slip faulting the fraction may be substantially higher. In Managua (1972), for example, recorded near the epicenter were values of

$\alpha_{h(peak)}=0.35\ g$ and $\alpha_{v(peak)}=0.28\ g$. Hall and Newmark (1977) recommended taking design motions in the vertical direction as two thirds those in the horizontal direction. These relationships generally are applicable where recorded strong motion data are not available.

"Effective" peak acceleration (a fraction of the peak) is often selected for design because very high peaks are frequently of short duration and have little effect on a structure. Low-magnitude events, of the order of $M=4.5$, can have peaks of 0.6 g, but these large accelerations usually occur only as two or three high-frequency peaks, which are probably S wave arrivals, and carry little energy.

Selection of values is described in Section 3.4.6.

Frequency Content

Significance

Maximum accelerations occur when the ground-motion period (frequency) approaches or equals the period of the structure and resonance occurs. Ground-motion amplitude decreases with distance by geometrical spreading and frictional dissipation (attenuation). The high frequencies (shorter periods) are close to the source; at distances of the order of 100 km it is the longer vibrational periods of the Rayleigh waves (1–3 s) that cause ground shaking.

Building periods are usually in the low to intermediate frequency ranges, and buildings are therefore subject to resonance from the long-period waves. Tall buildings are caused to sway, and, when in close proximity, to beat against each other as in Los Angeles during the Kern County event of 1952, where the focus was 125 km distant, and in Mexico City in 1957, where the focus was 300 km distant. In both cases, old and weak, but smaller structures, did not suffer damage.

Design Approach (Hudson, 1972)

In *high-frequency systems* (>5 Hz), the relationship between horizontal ground accelerations and lateral forces on a structure governs design.

Intermediate (≈1 Hz) and *low frequency* (<0.2 Hz, long periods) systems include most buildings and engineering works. Ground accelerations alone are not considered a good approximation of the actual lateral forces. In addition to peak ground accelerations, the maximum ground velocity for intermediate frequencies and the maximum ground displacement for low frequencies should be specified.

Duration

Duration, a measure of the number of cycles, is associated with fatigue in structures and has a major effect on the amount and degree of damage.

Length of faulting is considered to have a strong effect.

Bracketed durations (Section 3.2.7), prepared from strong ground-motion records, are often specified for design (see Section 3.4.6).

Comparisons of Acceleration, Frequency, and Duration

- *Parkfield* (1966, $M=5.5$): High peak α (0.5 g) but high frequency and very short duration caused little damage.
- *Mexico City* (1957, $M=7.5$): Lower peak α (0.01 to 0.1 g) had lower frequencies and a longer duration, causing complete collapse of multistory buildings in a geologic basin with weak soils.

- *Anchorage* (1964, $M=8.6$): High-magnitude event was similar to many historical events except for unusual duration of 3 min, which resulted in the liquefaction failure of many natural, previously stable, slopes.

Soil Condition Effects

Ground Amplification Factor

Bedrock excitation accelerations generally increase in magnitude as soil thickness increases and soil stiffness decreases. The maximum ground amplification factor generally ranges between 1 and 2 for strong motions, and is a function of period (Section 3.2.6).

Effects on Frequency

Local soil conditions filter the motion so as to amplify those frequencies that are at or near the fundamental frequency of the soil profile (Whitman and Protonotarios, 1977), but frequencies are diminished by attenuation.

Various earthquakes may have the same peak acceleration but if they occur with differing periods, the ground response will differ and structural damage may be selective. For example, in San Francisco (1957), in stiff soils, acceleration peaks occurred at low values of the fundamental period (0.4 to 0.5 s); therefore, maximum accelerations would tend to be induced in relatively stiff structures 5 to 6 stories in height, rather than in high-rise buildings. In deep deposits of soft soils, however, peak acceleration occurred at intermediate values of the fundamental period (1.5 to 2.6 s), which would induce maximum acceleration in multistory buildings of 20 to 30 stories, leaving lower, stiffer buildings unaffected (Seed, 1975).

Other Factors

- *Depth effects*: accelerations at foundation level can be substantially lower than at the surface, as discussed in Section 3.2.6.
- Subsidence and liquefaction are discussed in Section 3.3.3.
- Slope failures are discussed in Section 3.3.4.

Microzonation Maps

Geologic conditions and ground response factors presented as microzonation maps have been prepared for a few urban locations. They are useful for planning and preliminary design, and emphasize hazardous areas.

3.4.4 Response Spectra

Description

A *response spectrum* is a plot of the maximum values of acceleration, velocity, and displacement response of an infinite series of single-degree-of-freedom systems subject to a time-dependent dynamic excitation, such as but not limited to ground motion. The maximum response values are expressed as a function of undamped natural period for a given damping. Approximate response spectrum acceleration, velocity, and displacement values may be calculated from each other assuming a sinusoidal relationship. The calculated values are sometimes referred to as pseudo-acceleration, pseudo-relative velocity, or pseudo-relative displacement response spectrum values (USACE, 1999). An example of a response spectrum showing maximum displacements, maximum pseudo-velocities, and maximum pseudo-accelerations presented on a logarithmic tripartite graph is given in Figure 3.43.

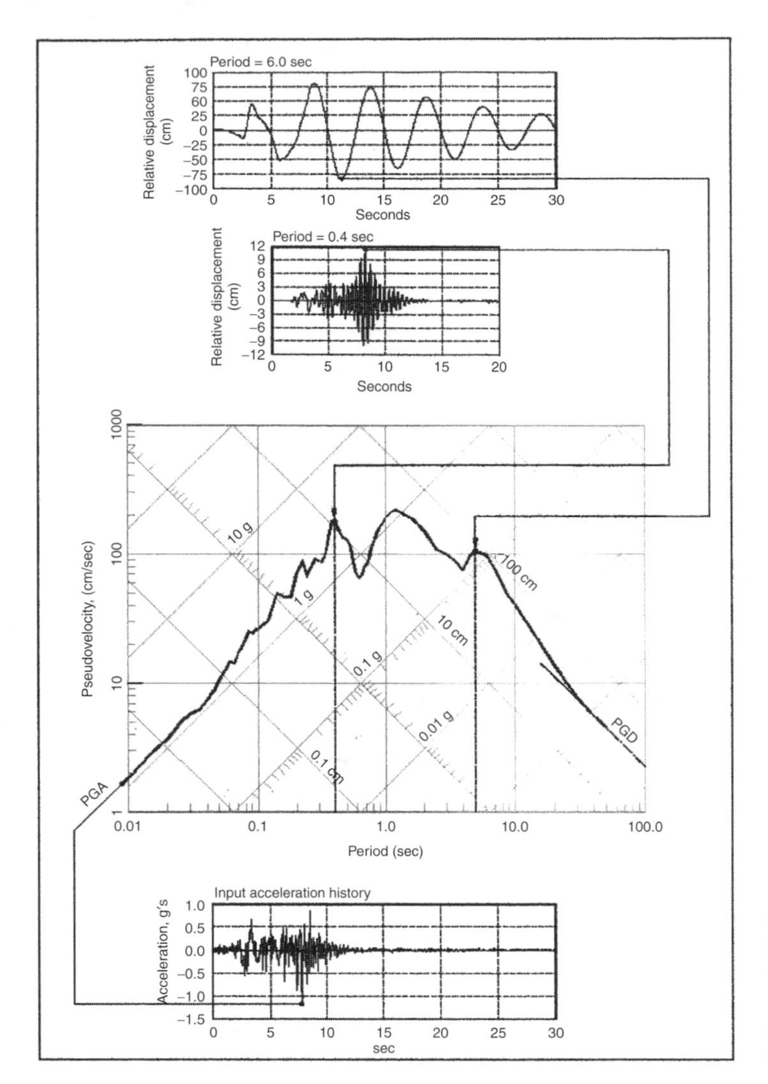

FIGURE 3.43
Construction of tripartite elastic design response spectrum. (From USACE, U.S. Army Corps of Engineers, ER 1110–2–6050, June 30, 1999. With permission.)

Response spectra are typically used to illustrate the characteristics of earthquake shaking at a site. There are two forms of response spectra, standard or normalized and site-specific. *Site-specific response spectra* are usually plots of accelerations *g* vs. a range in period for a specific percent damping. An example is given in Figure 3.18, a family of curves for the Mexico City event of 1985 (see discussion Section 3.2.6).

Standard or normalized response spectra plots are provided for particular areas based on strong motion data. For example, the USGS has prepared a series of maps for California and Nevada, and the Continental United States, showing pseudo-acceleration spectral response for 5% damping, for various periods, and 10% probability of exceedance of 50 and 250 years (USACE, 1995). An example is given in the Figure 3.44. Such maps are included in the 2000 International Building Code.

Applications

Response spectra are used as input in dynamic analysis of linear elastic systems, i.e., they are a convenient means of evaluating the maximum lateral forces developed in structures subjected to a given base motion. If the structure behaves as a single-degree-of-freedom system,

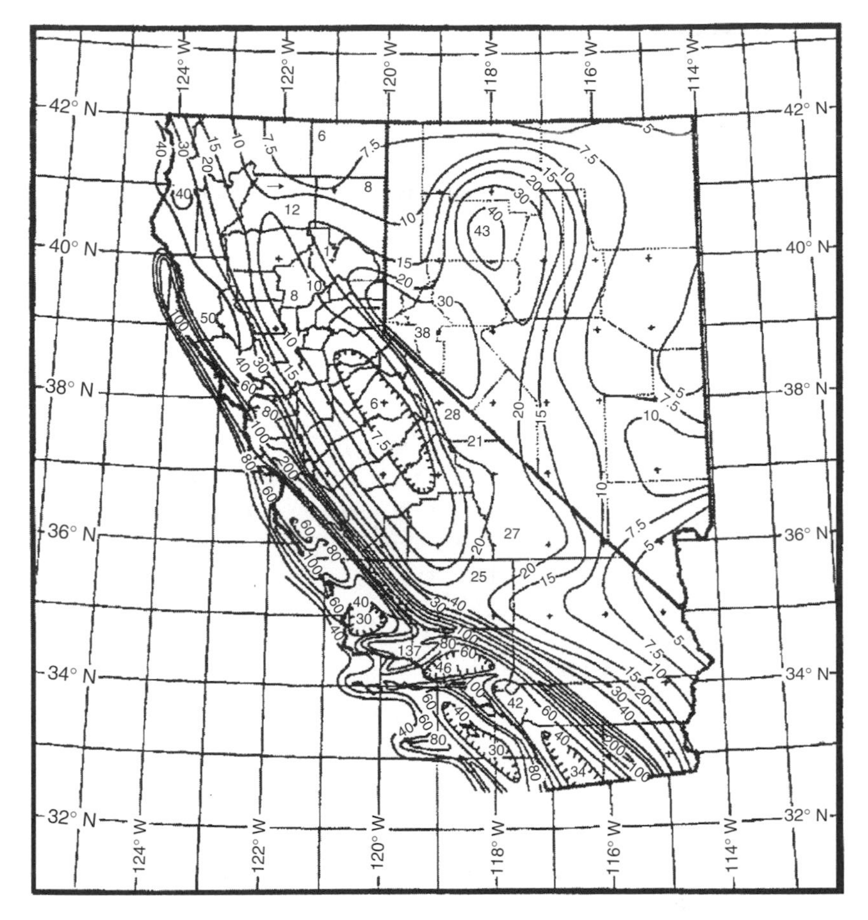

FIGURE 3.44
1991 USGS map of the 5% damped, 1.0 sec pseudoacceleration spectral response, expressed in percent of the acceleration of gravity, with a 10% probability of exceedence in 50 years. (From USACE, U.S. Army Corps of Engineers, ER 1110-2-1806, July 31, 1995. With permission.)

the maximum acceleration and thus the maximum inertia force may be determined directly from the acceleration response spectrum if the fundamental period of the structure is known.

They are also useful for comparing the response of structures in a given earthquake where soil conditions vary, or for comparing a number of earthquakes in the same area, as a function of soil conditions and natural periods.

References

With the availability of strong motion data, the methodology used to develop response spectra has changed significantly in complexity since the first edition of this book in 1984. The reader who wants more detail is referred to Scawthorn (2003) and USACE (1995, 1999). Response spectra have been developed by numerous investigators who are listed in Table 3-2 of EM 1110-2-6050 (USAEC, 1999). The table is a plot of seven investigator references vs. site condition (soil and rock) for western U.S. shallow earthquakes, eastern U.S. earthquakes, and subduction zones (plate edge) earthquakes.

Procedures to prepare response spectra are covered in Section 1615 of the IBC (2000). Two procedures are given: (1) obtaining ground motion data from maps such as Figure 3.44, or (2) a site-specific study.

3.4.5 Seismic Hazard Analysis

Method Selection

General

The method selected for dynamic analysis varies from the relatively simple conventional approach to comprehensive analytical procedures. The selection depends upon the degree of risk and hazard.

Structure Purpose and Type

Pertains to failure consequences (the risk).

- *Conventional structures*, moderate risk, include industrial plants, moderate-height buildings, fossil-fuel power plants, moderate-height dams, etc.
- *Lifeline structures*, moderate to high risk, include roadways, railways, tunnels, canals, pipelines (gas and liquid fuel, water, and sewage), electric power, etc.
- *Critical public structures*, high risk, include schools and hospitals.
- *Unconventional structures*, high to very high risk, include nuclear power plants, 50-story buildings, long suspension bridges, large dams, offshore drilling platforms, etc.

Earthquake Occurrence and Magnitude

Pertains to the *hazard*.

- *High hazard*: Frequent occurrence of high magnitude events (I>VI) or occasional occurrence of high-magnitude events (I>VIII). Zones 3 and 4 in Figure 3.13.
- *Moderate hazard:* Frequent occurrence of I=VI and occasional occurrence of I=VII. Zones 2A and 2B in Figure 3.13.
- *Low hazard*: Generally areas of no activity, or events seldom exceed I=V, and activity is normally low. Zones 0 and 1 in Figure 3.13.

Approach

Conventional or standard design is applied for conventional structures in low-hazard areas, and considers only *g* forces.

Comprehensive or site-specific design is applied to unconventional structures in all areas and to conventional structures, critical public structures, and lifelines in high-hazard areas. The approach considers all site response factors in the development of the design earthquake (see Section 3.4.6), which are employed in more comprehensive analysis.

Standard Approach

Generally standard studies use preliminary values of ground motion (seismic coefficient) obtained from published seismic zone maps (Figures 3.13 through 3.44), a preliminary structural analysis, and a simplified assessment of soil liquefaction and deformation. Standard studies may be satisfactory for final design and evaluation in seismic zones 1 or 2A (Figure 3.13), and are used to set the scope of site-specific studies.

Comprehensive Analytical Methods

General

The combined influence of ground accelerations, their frequency contents, and, to some extent, ground shaking duration, in relation to the period and damping of the structure, are considered in comprehensive analytical methods. Input can be from actual strong-motion records from simulated earthquake motion, or from response spectra.

Analysis considers the structure as being linear-elastic and having single or multiple degrees of freedom. Ground shaking causes base excitation, which is established by equations expressing the time-dependent (dynamic) force in terms of the structure's characteristics (mass, stiffness, and damping factors) as related to time-dependent acceleration, velocity, and displacement. A dynamic mathematical model is developed representing the structure and all of its elements, closely simulating the interaction effect of components on each other and the response of each to the dynamic forces.

Deterministic Seismic Hazard Analysis (DSHA)

This site-specific approach uses the known seismic sources (usually faults) sufficiently near the site and available historical seismic and geologic data to prepare models of ground motion at the site. One or more earthquakes are specified by magnitude and location, and are usually assumed to occur on the portion of the source closest to the site. The ground motions are estimated deterministically, given the magnitude, source-to-site distance, and site conditions (USACE, 1995).

Probabilistic Seismic Hazard Analysis (PSHA)

This approach uses the elements of the DSHA and adds an assessment of the potential for ground motions during the specified time period. The probability or frequency of occurrence of different magnitude earthquakes on each significant seismic source and inherent uncertainties are directly accounted for in analysis. Ground motions are selected based on the probability of exceedance of a given magnitude during the service life of the structure or for a given return period (USACE, 1995).

For both the DSHA and PSHA methods, site-specific ground motion studies are required to provide magnitude, duration, and site-specific values for the PGA, PGV, PGD, and design response spectra and time histories in both the horizontal and vertical directions at the ground surface or on a rock outcrop. Studies should also consider Soil Structure Interaction (SSI) effects which may reduce ground motions at the base of the structure.

Time-History Analysis

The response spectrum mode superposition fully accounts for the multimode dynamic behavior of the structure, but is limited to the linear elastic range of behavior and provides only maximum values of the response quantities. The time-history method is used to compute deformations, stresses, and section forces more accurately considering the time-dependent nature of the dynamic response to earthquake ground motion (USAEC, 2003).

The basic data input may be ground motion from an actual strong-motion record, or simulated earthquake motion. The corresponding response in each configuration of the vibrating system (mode) is calculated as a function of time. The total response, obtained by summing all significant modes, can be evaluated for any desired instant.

Seismic Analysis Progression

The USACE (1995) provides a summary of the method of seismic analysis applicable for the various phases of investigation for the seismic zones given in Figure 3.13. These relationships are given in Table 3.13.

TABLE 3.13
Seismic Analysis Progression

Zone	Project Stage				
	Reconnaissance		Feasibility		Design Memo[a]
0 and 1	E	→	SCM	→	RS[b]
2A and 2B	E	→	SCM	→	RS
	SCM[b]	→	RS[b]	→	TH[c]
	SCM	→	RS	→	TH
3 and 4	SCM	→	RS	→	RS[d]
	RS[b]	→	TH[c]	→	TH[c]

Note: E = Experience of the structural design engineer; SCM = Seismic coefficient method of analysis; RS = Response spectrum analysis; TH = Time-history analysis.

[a] If the project proceeds directly from feasibility to plans and specifications stage, a seismic design memorandum will be required for all projects in zones 3 and 4, and projects for which a TH analysis is required.

[b] Seismic loading condition controls design of an unprecedented structure or unusual configuration or adverse foundation conditions.

[c] Seismic loading controls the design requiring linear or nonlinear time-history analysis.

[d] RS may be used in seismic zones 3 and 4 for the feasibility and design memorandum phases of project development only if it can be demonstrated that phenomena sensitive to frequency content (such as soil–structure interaction and structure–reservoir interaction) can be adequately modeled in an RS.

Source: From USACE, U.S. Army Corps. of Engineers, ER 1110-2-1806, July 31, 1995. With permission.

3.4.6 The Design Earthquake

Definitions

General

The design earthquake is normally defined as the specification of the ground motion as a basis for design criteria to provide resistance to a moderate earthquake without damage, and to provide resistance to a major event without collapse.

Standard Structures (Uniform Building Code, 1997)

- *Design basis earthquake* (*DBE*): The building should not collapse in an earthquake with a 10% probability of exceedance in 50 years.
- *Operating basis earthquake* (*OBE*): The building should not collapse in an earthquake with a 50% probability of exceedance in 50 years.

Standard Structures (USACE, 1999)

- *Operating basis earthquake* (*OBE*) is an earthquake with a 50% probability of exceedance during the service life of the structure, which performs with little or no damage and without interruption or loss of service.
- *Maximum design earthquake* (*MDE*) is the maximum level of ground motion for which a structure is designed, and the structure performs without catastrophic failure, although severe damage or economic loss may be tolerated.

Nuclear Power Plants

Two loading levels are given by the U.S. Nuclear Regulatory Commission (USNRC):

- *Operating-basis earthquake* (*OBE*) is the vibratory ground motion through which the safety features of the plant must remain functional while the plant remains in operation. Some portions of the plant needed for power generation but not required for safe shutdown or radiation protection, however, need not be designed to resist the OBE and conceivably the plant could cease to produce power if subjected to an event of OBE magnitude. Category 1 structures are those critical to safe operation and shutdown and include the reactor containment; the auxiliary, fuel handling, radioactive waste, and control buildings; and the intake screen house at the cooling water source. The OBE is often substantially lower than the SSE.

- *Safe-shutdown earthquake* (*SSE*) is the largest vibratory motion that could conceivably occur at any time in the future. The operator must be able to shut the plant down safely after such an event, even if some of the plant components are damaged.

Approaches to Selection

Standard Approach

Conventional structures in areas of low seismic hazard are based on selection of values for *g*, and normally such values are obtained from existing publications such as national or local building codes, or seismic risk and probability maps such as Figure 3.14.

Alternatively, if there are no published criteria, as may be the situation in many countries, *g* may be estimated from records of intensity or magnitude by conversion, using relationships such as Figure 3.13, Table 3.14, or Equation 3.10. In unfamiliar areas without building codes, if the hazard degree is in doubt, then a *site response study* is undertaken.

Comprehensive Approach

Design of unconventional structures or structures in high hazard zones (Section 3.4.5) is based on the selection of values for peak effective horizontal and vertical *g*, frequency content, and duration.

Existing data are collected and reviewed within a radius of about 200 mi (320 km) (NRC, 1997) from the proposed structure, including catalogs of intensity reports and event magnitudes, strong-motion records, geologic maps, known faults, and other information, and a site response study is performed.

TABLE 3.14

Design Seismic Horizontal Ground Motions[a]

	Design Acceleration (gravities *g*)		Design Velocity (in./sec)	
Magnitude[b]	**Ground Motion[c]**	**Structures[d]**	**Ground Motion[c]**	**Structures[d]**
8.0	0.60	0.33	29	16
7.5	0.45	0.22	22	11
7.0	0.30	0.15	14	7
5.5	0.12	0.10	6	5

[a] From Hall, W. J. and Newmark, N. M., *Proceedings of ASCE*, New York, 1977, pp. 18–34. With permission.

[b] *Magnitudes* are considered as the design maximum earthquake (were developed for four seismic zones along the trans-Alaska pipeline).

[c] *Ground motion:* Peak values which may affect slope stability, liquefaction of cohesionless materials, or apply strains to underground piping.

[d] *Structural design:* Peak values used for design of structures or other facilities Since they account for structural features and response, and soil–structure interaction, they are generally less than those used to define soil response.

Site Response Study

General

The following applies to: (1) areas where seismic coefficient data (Figures 3.14, 3.15, and 3.44) are not available, or (2) the reconnaissance phase of a site-specific study.

Historical Seismicity

Catalogs of worldwide earthquakes are available from the USGS Earthquake Hazards Program, or from the Advanced National Seismic System (ANSS) hosted by the Northern California Earthquake Data Center. Information is provided in terms of intensity and magnitude, and in some cases, the magnitude vs. the number of times exceeded for a particular time interval.

Depending on the region and information available, the first step in a site response study is to estimate site intensity I from an analysis of earthquake history data. All known events of I=IV or greater, occurring within 320 km of the site (NRC requirement), are located and zoned by intensity as a seismicity map, such as given in Figure 3.11. Alternatively, relationships between various magnitudes ($M>4.0$) and exceedance for various earthquake locations within some site distance are prepared. From this, the distance of earthquakes of various magnitudes from the site are determined. For example, one may find that an event of $M=4.2$ occurred 3 mi from the site and the largest event of $M=7.5$ occurred 60 km from the site.

Recurrence analysis (Section 3.2.8) is performed to determine the probable return of events of various magnitudes to locations where they have occurred in the past (the source, usually given as the epicenter). Events of significant magnitudes with return periods of 50 or 100 years are usually selected, depending on the importance and type of the structure. "Engineering Lifetimes" for various types of constructions are given in Table 3.15. Events of the highest magnitude and those with a recurrence interval closest to the economic life of the structure are of major interest. Judgment is required to evaluate the results of the recurrence analysis. If the maximum intensities of some of the historical data are overestimated, the results might indicate that either the region is overdue for the occurrence of a damaging shock, or that there is a regional change toward a lower level of intensity.

Site intensity I_s, is estimated by the application of attenuation laws and relationships (see Section 3.2.5) such as Figure 3.16, or it is imposed on a "capable" fault and then attenuated to the site.

Geologic Study of Fault Structures

Geologic study is performed to locate fault structures and their lengths, and to identify capable faults (Section 3.3.1). Fault systems are correlated with intensity distributions.

TABLE 3.15

Engineering Lifetimes

Type of Construction	Lifetime (years)
Nuclear power plants	40–80
Buildings and pipelines	50
Bridges, tunnels, flood-control structures, and navigation locks	100
Dams	100–150
Solid-waste disposal in landfills	250
Repositories for hazardous nuclear wastes	10,000

From *Civil Engineering*, ASCE, November 1993.

Minimum fault lengths in terms of site distance requiring detailed study are given in Table 3.16. The potential magnitude may be estimated from the length of the capable fault (or faults) as given in Figure 3.27 or Equation 3.18. Assumptions regarding rupture length have varied by practitioners from one-half to one-third of the total (Adair, 1979).

When the information is available, all faults within some specified distance from the site should be identified in terms of distance, maximum magnitude of events, and fault type. Some relationships between construction damage for maximum source to site distances in terms of site acceleration, magnitude, and intensity are given in Table 3.17. Fault type is important since it has been found that for vertical or dipping faults the direction of rupture propagation in the near field can cause significant differences in the level of shaking for different orientations relative to the fault's strike (Section 3.3.1, Dip-Slip Displacement).

Control width: The NRC (1997) requires an evaluation for a specific fault of potential rupture width (Equation 3.19) and potential rupture area (Equation 3.20). Note that Equations 3.18 through 3.20 were developed for sites in the western United States. The *control width* of a fault is defined (NRC, 2003) as the maximum width of the zone containing mapped fault traces. These include all faults that can be reasonably inferred to have experienced differential movement during Quaternary times and that can join or reasonably be inferred to join the main fault trace, measured within 10 mi along the fault's trend in both directions from the point of nearest approach to the site, as shown in Figure 3.45. The control width requires detailed investigation for a specific Nuclear Power Plant location. The zone width requiring detailed study vs. the potential earthquake magnitude is given in Table 3.18.

TABLE 3.16

Minimum Fault Length to Be Considered in Establishing Safe-Shutdown Earthquake[a]

Distance from Site (mi)	Minimum Length (mi)
0–20	1
Greater than 20–50	5
Greater than 50–100	10
Greater than 100–150	20
Greater than 150–200	40

[a] From NRC, Appendix A to Part 100, U.S. Nuclear Regulatory Commission, 2003.

TABLE 3.17

Source to Site Distance for Earthquake Damage

Damage to Construction	Minimum Site Acceleration (g)	Earthquake: Richter Magnitude	Earthquake: Modified Mercalli Intensity	Maximum Distance Earthquake Source to Site (km)
Stable foundation	0.15	6.0	VIII	20
	0.15	7.0	X	32
	0.15	8.0	XI	50
Soil liquefaction, permanent ground displacement	0.10	5.3	VII	1
	0.10	6.0	VIII	10
	0.10	7.0	X	50
	0.10	8.0	XI	150
Seismic-wave amplification in soft soil	0.05	7.0	X	230
	0.05	8.0	XI	400

Maximum distance for damage to good construction (mean excitation) in western United States.
From *Civil Engineering*, ASCE, November 1993.

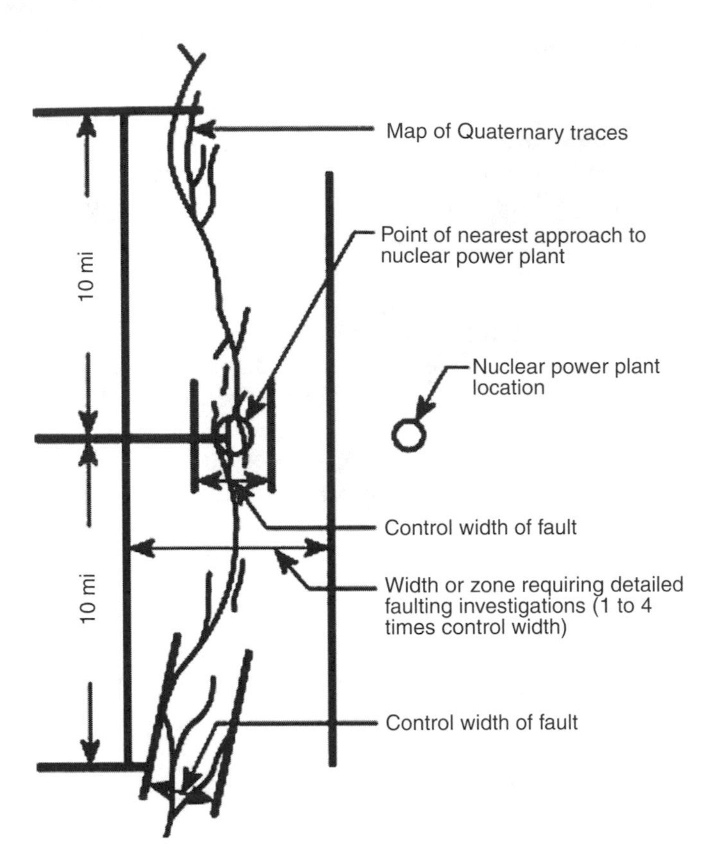

FIGURE 3.45
Diagrammatic illustration of delineation of width of zone requiring detailed faulting investigations for the specific Nuclear Power Plant location. (From NRC, Appendix A to Part 100, 2003. With permission.)

TABLE 3.18
Determination of Zone Requiring Detailed Faulting Investigation[a]

Magnitude of Earthquake	Width of Zone of Detailed Study
Less than 5.5	1 × control width
5.5–6.4	2 × control width
6.5–7.5	3 × control width
Greater than 7.5	4 × control width

[a] From NRC, Appendix A to Part 100, U.S. Nuclear Regulatory Commission, 2003.

It is assumed that the focus, whether the intensity is selected from the recurrence analysis or is based on possible rupture length, will be located at the closest point on the capable fault. This is considered a reasonable assumption for shallow-focus events.

Attenuation: Peak acceleration for rock excitation at the site, in terms of g, is estimated by *attenuation* from the fault (Section 3.3.1). One relationship is given in Figure 3.28. Attenuation relationships have been developed by numerous investigators who are listed in Table 3-1 of EM 1110-2-6050 (USAEC, 1999). The table is a plot of 16 investigator references vs. site condition (soil and rock) for western U.S. shallow earthquakes, eastern U.S. earthquakes, and subduction zone (plate edge) earthquakes.

The earthquake considered in planning the ultimate design is usually the largest that might occur at the closest approach to the site of any capable fault. Judgment is used in the selection of the design event; main considerations are recurrence probability and magnitude.

Site intensity, magnitudes, or estimates of g are thus found for rock excitation. Evaluations obtained by these foregoing procedures are, at best, only approximations applicable to shallow-focus shocks.

Selection of Ground Motion for Rock Sites

Strong-motion records or *response spectra* for an earthquake of the design magnitude for various ground conditions in the site area would provide the most reliable data on acceleration and frequency content, but as yet such data are not available for many areas of the world. Response spectra are frequently being updated and there are a number of approaches (Section 3.4.4).

In many cases it is necessary to estimate site ground motion from correlations.

Horizontal acceleration for rock excitation may be estimated roughly from intensity or magnitude values such as given in Figure 3.13 or Equation 3.10, or from Table 3.14, developed from an evaluation of strong-motion records.

Vertical acceleration for rock excitation has been estimated as 0.5 PHGA or higher if the site is close to a capable fault. Hall and Newmark (1977) recommend using two thirds of the values given in Table 3.14.

Frequency content is estimated from Figure 3.30, giving the predominant period vs. M vs. distance from the causative fault, or estimated from Table 3.14, giving design velocity vs. M. For long-distance earthquakes, of the order of 100 km or more, the possibility of sway in high-rise structures due to long-period waves is considered.

Duration of shaking is estimated on the basis of the bracketed duration in terms of distance and magnitude from Table 3.5, which provides these data for an acceleration $>0.05\,g$ and a frequency >2 Hz, or is developed from strong-motion records. Consideration is given to the fact that the duration of large earthquakes depends largely upon the length of faulting.

Selection of Ground Motion for Soil Sites

Soil conditions usually influence acceleration by amplification. Maximum accelerations from the same earthquakes occur at different periods for different soil types and thicknesses. Damage is selective and varies with building height and period and other factors (see Section 3.4.3). *Soil amplification factor* may be estimated from Figure 3.20, which gives general relationships between acceleration in terms of g, period, and foundation conditions. Attenuation relationships and response spectra developed by numerous investigators for soil sites are listed in Tables 3-1 and 3-2 of EM 1110-2-6050 (USAEC, 1999). The alternative procedure for evaluating soil response is by SSI analysis employing dynamic soil properties (see Section 3.4.7).

The prerequisite for selecting ground-motion relationships for soil sites is a subsurface investigation (Section 3.5.3). *In situ* and laboratory testing is performed to obtain data on the soils for evaluations of the liquefaction potential, and for the amplification factor to determine "g" for SSI analysis, and dynamic soil properties for the SSI analysis.

Synopsis

Site acceleration can be given for design in several ways:

- Maximum peak or effective peak acceleration.
- Acceleration at a given period.
- Real or synthetic time motion that provides for structural periods and damping.
- Continuous spectrum of time motion based on actual recorded events.

3.4.7 Soil-Structure Interaction (SSI) Analysis

General

The *purpose* of a SSI analysis is to evaluate a coupled bedrock–soil–structure system including resonance and feedback effects, for foundations on or below the surface.

Feedback from Structural Oscillations

For relatively light structures founded on rock or strong soils, the influence of the structure is minimal and the structural model excitation is essentially the same as for the prescribed ground motion. The situation is different for massive structures on strong soils and conventional structures on deep, weak deposits. Feedback of structural oscillations to the underlying soils, in these cases, may significantly affect the motion at the soil–structure interface, which in turn may result in amplification or reduction of the structural response. Deformation of the soil formation may also be caused by the feedback from the horizontal, vertical, or rotational oscillatory motion of the structure. The problem is complicated by founding below ground level, the usual procedure for most heavy structures.

Soil Classes in Seismic Loadings

Stable soils undergo plastic and elastic deformations but will dampen seismic motion, will maintain some characteristic strength level, and are amenable to SSI analysis. Both the dynamic input to the soil from the excitation of the underlying rock (which basically applies a shearing stress to the soils), as well as the feedback from structural oscillation, should be considered in dynamic analysis, for which dynamic soil properties (shear modulus and damping ratio) should be used (see Section 3.3.2).

Unstable soils are subject to a sudden and essentially complete strength loss by liquefaction, or sudden compression resulting in subsidence, and are not readily considered in SSI analysis.

Half-Space Analysis

The Model

An early approach to SSI analysis was the *half-space analysis* (Seed et al., 1975). The soil effects on structural response are represented by a series of springs and dashpots (energy adsorbers) in a theoretical half-space surrounding the structure as shown in Figure 3.46. The approach has limitations when applied to buried structures and is best used to analyze surface structures.

Ground Motions

For the problem illustrated, the horizontal earthquake motion was specified at the ground surface in the *free field* (a location where interaction between soil and structure is not occurring). It was assumed that the ground surface motions for the relatively thin soil deposit were the result of vertically propagating shear waves and the maximum accelerations and corresponding time histories were found for other depths down to the bedrock surface. A maximum acceleration at the foundation base level was selected. (The procedure is described in Schnabel et al., 1971.)

Analysis

Representative values for the spring constants were computed and an analysis of structural response made for damping ratios of 7 and 15%, which led to values of maximum acceleration at the base of the structure of 0.38 and 0.32 g, respectively.

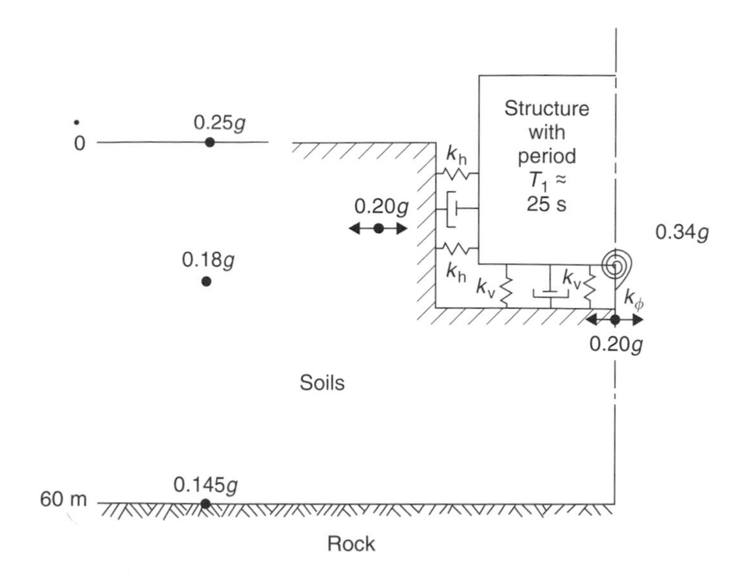

FIGURE 3.46
Soil–structure interaction model for half-space analysis. (From Seed, H. B. et al., Report No. EERC 75–25, Earthquake Engineering Center, University of California, Berkeley, August 1975. With permission.)

Limitations

As of 1974, the approach does not consider material damping, can only be applied to one- or two-layer soil systems, and provides no means for determining strains induced in the soils. These and other limiting factors are summarized in Table 3.19. The strains induced in the soils greatly affect soil deformation moduli G used in the determination of the spring constant.

Finite-Element Method of Analysis

Soil Profile Presentation

The model idealizes the soil continuum as a system of finite elements interconnected at a finite number of nodal points (Figure 3.47). Either triangular or rectangular elements can be used, depending upon the geometry of the conditions being modeled.

In most cases, the soils are considered to be equivalent linear-elastic materials. Soil response is described by formulating stiffness and mass matrices, and a nodal solution or a time-marching integration is effected, depending upon the capability of the particular computer program employed. Response–time histories of displacement, velocity, and acceleration can be computed for each nodal point. Soil characteristics required include shear modulus, Poisson's ratio, soil unit weight, and damping coefficients. The variations of the shear moduli and damping coefficients with strain are considered.

Ground Motion

Control motion is based on actual earthquake data, or on synthetic records, to obtain vertical and horizontal excitation in terms of g, and can be specified as located in the free field at the surface, or at foundation level. A wide variety of ground response spectra has been specified for the design of major facilities, such as nuclear power plants, major bridges, and other critical structures such as LNG storage and processing plants. For nuclear power plants the majority of the design ground response spectra is usually based on a statistical analysis of earthquakes of different magnitudes and site distances.

Design spectra construction is usually based on statistical analysis of recorded motions, frequently for a 50 to 84% nonexceedence probability. Figure 3.48 compares several *site-independent* ground response spectra used in the design or evaluation process in terms

TABLE 3.19
Comparison of Half-Space Analysis with Finite Element Analysis[a]

Consideration	Half-Space Theory or Interaction Springs	Finite Element Analysis
Deformability of soil profile and variation of accelerations with depth	Usually assumes that accelerations are constant with depth	Can take account of deformations of profile and variability of accelerations with depths
Characteristics of motion below the base of structure	Usually assumes that the motions below the base of the structure (and usually around structure) are the same as those in the free field	Can readily take into account the influence of interaction on the characteristics of the motions below the base of the structure
Determination of soil motions adjacent to structure	Provides no means for determining motions adjacent to structure	Provides means for determining motions adjacent to the structure
Determination of soil deformation characteristics	Characteristics can only be approximated	Characteristics can be determined on rational basis
Determination of damping effects	Damping can only be estimated	Damping can be appropriately characterized and considered in analysis
Effects of adjacent structures	Effects cannot be considered	Effects can readily be evaluated
Inclusion of high-frequency effects	Effects are appropriately included	Effects may well be masked by computational errors, including: (1) the use of a coarse mesh, (2) the use of Rayleigh damping, and (3) the use of too few modes
Lateral extent of model	Not a factor in analysis	Must be large enough to provide required boundary conditions
Three-dimensional configuration	Can be considered in analysis	Must be represented by two-dimensional model

[a] From Seed, H. B. et al., Report No. EERC. 75–25, Earthquake Engineering Center, University of California, Berkeley, August 1975. With permission.

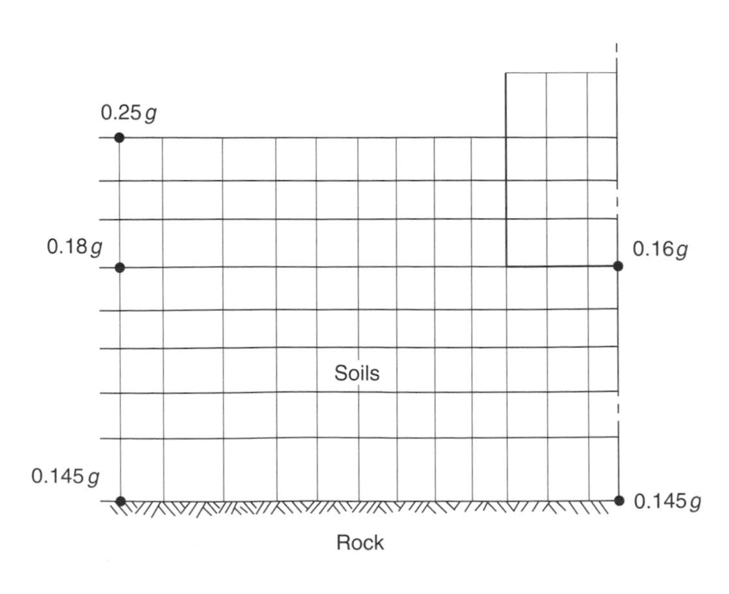

FIGURE 3.47
Finite–element model of soil–structure system of Figure 3.46. (From Seed, H. B. et al., Report No. EERC 75–25, Earthquake Engineering Center, University of California, Berkeley, August 1975. With permission.)

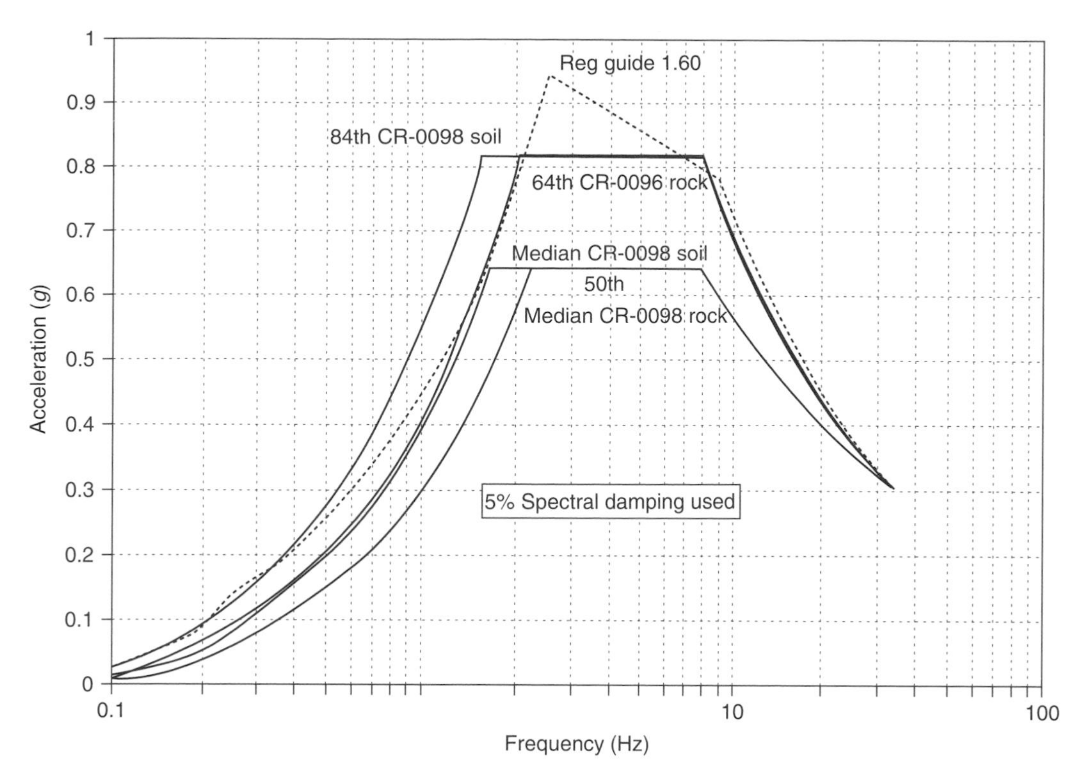

FIGURE 3.48
Examples of aggregated site-independent ground motion response spectra. (From Johnson, J. J., *Earthquake Engineering Handbook,* Chen, W. and Scawthorm, C., Eds., CRC Press, Boca Raton, Florida, 2003, Chap. 10. With permission. After U.S. NRC NUREG-0098.)

of acceleration vs. frequency. The frequency content of the motion is one of the most important aspects of the free-field motion as it affects structural response. In the absence of site-specific response spectra it has been used to define design criteria in the United States and in other countries (Johnson, 2003).

General Procedure

The procedure is illustrated by a sample analysis of major structures for a nuclear power plant presented by Idriss and Sadigh (1976), as shown in Figure 3.49. Ground motion was specified at the foundation level.

Step 1: Because the control motion is specified typically at some point below the surface in the free field, a deconvolution analysis is necessary to determine compatible base-rock motions. These are the motions that must develop in an underlying rock formation to produce the specified motions at the control point. SHAKE is a widely used computer program for computing one-dimensional seismic response of horizontally layered soil deposits based on the equivalent-linear method. It is used in the deconvolution analysis, in which the most important assumptions are

- The site response is dominated by shear shaking from below, and all other modes of seismic energy are neglected.
- The shear shaking is undirectional and the site responds with a state of plain strain.

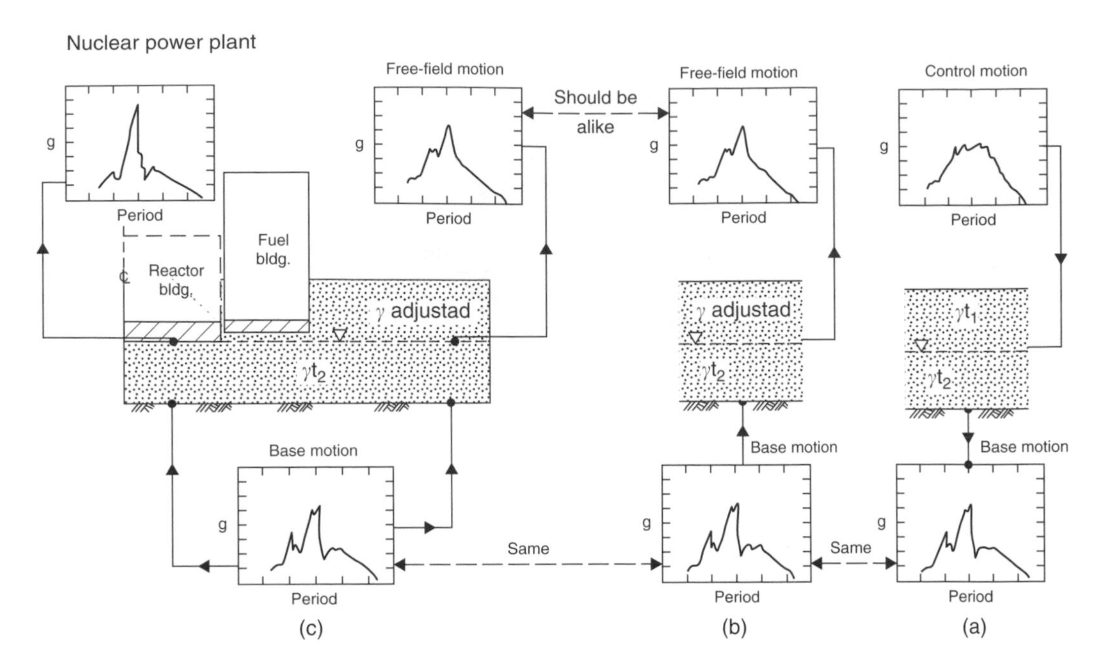

FIGURE 3.49
Representation of general analysis procedure used in soil–structure interaction analysis: (a) soil deposit model used in deconvolution analysis to obtain compatible motion of base rock; (b) soil deposit representing free field with modified soil properties (resulting from turbine building); (c) finite–element model of soil–structure system. (After Idriss, I. M. and Sadigh, K., *Proc. ASCE, J. Geotech. Eng. Div.*, 102, 663–682, 1976.)

- The stress–strain trajectories are cyclic.
- There are no residual displacements.
- There is no soil liquefaction.

Step 2: Base-rock motion from Step 1 is then used for analysis of the soil–structure system, leading to an evaluation of the motions at any selected points such as the base of the structure, operating floor, etc.

A number of other computer programs are available that allow for nonlinear dynamic analysis, as well as consideration of excess pore pressures and the presence of structural elements. PLAXIS and SASSI 2000 are based on the finite-element method (FEM).

Conclusions

Comparison with Half-Space Analysis

In the FEM model illustrated in Figure 3.47 of the problem illustrated in Figure 3.46, Seed et al. (1975) found that the maximum acceleration at the base of the structure would be only 0.16 *g*, or roughly half that found by the half-space analysis.

Damping was appropriately characterized by soil modulus values compatible with the strains that developed in the different elements representing the soil deposit. A comparison between the FEM and the half-space approach for buried structures is summarized in Table 3.19.

From the study comparing half-space analysis with the FEM, Seed et al. (1975) concluded that there is still much to be learned about SSI that even sophisticated analyses do not have the capability to incorporate all aspects of reality, and that considerable judgment is required in evaluating the results obtained in any analysis.

In addition to the limitations of the analytical procedures, there are uncertainties in measuring soil properties and ground-motion characteristics. Thus the problem is not only extremely complicated, but it is not well defined. The result is necessary conservatism in the design of important structures.

3.5 Investigation: Important Structures in High-Hazard Areas

3.5.1 Introduction

Purpose

This chapter tends to follow procedures and guidelines employed in studies for nuclear power plants in the United States, but is intended as a conservative guide for important structures located in any high-hazard seismic region. References are NRC (1979, 1997, 2003) and USAEC (1995, 1999).

To establish the design earthquake a "probabilistic seismic hazard analysis" (PSHA) is performed. The methodology quantifies the hazard at a site from all earthquakes of all possible magnitudes, at all significant distances from the site of interest, as a probability by taking into account their frequency of occurrence (Thenhaus and Campbell, 2003). The elements of the PSHA are given in Figure 3.50. One of the elements, "The Logic Tree," is illustrated in Figure 3.51. It provides a systematic approach to evaluating a region with a number of seismic sources and an earthquake history.

Objectives and Scope

Investigation objective is basically safe and economical construction, which requires:

- Identification and treatment of geologic hazards (avoid, reduce, or eliminate).
- Estimation of the design earthquake.
- Establishment of foundation design criteria.
- Evaluation of structural response to dynamic forces.

Study scopes range from simple to complex, depending upon several major factors, including:

- Importance of structure and the degree of risk (see Section 3.4.5).
- Regional seismicity (the degree of hazard) and adequacy of available data.
- Physiographic conditions (mountains, coastline, plains, etc.).
- Regional and local geology (hazards, rock types and structure, soil types, and characteristics, and groundwater conditions).

An *investigative team* to accomplish the objectives includes geologists, seismologists, geophysicists, geotechnical engineers, and structural engineers.

3.5.2 Preliminary Phase

Purpose

During the preliminary phase, existing data on regional and local seismicity and the natural environment are collected and reviewed either to provide a database for a site selection or a feasibility review of a previously selected site.

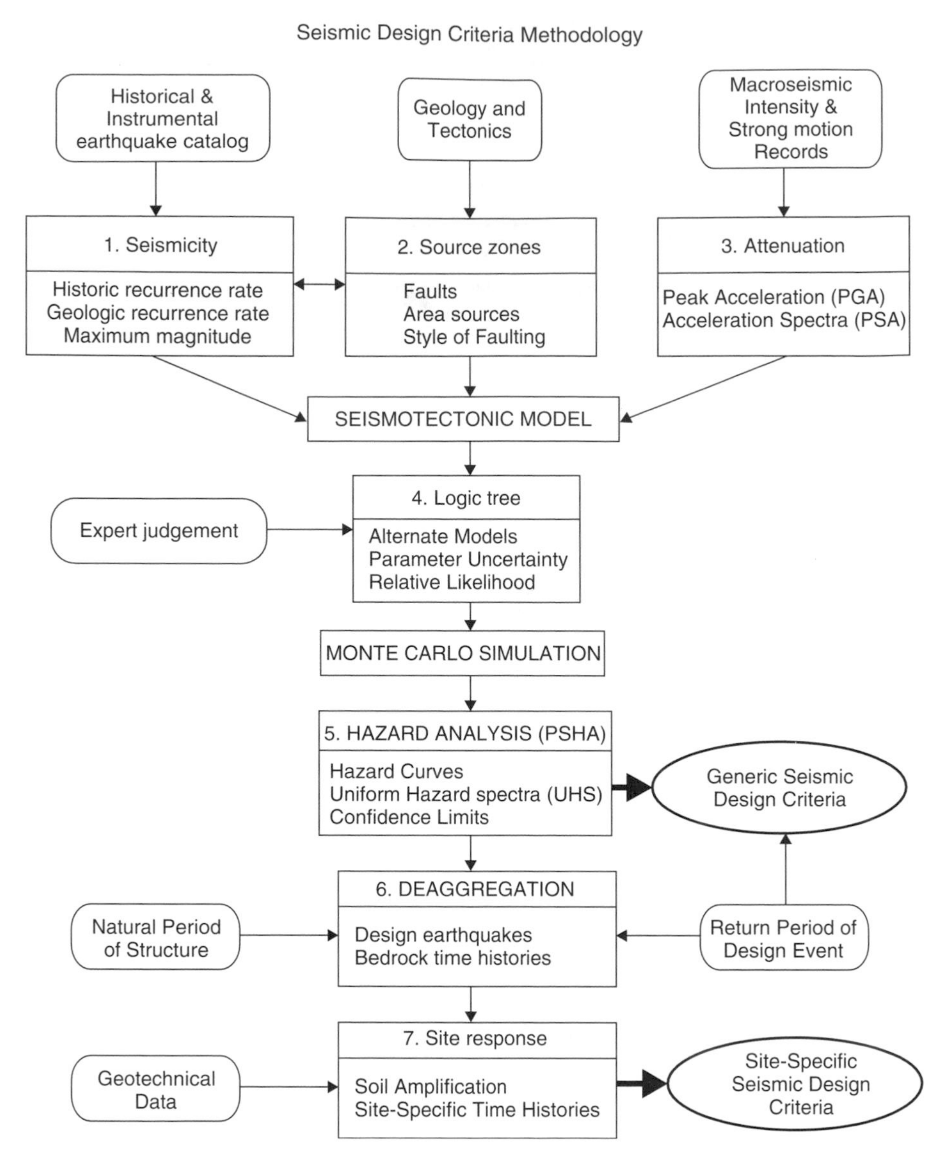

FIGURE 3.50
Flowchart showing the elements of the probabilistic seismic hazard methodology (PSHA). Since uncertainty is inherent in the earthquake process, the parameters of the seismotonic model are systematically varied via logic trees (Figure 3.51), Monte Carlo simulation, and other techniques. (From Thenhaus, P. C. and Campbell, K. W., *Earthquake Engineering Handbook*, Chen, W. and Scawthorn, C., Eds., CRC Press, Boca Raton, FL, 2003, Chap. 4. With permission.)

Seismicity

A "Site Response Study" is conducted as described in Section 3.4.6. *Study scope* depends upon the current regional seismic activity, the historical activity, and the completeness and type of historical records available.

Existing data review is the first step in evaluating the hazard degree. World *seismicity maps* are useful for projects in any Country for an overview to establish the general site location in relation to plate edges. *Seismic risk maps* or *microzonation maps* provide data on the hazard degree and ground response. *National* and *local building codes* provide information on

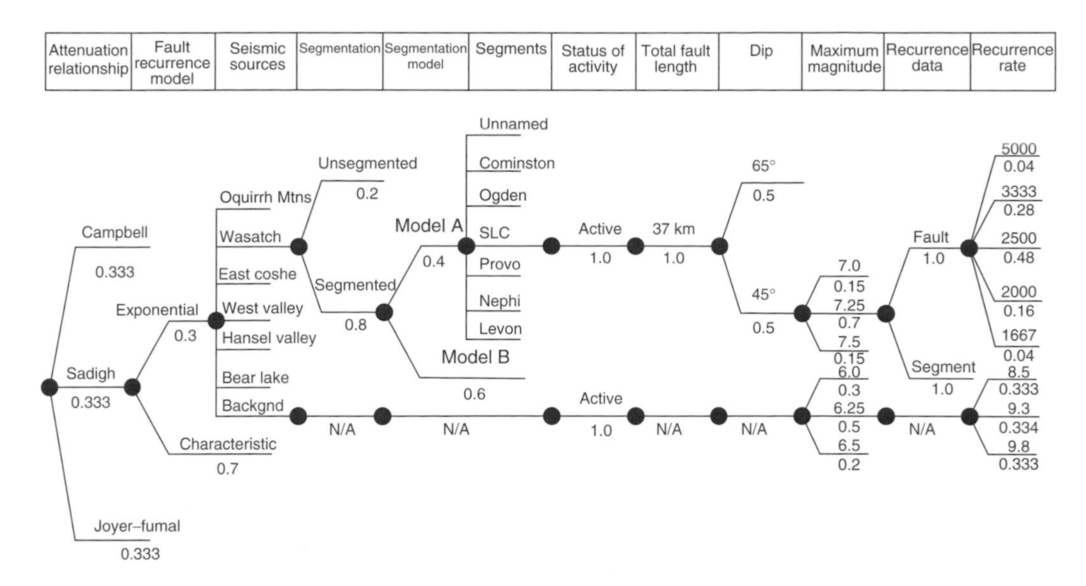

FIGURE 3.51
Example of logic tree simulation for seismic forces to assess regional earthquake hazards and risk along the Wasatch Fault, Utah. (From Youngs, R. R. et al., *Assessment of Regional Earthquake Hazards and Risk along the Wasatch Fault, Utah,* Gori, P. L. and Hays, W. W., Eds., USGS Open File Rep. 87–585, 1987, pp. M1–M110. With permission.)

design criteria. A guide to international seismic codes is available from the *National Center for Earthquake Engineering Research* (NCEER), in Buffalo, New York.

Data adequacy is evaluated. Codes, strong-ground-motion records, response spectra, recurrence studies, microzonation maps, etc. may provide information adequate for design, or a suitable design earthquake or other design basis may have already been established for the area. The possible impact of a nearby reservoir on induced seismicity should be considered (Section 3.2.1).

The Natural Environment

Purpose and Scope

The purpose of a review of existing data is the identification of important natural factors affecting the suitability of the site location from the viewpoint of hazard degree and the anticipation of potential foundation problems. Items of interest are physiography, climate, and geology. In addition to the collection and review of existing data, new data may be generated during the preliminary phase by terrain analysis techniques and field reconnaissance. Data are also collected for other environmental aspects including flora and fauna, but are not included in the scope of this discussion.

Physiography

Information on physiography is obtained from topographic maps and remote-sensing imagery. The importance of physiography lies in its direct relationship in many cases to the geologic hazards, and with respect to seismicity it influences attenuation (Section 3.2.6), although the relationships are not well defined. Some very general associations may be established:

- *Mountainous regions*: Characteristic features are slope failures, variable regional geology, intense surface fault systems, and, in some locations, volcanoes.

Intensity distribution may be reduced and highly modified, as in the western United States.

- *Great Plains and other large areas of reduced relief*: Regional geology is likely to be more uniform, but limestone and potential collapse conditions more prevalent, fault systems less pronounced and identifiable, and intensity distribution over very large areas relatively uniform, as in the central United States (see Figure 3.18).
- *River valleys* are possible flood areas with potentially liquefiable soils in the floodway, pastoral, and estuarine zones; relatively poor foundation conditions; and fault systems less pronounced and identifiable in broad, mature valleys.
- *Coastal areas* exhibit great geological variations, but may contain liquefiable or otherwise unstable soils, and, depending upon location and configuration, may be subjected to the tsunami hazard.

Climate

Climate relates to hazards and geologic conditions, as an aid or as a requirement for predictions. Some typical associations are the evaluation of the flood hazard, the type and depth of residual soil development, and the potential for slope failures, collapsing soils, and expansive soils.

Regional and Local Geologic Conditions

Regional conditions: include information on rock types and structures (faults, floods, etc.), and the hazards of slope failures, ground subsidence or collapse from fluid or solid extraction or from natural causes, regional warping and tilting, and volcanoes.

Local conditions: include information on physiography and geology. Evaluation of conditions may result in recommendations to abandon the site and select another location if the constraints are judged to be too severe. The most severe constraints are local active faulting and warping, high liquefaction potential, large-scale unstable or potentially unstable slopes, volcanism, ground collapse, and tsunami potential.

3.5.3 Detailed Study of Regional and Local Geologic Conditions

Fault Studies

Prepare Geologic Structure Map

Data from the literature and terrain analysis techniques are used to prepare a map showing all tectonic structures, including lineaments, within 200 mi of the site, with the primary objective of locating all faults with the length–distance relationships given in Table 3.16. If a known active or capable fault is within the 200 mi radius, and is a major fault, such as the San Andreas Fault, then it is considered as the limit of the study area. The geologic structure map is overlaid with the seismicity map and events correlated with lineaments to identify potentially capable faults.

The NRC (2003) provides guidance in determining which faults may be of significance in evaluating the safe-shutdown earthquake. In general, either capable or noncapable faults with lengths less than those indicated in Table 3.16 need not be considered.

Investigate for Capable Faults

Investigations start near the site and extend outward to locate the design capable fault. Judgment is required to evaluate the significance of lineament or fault length, site proximity, faulting evidence, etc. with regard to the necessity of a detailed investigation to evaluate the capability of a particular fault. Fault characteristics and identification are described in Section 3.3.2. Study procedures are given in Section 3.4.6, "Geologic Study."

Detailed reconnaissance and remotely sensed *imagery interpretation* are performed for each lineament of interest to identify faulting evidence, fault type, past displacements, and apparent activity on the basis of observations.

Fault zone extent and control width: Fault traces near the plant site are mapped along the fault for a distance of 10 mi in both directions from the point of its nearest approach to the site, as shown in Figure 3.45. Because surface faulting may have occurred beyond the limit of the mapped fault traces, detailed faulting investigation of a zone beyond the control width is required. The width of this zone depends upon the largest potential earthquake related to the fault as given in Table 3.18.

Explorations are made of candidate faults with geophysical surveys, vertical and angle borings (including coring, sampling, and sensing with nuclear probes), and trenches to determine fault existence, zone width, and geometric attitude.

Quaternary dating: Numerous methods have been developed to date geologic formations. *Radiometric dating* (Appendix A.4 and Table 3.8), performed to evaluate fault activity as exposed in test trenches, is still the most popular.

Investigate Other Major Hazards

- Flood potential
- Slope stability (see Sections 1.5 and 3.3.4)
- Ground subsidence and collapse, faulting, and induced seismic activity from fluid extraction (see Section 2.2)
- Ground subsidence and collapse from subsurface mining (see Section 2.3.5)
- Ground collapse from failure of cavities in soluble rock (see Section 2.4.3)
- Liquefaction and subsidence potential (see Section 3.3.3)

Site Soil and Foundation Studies

Objectives

Soil formations: Determine stratigraphy and soil types, identify the potential for ground compression and heave, and measure static and dynamic strength, deformation properties, and permeability.

Rock formations: Determine stratigraphy and rock types, and identify the degree and extent of weathering and distribution and the nature of discontinuities. Measure pertinent engineering properties.

Groundwater conditions: Locate the static, perched, and artesian conditions, and determine water chemistry. Evaluate susceptibility to changes with time and weather, or other transient conditions.

Explorations

Terrain analysis and field reconnaissance are performed to provide data for a detailed geologic map of surface conditions.

Stratigraphy is investigated with geophysics, test borings, trenches, pits, etc.

Samples for identification and laboratory testing are obtained from borings, trenches, and pits.

Property Measurements

Static and dynamic properties are measured *in situ* and in the laboratory. Of particular interest are cyclic shear moduli and damping ratios for SSI studies. The very low strains of *in situ* direct seismic tests should be correlated with laboratory results to obtain simulations with earthquake strains.

Instrumentation

Piezometers are installed as standard procedure to monitor groundwater fluctuations. Other instrumentation may be installed to monitor natural slopes, areas of ground collapse or subsidence, or active or potentially active faults, depending upon the time available for the study.

3.5.4 Evaluation and Analysis

Ground-Motion Prediction

Select Design Earthquake for Rock Excitation

The design earthquake is obtained with the procedures given in Section 3.4.6 and is designated as the Safe Shutdown Earthquake (SSE) or the Operating Basis Earthquake (OBE) depending upon the ground motion criteria selected.

Soil Profile Effects (Section 3.4.3)

Rock motion must be converted into soil motion either at the surface or at foundation level. This is normally accomplished either by applying amplification factors, or by SSI analysis based on measured soil dynamic properties. Both frequency and duration should be considered. Ground amplification factors vary with soil type, thickness, and rigid-boundary conditions (i.e., bedrock surface configuration). Soil conditions usually lead to amplification, but at times attenuation occurs; maximum accelerations occur at different periods for different soil types and thicknesses during the same earthquake. Damage can be selective and varies with building rigidity, height, and period, among other factors, such as construction type and quality.

Foundation Design Criteria and Structural Response

Foundation Evaluation and Selection

Evaluations are made of potentially unstable soils to determine the possibility of subsidence, liquefaction, or permanent reduction in strength when they are subjected to dynamic shear forces. The evaluation includes the possibility of developing high shearing stresses in soft clays, which could lead to the rupture of deep foundations, and the possibility of embankment failures. (This evaluation should be made as early as feasible during the investigation, since it may show that the site should be abandoned.)

Suitable foundation types are selected for the structure, or structures, from evaluations of the mechanical properties of the soils and other factors.

Seismic Input for Design Response

A seismic design analysis can vary from a relatively simple one in which only a horizontal acceleration force is applied to the structure, to a complex one in which all elements of ground motion are considered.

Complex analysis considers the combined influence of the amplitude of ground motions, their frequency contents, and to some extent duration of shaking in terms of period and damping. A design response spectrum may be applied that represents an average of several appropriate spectra developed into a design envelope. The influence of the frequency content is important, and some investigators recommend that instead of scaling the entire response spectra to ground accelerations, a better overall picture of ground response is obtained by specifying in addition the maximum ground velocity for intermediate frequencies and the maximum ground displacement for the lower (long-period) frequencies. Magnitudes for design spectra can be determined from equations that estimate amplitudes directly, instead of scaling them from the estimates of peak ground velocity. Alternatively, complete input motion can be obtained from actual accelerograph records or produced synthetically from given base-rock motion for a particular site.

Seismic Hazard Analysis (Section 3.4.5)

For foundations supported on rock or strong soils, response is evaluated by time-motion methods or response-spectrum analysis. For structures founded on relatively deep and weak soils, the feedback from structural oscillations is evaluated by SSI analysis or soil amplification factors. The base-rock motion, given in terms of g varying with period, is used to evaluate the motions at any point, such as the structure base, employing values for the dynamic shear modulus and damping ratios of the soils. An alternative procedure is to specify a control or input motion at some point in the "free field" from which comparable rock motions are determined.

3.5.5 Limitations in the Present State of the Art

General

Limitations of knowledge regarding earthquakes as they pertain to engineering are severe and require the application of considerable judgment based on experience as well as generous safety factors on any project. The limitations are well recognized, and substantial effort is being applied by various disciplines involved with earthquake engineering to improve the various ignorance factors, some of the more significant of which are given below. Unfortunately, capabilities in rigorous mathematical analysis appear to be far in advance of the capabilities of generating accurate and representative data.

Earthquake Characteristics

Focal-Depth Effects

Procedures to define the design earthquake are considered approximately valid for the continental western United States, where modern earthquakes are generally shallow and fault-related. The applicability to intermediate and deep-focus events, which usually are not associated with surface faulting, is not well defined.

Occurrence Prediction

Seismicity data in many locations are meager, usually based on "felt" reports, and cover only a relatively brief historical period (200 years or so in the United States) in comparison with "recent" geologic time of, say, about 10,000 years from the last glacial age. Seismic activity has been found to be cyclic in many areas, with cycle lengths longer than the time interval of data in some areas. (How can it be known when essentially singular events such as New Madrid or Charleston may return, or may occur in some other region with "historically" low activity?)

Felt reports are based on the response of people and structures; therefore, they depend on development and demography for registry. As the world population increases, so does the incidence of "felt" earthquakes. Seismograph stations have only been in operation for about 100 years, and only since World War II has there been a substantial number of installations in many areas.

Relationships between intensity and magnitude are very general; therefore, conversion of I from felt reports to M or g is, at best, approximations. Recurrence predictions of M and the time interval must be considered as broad approximations for most locations.

Attenuation

Attenuation, an important element in evaluating data from an existing event, depends on many variables such as topography, geology, fault rupture length, and focal depth, and can undergo extreme variation in a given area, as related to both distribution and areal extent. Relationships have not been well defined and estimates must be considered only as broad approximations for most areas.

Ground Conditions and Response

Fault Identification and Capability Evaluation

Faulting does not always extend to the surface; therefore, positive identification can be difficult. Estimation of capability has many uncertainties, and the knowledge of causes and the basis for anticipation of activity are not well established. Faults, considered as dead, and not studied thoroughly, can suddenly become active.

Relationships between fault rupture length and event magnitude are based on relatively few data and restricted geographic areas. Relationships between percent of rupture length and M, or between magnitude propagation (stress drop) and duration, have barely been addressed. Relationships between fault displacement and M, especially for low-intensity events, need study. It is not unusual to find structures in a risky condition with foundations bearing on rock located over a fault in areas of low seismicity.

Ground Response

Relatively few events have been recorded on strong ground-motion instruments and the greater majority of these are from the western United States, and a few other seismically active countries, such as Japan. Foundation and topographic conditions for instrument locations vary and require consideration when accelerograms are evaluated. Relationships among peak and effective horizontal acceleration, vertical acceleration, frequency content, and duration are not established for many conditions of geology and topography, including effects of distance and foundation depth.

Effects of soil conditions have not been well defined; although soil is generally considered to amplify rock excitation, cases of attenuation have been reported. Relationships among soil type, depth, layering effects, bedrock boundary configuration, and response characteristics of acceleration and frequency content are not well established.

Dynamic Soil Properties

Dynamic shear moduli and damping ratios can be measured with acceptable accuracy, primarily in cyclic simple-shear testing, but basically only for those soils in which high-quality undisturbed sampling is possible. Evaluations of other soils must rely upon estimates of properties from various correlation procedures.

Prediction

It is still not possible to predict when an earthquake will occur, but we do know the general location of where they are likely based on historical events. A better understanding of the relationships between surface and fault movements and earthquake occurrences should be obtained by the monitoring data from Space-based instruments such as GPS navigation system and Interferometric Synthetic Aperture Radar (InSAR) (Section 3.2.8).

Instrumentation installed to monitor movements near and on faults will provide valuable data, as will the deep borehole observatory proposed for the San Andreas Fault near Parkfield (Section 3.2.8). They will provide data for specific locations, however, whereas GPS stations can be installed on numerous faults and InSAR obtains regional data on a periodic basis.

References

Adair, M. J., Geologic evaluation of a site for a nuclear power plant, in *Reviews in Engineering Geology*, Hatheway, A. W. and McClure, C. R. Jr., Eds., Vol. IV, Geology in the Siting of Nuclear Power Plants, Geological Society of America, 1979, pp. 27–39.

Aggarwal, Y. P. and Sykes, L. R., Earthquakes, faults and nuclear power plants in southern New York and Northern New Jersey, *Science*, American Association for Advancement of Science, Vol. 200, 1978, pp. 425–429.

Algermissen, S. T. and Perkins, D. M., A Probabalistic Estimate of Maximum Acceleration in Rock in the Contiguous United States, U.S. Geological Survey, Open File Report 76–416, 1976.

Bollinger, G. A., Seismicity and crustal uplift in the southeastern United States, *Am. J. Sci.*, 273, 396–408, 1973.

Bollinger, G. A., The seismic regime in a minor earthquake zone, *Proc. ASCE Numer. Methods Geomech.*, 2, 917–937, 1976.

Bolt, B. A., Duration of Strong Ground Motion, *Proceedings of the 5th World Conference on Earthquake Engineering*, Rome, 1973.

Bolt, B. A., *Earthquakes: A Primer*, W. H. Freeman & Co., San Francisco, 1978.

Bolt, B. A., Horn, W. L., Macdonald, G. A., and Scott, R. F., *Geological Hazards*, Springer-Verlag, New York, 1975.

Bonilla, M. G., Surface faulting and related effects, in *Earthquake Engineering*, Weigel, R. L., Ed., Prentice-Hall Inc., Englewood Cliffs, NJ, 1970, Chap. 3.

Brandes, H.G., Geotechnical and foundation aspects, in *Earthquake Engineering Handbook*, Chen, W. and Scawthorn, C., Eds., CRC Press, Boca Raton, FL, 2003, Chap. 7.

Bray, J. D., Geotechnical earthquake engineering, in *The Civil Engineering Handbook*, CRC Press, Boca Raton, FL, 1995, Chap. 4.

Campbell, K.W., Engineering models of strong ground motion, in *Earthquake Engineering Handbook*, Chen, W. and Scawthorn, C., Eds., CRC Press, Boca Raton, FL, 2003, Chap. 5.

Castro, G., Liquefaction and cyclic mobility of saturated sands, *Proc. ASCE J. Geotech. Eng. Div.*, 101, 1975.

Christian, J. T., Borjeson, R. W., and Tringale, P. T., Probabalistic evaluation of OBE for nuclear power plant, *Proc. ASCE J. Geotech. Eng. Div.*, 104, 907–919, 1978.

Christian, J. T. and Swiger, W. F., Statistics of liquefaction and SPT results *Proc. ASCE J. Geotech. Eng. Div.*, 101, 1975.

Clough, R. W., Earthquake response of structures, in *Earthquake Engineering*, Weigel, R. L., Ed., Prentice-Hall, Inc., Englewood, NJ, 1970, Chap. 12.

Cluff, L. S., Hansen, W. R., Taylor, C. L., Weaver, K. D., et al., Site Evaluation in Seismically Active Regions–Interdisciplinary Approach, Proceedings International Conference on Microzonation, Seattle, October, Vol. II, 1972, pp. 957–987.

D'Appolonia, E., Dynamic loadings, *Proc. ASCE J. Soil Mech. Found. Eng.* Div., 95, 49, 1970.

Dietz, R. S. and Holden, J. C., Reconstruction of Pangaea; breakup and dispersion of the continents; Permian to the present, *J. Geophys. Res.*, 75, 4939–4959, 1970.

Donovan, N. C. and Bornstein, A. E., Uncertainties in seismic risk procedures, *Proc. ASCE J. Geotech. Eng. Div.*, 869–887, 1978.

ENR, Reservoir Filling Linked to Quake, *Engineering News-Record*, Dec. 18, 1975.

ENR, Geothermal Plant Threatened by Volcanoes, *Engineering News-Record*, Dec. 9, 1976, p. 11.

ENR, Quakes Nudge Palmdale Bulge, *Engineering News-Record*, Mar. 22, 1979, p. 3.

Environmental Science Services Administration, Studies in Seismicity and Earthquake Damage Statistics, Appendix B, U.S. Dept. of Commerce, Coast and Geodetic Survey, 1969.

Erdik, M. and Durukal, E., Simulation modeling of strong ground motion, in *Earthquake Engineering Handbook*, Chen, W. and Scawthorn, C., Eds., CRC Press, Boca Raton, FL, 2003.

Esteva, L. and Rosenblueth, E., Espectos de temblores a distancias moderadas y grandes, *Bol. Soc. Mexicano Ing. Sismica*, 2, 1–18, 1969.

Faccioli, E. and Resendiz, D., Soil dynamics: behavior including liquefaction, in *Seismic Risk and Engineering Decisions*, Lomnitz and Rosenblueth, Eds., Elsevier Scientific Publishing Co., New York, 1976, pp. 71–140, Chap. 4.

Finn, W. D., Soil Dynamics-Liquefaction of Sands, *Proceeding of the International Conference on Microzonation*, Seattle, November, Vol. I, 1972, pp. 87–112.

Geotimes, Earthquakes in and Near the Northeastern U.S., 1638–1998, Excerpted from USGS Fact Sheet FS-006–01, 2001.

Guttenberg, B. and Richter, C. F., *Seismicity of the Earth and Associated Phenomena*, Princeton University Press, Princeton, NJ, 1954.

Hall, W. J. and Newmark, N. M., Seismic design criteria for pipelines and facilities, in *The Current State of Knowledge of Lifeline Earthquake Engineering, Proceeding of the ASCE*, New York, 1977, pp. 18–34.

Hamilton, R. M., Earthquake Hazards Reduction Program-Fiscal Year 1978 Studies Supported by the U.S. Geological Survey, Geological Survey Circular 780, U.S. Dept of the Interior, 1978.

Hardin, B. O. and Dmevich, V. P., Shear modulus and damping in soils: Measurement and parameter effects, *Proc. ASCE J. Soil Mech. Found. Eng. Div.*, 98, 603–624, 1972.

Hardin, B. O. and Dmevich, V. P., Shear modulus and damping in soils: design equations and curves, *Proc. ASCE J. Soil Mech. Found. Eng. Div.*, 98, 667–692, 1972.

Hodgson, J. H., *Earthquakes and Earth Structures*, Prentice-Hall Inc., Englewood Cliffs, 1964.

Housner, G. W., Intensity of Ground Shaking Near the Causative Fault, *Proceedings of the 3rd World Conference on Earthquake Engineering*, New Zealand, Vol. 1, 1965.

Housner, G. W., Strong ground motion, in *Earthquake Engineering*, Prentice-Hall Inc., Englewood Cliffs, NJ, 1970, chap. 4.

Hudson, D. E., Strong Motion Seismology, *Proceedings of the International Conference on Microzonation*, Seattle, October, Vol. I, 1972, pp. 29–60.

Hunt, R.E., *Geotechnical Engineering Investigation Manual*, McGraw-Hill Book Co., New York, 1984.

ICOLD, *A Review of Earthquake Resistant Design of Dams*, Bull. 27, *International Committee on Large Dams*, March 1975.

Idriss, I. M., Earthquake Ground Motions at Soft Soil Sites, *Proceedings of the 2nd International Conference on Recent Advances in Geotechniques and Engineering and Soil Dynamics*, St. Louis, MO, II, 2265–2273, 1991.

Idriss, I. M. and Sadigh, K., Seismic SSI of nuclear power plant structures, *Proc. ASCE J. Geotech. Eng. Div.*, 102, 663–682, 1976.

Johnson, J. J., Soil Structure Interaction, in *Earthquake Engineering Handbook*, Chen, W. and Scawthorn, C., Eds., CRC Press, Boca Raton, FL, 2003, chap. 10.

Kerr, R. A., Tidal waves: a new method suggested to improve prediction, *Science*, 200, 521–522, 1978.

Lawson, A. C. et al., The California Earthquake of April 18, 1986, Carnegie Inst. of Washington, 1908; 2 vols and atlas.

Leeds, D. J., The Design Earthquake, in *Geology, Seismicity and Environmental Impact*, Special Publication Association of Engineering Geology, Los Angeles, CA, 1973.

Leet, L. D., *Vibrations from Blasting Rock*, Harvard University Press, Cambridge, MA, 1960.

Leggett, R. F., *Cities and Geology*, McGraw-Hill Book Co., New York, 1973.

Liao, S. and Whitman, R.V., Overburden correction factors for SPT in sand, *J. Geotech. Eng., ASCE*, 112, 373–377, 1986.

Lomnitz, C., *Global Tectonics and Earthquake Risk*, Elsevier Scientific Pub. Co., Amsterdam, 1974.

Makdisi, F. I. and Seed, H. B., Simplified procedure for estimating dam and embankment earthquake-induced deformations, *Proc. ASCE J. Geotech. Eng. Div.*, 104, 849–868, 1978.

Munfakh, G., Kavazanjian, E., Matasovic, N., Hadj-Hamou, T., and Wang, J., Ground Motion Characterization, in Geotechnical Earthquake Engineering Reference Manual, Report No. FHWA-HI-99–012, FHA, Arlington, VA, 1998, Chap. 4.

Murphy, L. M. and Cloud, W. K., United States Earthquakes, 1952, U.S. Dept. of Commerce, *Coast and Geodetic Survey, Serial No. 773*, U.S. Govt. Printing Office, 1954.

Neumann, F., Principles Underlying the Interpretation of Seismograms, Spec. Pub. No. 264 (revised edition), ESSA, Coast and Geodetic Survey, U. S. Govt. Printing Office, Washington, DC, 1966.

NRC, Site Investigations for Foundations of Power Plants; U.S. Nuclear Regulatory Commission Regulatory Guide 1.132, 1979.

NRC, Identification and Characterization of Seismic Sources and Determination of Safe Shutdown Earthquake Ground Motion, U.S. Nuclear Regulatory Commission Regulatory Guide 1.165, Draft issued as DG-1032, March 1979.

NRC, Seismic and Geologic Siting Criteria for Nuclear Power Plants, Appendix A to Part 100, U.S. Nuclear Regulatory Commission, 2003.

Nuttli, O. W., Seismicity of the Central United States, in *Reviews in Engineering Geology,* Vol. IV, Geology in the Siting of Nuclear Power Plants, Geological Society of America, Boulder, CO, 1979, pp. 67–93.

Okumura, K., Yoshioka, T., and Kusfu, I., Surface Faulting on the North Anatolian Fault, U.S.G.S. Open File Report, 94–568, 1993, pp. 143–144.

Rainer, H., Are there connections between earthquakes and the frequency of rock bursts in the mine at Blieburg? *J. Int. Soc. Rock Mech.*, 6, 1974.

Raisz, E., *Map of the Landforms of the United States*, 4th ed., Institute of Geographical Exploration, Harvard University, Cambridge, MA, 1946.

Raleigh, C. B., Healy, J. H., and Bredehoeft, J. D., Faulting and Crustal Stress at Rangely, Colorado, Geophysical Monogram No. 16, Amer. Geophys. Union, Washington, DC, 1972.

Richter, C. F., *Elementary Seismology*, W. H. Freeman & Co., San Francisco, 1958.

Robertson, P. K. and Wride, C. E., Cyclic Liquefaction and its Evaluation Based on the SPT and the CPT, *Proceedings of the NCEER Workshop on Evaluation of Liquefaction Resistance of Soils*, Salt Lake City, UT, Multidisciplinary Center for Earthquake Engineering Research, Buffalo, NY, 1997.

Scawthorn, C., Earthquakes: seismogenesis, measurement and distribution, in *Earthquake Engineering Handbook*, Chen, W. and Scawthorn, C., Eds., CRC Press, Boca Raton, FL, 2003, Chap. 4.

Schnabel, P., Seed, H. B., and Lysmer, J., Modifications of Seismograph Records for Effects of Local Soil Conditions, Report No. EERC 71-8, Earthquake Engineering Research Center, University of California, Berkeley, December 1971.

Schwartz, D. P. and Coppersmith, K. J., Seismic hazards: new trends in analysis using geologic data, in *Active Tectonics*, National Academy Press, Washington, DC, pp. 215–230, 1986.

Seed, H. B., Soil problems and soil behavior, in *Earthquake Engineering,* Prentice-Hall Inc., Englewood Cliffs, NJ, 1970, Chap. 10.

Seed, H. B., Earthquake effects on soil–foundation systems, in *Foundation Engineering Handbook,* Winterkorn and Fang, Eds., Van Nostrand Reinhold Book Co., New York, 1975, Chap. 25.

Seed, H. B., Evaluation of Soil Liquefaction Effects on Level Ground During Earthquakes, State-of-the-Art-Paper, Liquefaction Problems in Geotechnical Engineering, ASCE Preprint 2752, New York, 1976, pp. 1–104.

Seed, H. B., Romo, M. P., Sun, J., Jaime, A., and Lysmer, J., Relationships between Soil C Conditions and Earthquake Ground Motions in Mexico City in the Earthquake of Sept. 19, 1985, Earthquake Engineering Research Center Report No. UCB/EERC-87/15, Univ. of California, Berkeley, 1987.

Seed, H. B. and Booker, J. R., Stabilization of potentially liquefiable sand deposits using gravel drains, *Proc. ASCE J. Geotech. Eng Div.*, 103, 757–768, 1977.

Seed, H. B., Mori, K., and Chan, C. K., Influence of Seismic History on the Liquefaction Characteristics of Sands, Report No. EERC 75–25, Earthquake Engineering Center, Univ. of California, Berkeley, August 1975.

Seed, H. B., Lee, K. L., and Idriss, I. M., Analysis of the Sheffield dam failure, *Proc. ASCE J. Soil Mech. Found. Eng. Div.*, 95, 1969.

Seed, H. B., Idriss, I. M., Lee, K. L., and Makdisi, F. I., Dynamic analysis of the slide in the lower San Fernando Dam during the earthquake of Feb. 9, 1971, *Proc. ASCE J. Geotech. Eng. Div.*, 101, 889–911, 1975.

Seed, H. B., Idriss, I. M., and Kiefer, F. W., Characteristics of rock motion during earthquakes, *Proc. ASCE J. Soil Mech. Found. Eng.* Div., 95, 1199–1218, 1969.

Seed, H. B., Lee, K. L., Idriss, I. M., and Makdisi, F. I., The slides in the San Fernando Dams during the earthquake of February 9, 1971, *Proc. ASCE J. Geotech. Eng. Div.*, 101, 1975.

Seed, H. B., Lysmer, J., and Hwang, R., Soil-structure interaction analysis for seismic response, *Proc. ASCE J. Geotech. Eng. Div.*, 101, 439–458, 1975.

Seed, H. B., Makdisi, F. I., and DeAlba, P., Performance of earth dams during earthquakes, *Proc. ASCE J. Geotech. Eng. Div.*, 101, 967–994, 1978.

Seed, H. B., Tokimatsu, K., Harder, L. F., and Chung, R. M., Influece of SPT procedures in soil liquefaction resistance evaluations, *Proc ASCE J. Geotech. Eng. Div*, 111, 1425–1445, 1985.

Shannon & Wilson Inc. and Agbabian-Jacobsen Associates, Soil Behavior Under Earthquake Loading Conditions, Report prepared for USAEC, Contract W-7405-eng.-26, 1971.

Synolakis, C., in *Earthquake Engineering Handbook,* Chen, W.-F. and Scawthorn, C., Eds., CRC Press, Boca Raton, FL, 2003, pp. 9–13.

Taylor, C. L. and Cluff, L. S., Fault Displacement and Ground Deformation Associated with Surface Faulting, *Proceedings of the ASCE The Current State of Knowledge of Lifeline Earthquake Engineering,* Specialty Conference University of California, Los Angeles, 1977, pp. 338–353.

Taylor, P. W. and Larkin, T. J., Seismic site response of nonlinear soil media, *Proc. ASCE J. Geotech. Eng. Div.*, 104, 369–383, 1978.

Thenhaus, P. C. and Campbell, K. W., Seismic hazard analysis, in *Earthquake Engineering Handbook,* Chen, W. and Scawthorn, C., Eds., CRC Press, Boca Raton, FL, 2003, Chap. 4.

USAEC, Soil Behavior under Earthquake Loading Conditions, National Technical Information Service TID-25953, U.S. Dept. of Commerce, Oak Ridge National Laboratory, Oak Ridge, TN, January 1972.

USACE, Earthquake Design and Evaluation for Civil Works Projects, U.S. Army Corps of Engineers, ER 1110–2–1806, July 31, 1995.

USACE, Response Spectra and Seismic Analysis for Concrete Hydraulic Structures, U.S. Army Corps of Engineers, ER 1110-2-6050, June 30, 1999.

USACE, Time-History Dynamic Analysis Of Concrete Hydraulic Structures, U.S. Army Corps of Engineers, ER 1110-2-6051, Dec. 22, 2003.

UBC, Uniform Building Code Response Spectra, *International Conference of Building Officials,* 1994.

USGS, USGS Earthquake Hazards Program 2003, Internet: earthquake.usgs.gov/docs.

Wallace, R. E., Discussion-nomograms for estimating components of fault displacements from measured height of fault scarp, *Bull. Assoc. Eng. Geol.*, 17, 39–45, 1980.

Wells, D. L. and Coppersmith, K. J., New empirical relationships among magnitude, rupture length, rupture width, rupture area and surface displacement, *Bull. Seismol. Soc. Am.*, 84, 1994.

Whitcomb, J. H., Garmany, J. D., and Anderson, R., Earthquake prediction: variation of seismic velocities before the San Fernando earthquake, *Science, 180,* 1973.

Whitman, R. V. and DePablo, P. O., Densification of Sand by Vertical Vibrations, *Proceedings of the 4th World Conference on Earthquake Engineering*, Santiago, Chile, 1969.

Whitman, R. V. and Protonotarios, J. N., Inelastic response to site-modified ground motions, *Proc. ASCE J. Geotech. Eng. Div.*, 103, 1037–1053, 1977.

Wylie, E. B. and Streeter, V. L., Characteristics Method for Liquefaction of Soils, *Proceedings of the ASCE, Numerical Methods in Geomechanics*, ASCE, New York, Vol. II, 1976, pp. 938–954.

Youd, T. L., Major Cause of Earthquake Damage is Ground Failure, *Civil Engineering*, ASCE, April 1978, pp. 47–51.

Youd, T. L. and Idriss, I. M., Eds., *Proceedings of the NCEER Workshop on Evaluation of Liquefaction Resistance of Soils,* Salt Lake Ciy, UT, NCEER Tech. Rep. NCEER-97-0022, Buffalo, NY, Jan. 5–6, 1997.

Youd, T. L. and Perkins, D. M., Mapping liquefaction-induced ground failure potential, *Proc. ASCE J. Geotech. Eng. Div.*, 104, 433–446, 1978.

Youngs, R. R., Swan, F. H., Powers, M. S., Schwartz, D. P., and Green, R. K., Probabilistic Analysis of Earthquake Ground Shaking along the Wasatch Front, Utah, in Gori, P. L. and Hays, W. W., Eds.,

Assessment of Regional Earthquake Hazards and Risk along the Wasatch Fault, Utah, U.S.G.S. Open File Rep. 87–585, 1987, pp. M1-M110.

Zeevaert, L., *Foundation Engineering for Difficult Soil Conditions,*Van Nostrand Reinhold Book Co., New York, 1972.

Further Reading

Ambraseys, N. N., On the Seismic Behavior of Earth Dams, *Proceedings of the 2nd International Conference on Earthquake Engineering*, Tokyo, July 1960.

Arango, I. and Dietrich, R. J., Soil and Earthquake Uncertainties on Site Response Studies, *International Conference on Microzonation*, Seattle, November 1972.

Bergstrom, R. N., Chu, S. L., and Small, R. J., Dynamic Analysis of Nuclear Power Plants for Seismic Loadings, Presentation Reprint, ASCE Annual Meeting, Chicago, October 1969.

Blazquez, R., Krizek, R. J., and Baiant, Z. P., Site factors controlling liquefaction, *Proc. ASCE J. Geotech. Eng. Div.*, 106, 785–802, 1980.

Bolt, B. A., Elastic waves in the vicinity of the earthquake source, in *Earthquake Engineering,* Weigel, R. L., Ed., Prentice-Hall, Inc., Englewood Cliffs, NJ, 1970a, Chap. 1.

Bolt, B. A., Causes of earthquakes, in *Earthquake Engineering,* Weigel, R. L., Ed., Prentice-Hall Inc., Englewood Cliffs, NJ, 1970b, Chap. 2.

Bolt, B. A., Seismicity, *Proceedings of the International conference on Microzonation,* Seattle, Vol. I., October 1972, pp. 13–28.

Bolt, B. A. and Hudson, D. E., Seismic instrumentation of dams, *Proc.ASCE J. Geotech. Eng. Div.*, 101, 1975.

Bonilla, M. G. and Buchanan, J. M., Interim Report on Worldwide Surface Faulting, U.S. Geological Survey, Open-File Report, 1970.

Cluff, L. S. and Brogan, G. E., Investigation and Evaluation of Fault Activity in the USA, *Proceedings of the 2nd International Congress,* International Association of Engineering Geologists, São Paulo, Vol. I, 1974.

Donovan, N. C., Bolt, B. A., and Whitman, R. V., Development of Expectancy Maps and Risk Analysis, Preprint 2805, ASCE Annual Convention and Exposition, Philadelphia, PA, September 1976.

Epply, R. A., Earthquake History of the United States, Part I, *Strong Earthquakes of the United States* (Exclusive of California, Nevada), U.S. Govt. Printing Office, Washington, DC, 1965.

Esteva, L., Seismicity, in *Seismic Risk and Engineering Decisions,* Lomnitz and Rosenblueth, Eds., Elsevier Scientific Publishing Co., New York, 1976, Chap. 6.

Fischer, J. A., North, E. D., and Singh, H., Selection of Seismic Design Parameters for a Nuclear Facility, *Proceedings of the International Conference on Microzonation*, Seattle, Vol. II, October 1972, pp. 755–770.

Haimson, B. C., Earthquake Related Stresses at Rangely, Colorado, *New Horizons in Rock Mechanics, Proceedings of the* ASCE, 14th Symposium on Rock Mechanics, University Park, PA, June 1972, 1973.

Hudson, D. E., Ground Motion Measurements in Earthquake Engineering, *Proceedings of the Symposium on Earthquake Engineering*, The University of British Columbia, Vancouver, BC, 1965.

Lamar, D. L., Merifield, P. M., and Proctor, R. J., Earthquake Recurrence Intervals on Major Faults in Southern California, in *Geology, Seismicity and Environmental Impact*, Spec. Pub., Assoc. of Eng. Geol., Los Angeles, CA, 1973.

Martin, G. M., Finn, W. D., and Seed, H. B., Fundamentals of liquefaction under cyclic loading, *Proc. ASCE J. Geotech. Eng. Div.*, 101, 1975.

Newmark, N. M. and Hall, W. J., Seismic Design Spectra for Trans-Alaska Pipeline, *Proceedings of the 5th World Conference on Earthquake Engineering*, Rome, Paper No. 60, 1973.

Panovko, Y., *Elements of the Applied Theory of Elastic Vibration*, Mir Publishers, Moscow, 1971.

Park, T. K. and Silver, M. L., Dynamic triaxial and simple shear behavior of sand, *Proc. ASCE J. Geotech. Eng. Div.*, 101, 1975.

Pensien, J., Soil-Pile Interaction, in *Earthquake Engineering*, Prentice-Hall Inc., Englewood Cliffs, NJ, 1970.

Pyke, R., Seed, H., and Chan, C. K., Settlement of sands under multidirectional shaking, *Proc. ASCE J. Geotech. Eng. Div.*, 101, 1975.

Scholz, C. H., Crustal movement in tectonic areas, *Tectonophysics*, 14, 1974.

Seed, H. B., Earth slope stability during earthquakes, in *Earthquake Engineering*, Prentice-Hall Inc., Englewood Cliffs, NJ, 1970b, Chap 15.

Seed, H. B. and Schnable, P. B., Soil and Geologic Effects on Site Response During Earthquakes, *International Conference on Microzonation*, Seattle, Washington, November 1972.

Sherard, J. L., Cluff, L. S., and Allen, C. R., Potentially active faults in dam foundations, *Geotechnique*, 24, 1974.

Sherif, M. A., Bostrom, R. C., and Ishibashi, I., Microzonation in Relation to Predominant Ground Frequency, Amplification and Other Engineering Considerations, *Proceedings of the 2nd International Congress International Association of Engineering Geologists*, São Paulo, Vol. 1, 1974, pp. 11–2.1 to 2.11.

Steinbrugge, K. V., Earthquake damage and structural performance in the U.S., in *Earthquake Engineering*, Prentice-Hall Inc., Englewood Cliffs, NJ, 1970, Chap. 9.

Wallace, R. E., Earthquake recurrence intervals on the San Andreas fault, *Bull. Geol. Soc. Am.*, 81, 2875–2890, 1970.

Wiegel, R. L., Tsunamis, in *Earthquake Engineering*, Prentice-Hall Inc., Englewood Cliffs, NJ, 1970, Chap. 11.

Wong, R. T., Seed, H. B., and Chan, C. K., Cyclic loading liquefaction of gravelly soils, *Proc. ASCE J. Geotech. Eng. Div.*, 101, 1975.

Zaslawsky, M. and Wight, L. H., Comparison of Bedrock and Surface Seismic Input for Nuclear Power Plants, *Proc. ASCE, Numerical Methods in Geomechanics*, Vol. II, ASCE, New York, pp. 991–1000.

Appendix

The Earth and Geologic History

A.1 Significance to the Engineer

To the engineer, the significance of geologic history lies in the fact that although surficial conditions of the Earth appear to be constants, they are not truly so, but rather are transient. Continuous, albeit barely perceptible changes are occurring because of warping, uplift, faulting, decomposition, erosion and deposition, and the melting of glaciers and ice caps. The melting contributes to crustal uplift and sea level changes. Climatic conditions are also transient and the direction of change is reversible.

It is important to be aware of these transient factors, which can invoke significant changes within relatively short time spans, such as a few years or several decades. They can impact significantly on conclusions drawn from statistical analysis for flood-control or seismic-design studies based on data that extend back only 50, 100, or 200 years, as well as for other geotechnical studies. To provide a general perspective, the Earth, global tectonics, and a brief history of North America are presented.

A.2 The Earth

A.2.1 General

Age has been determined to be approximately 4 1/2 billion years.

Origin is thought to be a molten mass, which subsequently began a cooling process that created a crust over a central core. Whether the cooling process is continuing is not known.

A.2.2 Cross Section

From seismological data, the Earth is considered to consist of four major zones: crust, mantle, and outer and inner cores.

Crust is a thin shell of rock averaging 30 to 40 km in thickness beneath the continents, but only 5 km thickness beneath the seafloors. The lower portions are a heavy basalt ($\gamma = 3$ t/m^3, 187 pcf) surrounding the entire globe, overlain by lighter masses of granite ($\gamma = 2.7$ t/m^3, 169 pcf) on the continents.

Mantle underlies the crust and is separated from it by the Moho (Mohorovicic discontinuity). Roughly 3000 km thick, the nature of the material is not known, but it is much denser than the crust and is believed to consist of molten iron and other heavy elements.

Outer core lacks rigidity and is probably fluid.

Inner core begins at 5000 km and is possibly solid ($\gamma \approx 12$ t/m^3, 750 pcf), but conditions are not truly known. The center is at 6400 km.

A.3 Global Tectonics

A.3.1 General

Since geologic time the Earth's surface has been undergoing constant change. Fractures occur from faulting that is hundreds of kilometers in length in places. Mountains are pushed up, then eroded away, and their detritus deposited in vast seas. The detritus is compressed, formed into rock and pushed up again to form new mountains, and the cycle is repeated. From time to time masses of molten rock well up from the mantle to form huge flows that cover the crust.

Tectonics refers to the broad geologic features of the continents and ocean basins as well as the forces responsible for their occurrence. The origins of these forces are not well understood, although it is apparent that the Earth's crust is in a state of overstress as evidenced by folding, faulting, and other mountain-building processes. Four general hypotheses have been developed to describe the sources of global tectonics (Hodgson, 1964; Zumberge and Nelson, 1972).

A.3.2 The Hypotheses

Contraction hypothesis assumes that the Earth is cooling, and because earthquakes do not occur below 700 km, the Earth is considered static below this depth, and is still hot and not cooling. The upper layer of the active zone, to a depth of about 100 km, has stopped cooling and shrinking. As the lower layer cools and contracts, it causes the upper layer to conform by buckling, which is the source of the surface stresses. This hypothesis is counter to the spreading seafloor or continental drift theory.

Convection-current hypothesis assumes that heat is being generated within the Earth by radioactive disintegration and that this heat causes convection currents that rise to the surface under the mid-ocean rifts, causing tension to create the rifts, then moves toward the continents with the thrust necessary to push up mountains, and finally descend again beneath the continents.

Expanding Earth hypothesis, the latest theory, holds that the Earth is expanding because of a decrease in the force of gravity, which is causing the original shell of granite to break up and spread apart, giving the appearance of continental drift.

Continental drift theory is currently the most popular, but is not new, and is supported by substantial evidence. Seismology has demonstrated that the continents are blocks of light granitic rocks "floating" on heavier basaltic rocks. It has been proposed that all of the continents were originally connected as one or two great land masses and at the *end of the Paleozoic era* (Permian period) they broke up and began to drift apart as illustrated in sequence in Figure A.1. The proponents of the theory have divided the earth into "plates" (Figure A.2) with each plate bounded by an earthquake zone as shown in Figure 3.1. (Note that more plates have been identified in Figure A.2 than are shown in Figure 3.1.)

Wherever plates move against each other, or a plate plunges into a deep ocean trench, such as that exists off the west coast of South America or the east coast of Japan, so that it slides beneath an adjacent plate (see Figure 3.2), there is high seismic activity. This

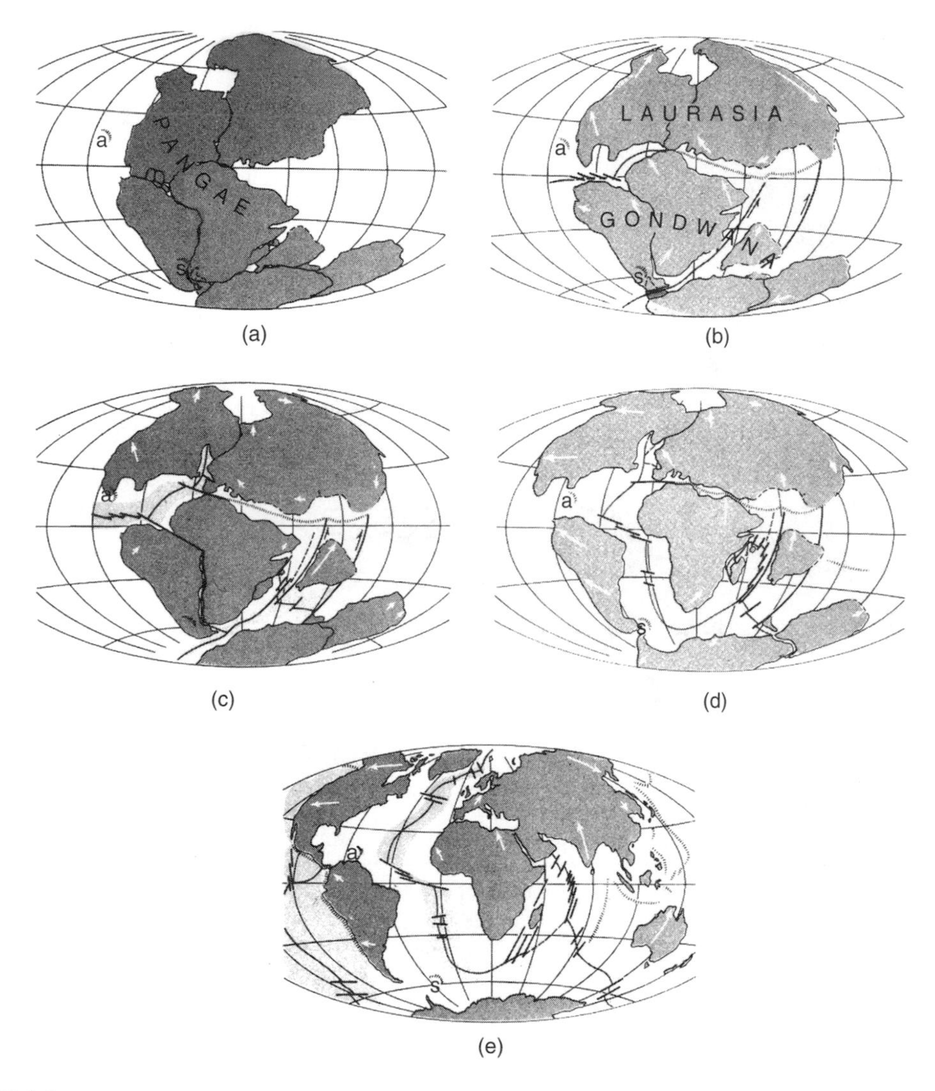

FIGURE A.1
The breakup and drifting apart of the original land mass, Pangaea: (a) Pangaea, the original continental land mass at the end of the permian, 225 million years ago; (b) Laurasia and Gondwana at the end of the Triassic, 180 million years ago; (c) positions at the end of the Jurassic, 135 million years ago (North and South America beginning to break away); (d) positions at the end of the Cretaceous, 65 million years ago; (e) positions of continents and the plate boundaries at the present. (From Dietz, R. S. and Holden, J. C., *J. Geophys. Res.*, 75, 4939–4956, 1970. With permission.)

concept is known as "plate tectonics" and appears to be compatible with the relatively new concept of *seafloor spreading*, as shown in Figure 3.2.

A.4 Geologic History

A.4.1 North America: Provides a General Illustration

The geologic time scale for North America is given in Table A.1, relating periods to typical formations. A brief geologic history of North America is described in Table A.2. These

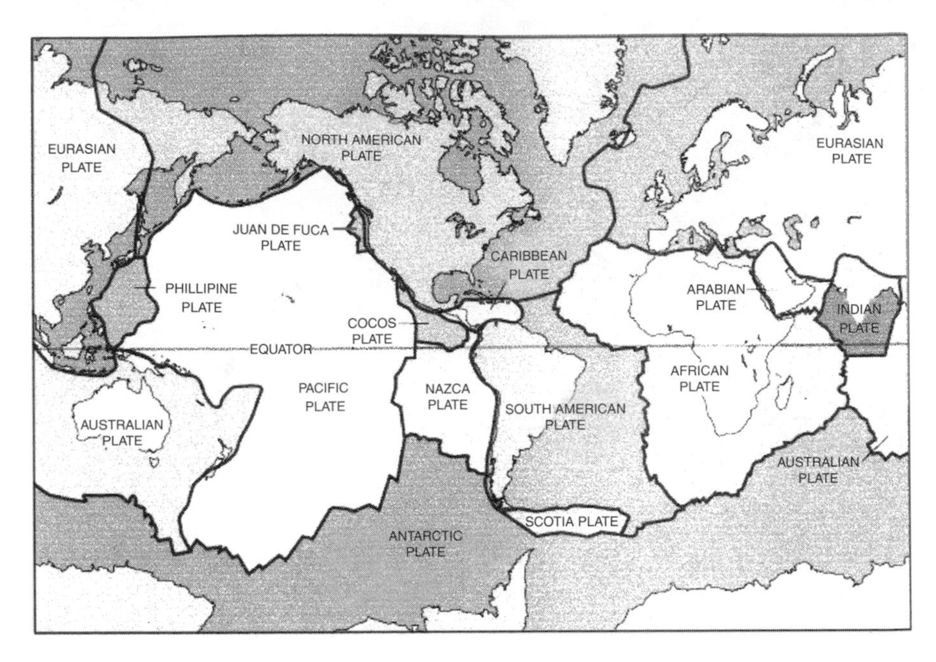

FIGURE A.2
Major tectonic plates of the world. (Courtesy of USGS, 2004.)

TABLE A.1
Geologic Time Scale and the Dominant Rock Types in North America

Era	Period	Epoch	Dominant Formations	Age (millions of years)
Cenozoic	Quaternary	Holocene	Modern soils	0.01
		Pleistocene	North American glaciation	2.5–3
	Neogene	Pliocene		7
		Miocene	"Unconsolidated" coastal-plain sediments	26
	Tertiary	Oligocene		37
	Paleogene	Eocene		54
		Paleocene		65
Mesozoic	Cretaceous		Overconsolidated clays and clay shales	135
	Jurassic		Various sedimentary rocks	180
	Triassic		Clastic sedimentary rocks with diabase intrusions	
Paleozoic	Permian		Fine-grained clastics, chemical precipitates, and evaporites. Continental glaciation in southern hemisphere	225
	Pennsylvanian		Shales and coal beds	280
	Mississippian	Carboniferous	Limestones in central United States. Sandstones and shales in east	310
	Devonian		Red sandstones and shales	400
	Silurian		Limestone, dolomite and evaporites, shales	435
	Ordovician		Limestone and dolomite, shales	500
	Cambrian		Limestone and dolomite in late Cambrian, sandstones and shales in early Cambrian	600
Precambrian	Precambrian		Igneous and metamorphic rocks	About 4.5 billion years

TABLE A.2

A Brief Geologic History of North America[a]

Period	Activity
Precambrian	Period of hundreds of millions of years during which the crust was formed and the continental land masses appeared
Cambrian	Two great troughs in the east and west filled with sediments ranging from detritus at the bottom, upward to limestones and dolomites, which later formed the Appalachians, the Rockies, and other mountain ranges
Ordovician	About 70% of North America was covered by shallow seas and great thicknesses of limestone and dolomite were deposited There was some volcanic activity and the eastern landmass, including the mountains of New England, started to rise (Taconic orogeny)
Silurian	Much of the east was inundated by a salty inland sea; the deposits ranged from detritus to limestone and dolomite, and in the northeast large deposits of evaporites accumulated in landlocked arms of the seas Volcanos were active in New Brunswick and Maine
Devonian	Eastern North America, from Canada to North Carolina, rose from the sea (Arcadian orogeny). The northern part of the Appalachian geosyncline received great thicknesses of detritus that eventually formed the Catskill Mountains In the west, the stable interior was inundated by marine waters and calcareous deposits accumulated In the east, limestone was metamorphosed to marble
Carboniferous	Large areas of the east became a great swamp which was repeatedly submerged by shallow seas. Forests grew, died, and were buried to become coal during the Pennsylvanian portion of the period
Permian	A period of violent geologic and climatic disturbances. Great wind-blown deserts covered much of the continent. Deposits in the west included evaporites and limestones The Appalachian Mountains were built in the east to reach as high as the modern Alps (Alleghanian orogeny) The continental drift theory (Section 3.2) considers that it was toward the end of the Permian that the continents began to drift apart
Triassic	The Appalachians began to erode and their sediments were deposited in the adjacent non-marine seas The land began to emerge toward the end of the period and volcanic activity resulted in sills and lava flows; faulting occurred during the Palisades orogeny
Jurassic	The Sierra Nevada Mountains, stretching from southern California to Alaska, were thrust up during the Nevadian disturbance
Cretaceous	The Rocky Mountains from Alaska to Central America rose out of a sediment-filled trough For the last time the sea inundated much of the continent and thick formations of clays were deposited along the east coast
Tertiary	The Columbia plateau and the Cascade Range rose, and the Rockies reached their present height Clays were deposited and shales formed along the continental coastal margins, reaching thicknesses of some 12 km in a modern syncline in the northern Gulf of Mexico that has been subsiding since the end of the Appalachian orogeny Extensive volcanic activity occurred in the northwest
Quaternary	During the Pleistocene epoch, four ice ages sent glaciers across the continent, which had a shape much like the present In the Holocene epoch (most recent), from 18,000 to 6,000 years ago, the last of the great ice sheets covering the continent melted and sea level rose almost 100 m Since then, sea level has remained almost constant, but the land continues to rebound from adjustment from the tremendous ice load. In the center of the uplifted region in northern Canada, the ground has risen 136 m in the last 6,000 years and is currently rising at the rate of about 2 cm/year (Walcott,1972) Evidence of ancient postglacial sea levels is given by raised beaches and marine deposits of late Quaternary found around the world. In Brazil, for example, Pleistocene sands and gravels are found along the coastline as high as 20 m above the present sea level

[a] The geologic history presented here contains the general concepts accepted for many decades, and still generally accepted. The major variances, as postulated by the continental drift concept, are that until the end of the Permian, Appalachia [a land mass along the U.S. east coast region] may have been part of the northwest coast of Africa [Figure 1*a*], and that the west coast of the present United States may have been an archipelago of volcanic islands known as Cascadia. (From Zumberge, J. H. and Nelson, C. A., *Elements of Geology*, 3rd ed., Wiley, New York, 1972. Reprinted with permission of Wiley.)

relationships apply in a general manner to many other parts of the world. Most of the periods are separated by major crustal disturbances (orogenies). Age determination is based on fossil identification (paleontology) and radiometric dating.

The classical concepts of the history of North America have been modified in conformity with the modern concept of the continental drift hypothesis. The most significant modification is the consideration that until the end of the Permian period, the east coast of the United States was connected to the northwest coast of Africa as shown in Figure A.1.

A.4.2 Radiometric Dating

Radiometric dating determines the age of a formation by measuring the decay rate of a radioactive element.

In radioactive elements, such as uranium, the number of atoms that decay during a given unit of time to form new stable elements is directly proportional to the number of atoms of the radiometric element of the sample. This decay rate is constant for the various radioactive elements and is given by the half-life of the element, i.e., the time required for any initial number of atoms to be reduced by one half. For example, when once-living organic matter is carbon-dated, the amount of radioactive carbon (carbon 14) remaining and the amount of ordinary carbon present are measured, and the age of a specimen is computed from a simple mathematical relationship. A general discussion on dating techniques can be found in Murphy et al. (1979). The various isotopes, effective dating range, and minerals and other materials that can be dated are given in Table A.3.

In engineering problems the most significant use of radiometric dating is for the dating of materials from fault zones to determine the age of most recent activity (see Section 3.3.1). The technique is also useful in dating soil formations underlying colluvial deposits as an indication as to when the slope failure occurred.

TABLE A.3
Some of the Principal Isotopes Used in Radiometric Dating

Isotope				
Parent	**Offspring**	**Parent Half-Life (years)**	**Effective Dating Range (years)**	**Material That Can Be Dated**
Uranium 238	Lead 206	4.5 billion	10 million to 4.6 billion	Zircon, uraninite, pitchblende
Uranium 235	Lead 207	710 million		
Potassium 40[a]	Argon 40 Calcium 40	1.3 billion	100,000 to 4.6 billion	Muscovite, biotite hornblende, intact volcanic rock
Rubidium 87	Strontium 87	47 billion	10 million to 4.6 billion	Muscovite, biotite, microcline, intact metamorphic rock
Carbon 14[a]	Nitrogen 14	5,730±30	100 to 50,000	Plant material: wood, peat charcoal, grain. Animal material: bone, tissue. Cloth, shell, stalactites, groundwater and seawater.

[a] Most commonly applied to fault studies: Carbon 14 for carbonaceous matter, or K–Ar for noncarbonaceous matter such as fault gouge.

References

Hodgson, J. H., *Earthquakes and Earth Structure*, Prentice-Hall Inc., Englewood Cliffs, NJ, 1964.
Murphy, P. J., Briedis, J., and Peck, J. H., Dating techniques in fault investigations, geology in the siting of nuclear power plants, in *Reviews in Engineering Geology IV*, The Geological Society of America, Hatheway, A. W. and McClure, C. R., Jr., Eds, Boulder, CO, 1979, 153–168.
Zumberge, J. H. and Nelson, C. A., *Elements of Geology*, 3rd ed., Wiley, New York, 1972.

Further Reading

Dunbar, C. O. and Waage, K. M., *Historical Geology*, 3rd ed., Wiley, New York, 1969.
Guttenberg, B. and Richter, C. F., *Seismicity of the Earth and Related Phenomenon*, Princeton University Press, Princeton, NJ, 1954.
Walcott, R. L., Late quaternary vertical movements in eastern North America: quantitative evidence of glacio-isostatic rebound, *Rev. Geophys. Space Phys.*, 10, 849–884, 1972.

Index

Page references given in *italics* refer to figures.
Page references given in **boldface** type refer to tables.

C

F

Q

R

Y

Z